W9-BMX-403

NUS
CO-OP

QUEUEING SYSTEMS

VOLUME I: THEORY

"Ah, 'All things come to those who wait.'
They come, but often come too late."

From Lady Mary M. Currie: *Tout Vient à Qui Sait Attendre* (1890)

QUEUEING SYSTEMS

VOLUME I: THEORY

Leonard Kleinrock

Professor
Computer Science Department
School of Engineering and Applied Science
University of California, Los Angeles

A Wiley-Interscience Publication

John Wiley & Sons

New York • Chichester • Brisbane • Toronto

Library of Congress Cataloging in Publication Data:

Kleinrock, Leonard.
Queueing systems.

"A Wiley-Interscience publication."
CONTENTS: v. 1. Theory.
1. Queueing theory. I. Title.

T57.9.K6 519.8'2 74-9846
ISBN 0-471-49110-1

30 29 28 27 26

TO STELLA

Preface

How much time did you waste waiting in line this week? It seems we cannot escape frequent delays, and they are getting progressively worse! In this text we study the phenomena of standing, waiting, and serving, and we call this study *queueing theory*.

Any system in which arrivals place demands upon a finite-capacity resource may be termed a queueing system. In particular, if the arrival times of these demands are unpredictable, or if the size of these demands is unpredictable, then conflicts for the use of the resource will arise and queues of waiting customers will form. The lengths of these queues depend upon two aspects of the flow pattern: first, they depend upon the *average rate* at which demands are placed upon the resource; and second, they depend upon the *statistical fluctuations* of this rate. Certainly, when the average rate exceeds the capacity, then the system breaks down and unbounded queues will begin to form; it is the effect of this average overload which then dominates the growth of queues. However, even if the average rate is less than the system capacity, then here, too, we have the formation of queues due to the statistical fluctuations and spurts of arrivals that may occur; the effect of these variations is greatly magnified when the average load approaches (but does not necessarily exceed) that of the system capacity. The simplicity of these queueing structures is deceptive, and in our studies we will often find ourselves in deep analytic waters. Fortunately, a familiar and fundamental law of science permeates our queueing investigations. This law is the conservation of flow, which states that the rate at which flow increases within a system is equal to the difference between the flow rate into and the flow rate out of that system. This observation permits us to write down the basic system equations for rather complex structures in a relatively easy fashion.

The purpose of this book, then, is to present the theory of queues at the first-year graduate level. It is assumed that the student has been exposed to a first course in probability theory; however, in Appendix II of this text we give a probability theory refresher and state the basic principles that we shall need. It is also helpful (but not necessary) if the student has had some exposure to transforms, although in this case we present a rather complete transform theory refresher in Appendix I. The student is advised to read both appendices before proceeding with the text itself. Whereas our material is presented in the language of mathematics, we do take great pains to give as informal a presentation as possible in order to strike a balance between the abstractions usually encountered in such a study and the basic need for understanding and applying these tools to practical systems. We feel that a satisfactory middle ground has been established that will neither offend the mathematician nor confound the practitioner. At times we have relaxed the rigor in proofs of uniqueness, existence, and convergence in order not to cloud the main thrust of a presentation. At such times the reader is referred to some of the other books on the subject. We have refrained from using the dull "theorem–proof" approach; rather, we lead the reader through a natural sequence of steps and together we "discover" the result. One finds that previous presentations of this material are usually either too elementary and limited or far too elegant and precise, and almost all of them badly neglect the applications; we feel that the need for a book such as this, which treads the boundary inbetween, is necessary and useful. This book was written over a period of five years while being used as course notes for a one-year (and later a two-quarter) sequence in queueing systems at the University of California, Los Angeles. The material was developed in the Computer Science Department within the School of Engineering and Applied Science and has been tested successfully in the most critical and unforgiving of all environments, namely, that of the graduate student. This text is appropriate not only for computer science departments, but also for departments of engineering, operations research, mathematics, and many others within science, business, management and planning schools.

In order to describe the contents of this text, we must first describe the very convenient shorthand notation that has been developed for the specification of queueing systems. It basically involves the three-part descriptor A/B/m that denotes an m-server queueing system, where A and B describe the interarrival time distribution and the service time distribution, respectively. A and B take on values from the following set of symbols whose interpretation is given in terms of distributions within parentheses: M (exponential); E_r (r-stage Erlangian); H_R (R-stage hyperexponential); D (deterministic); G (general). Occasionally, some other specially defined symbols are used. We sometimes need to specify the system's storage capacity (which we denote by K) or perhaps the size of the customer population (which we denote by M), and in these cases we adopt the five-part descriptor A/B/m/K/M; if either of these last two descriptors is absent, then we assume it takes on the value of infinity. Thus, for example, the system D/M/2/20 is a two-server system with constant (deterministic) interarrival times, with exponentially distributed service times, and with a system storage capacity of size 20.

This is Volume I (Theory) of a two-volume series, the second of which is devoted to computer applications of this theory. The text of Volume I (which consists of four parts) begins in Chapter 1 with an introduction to queueing systems, how they fit into the general scheme of systems of flow, and a discussion of how one specifies and evaluates the performance of a queueing system. Assuming a knowledge of (or after reviewing) the material in Appendices I and II, the reader may then proceed to Chapter 2, where he is warned to take care! Section 2.1 is essential and simple. However, Sections 2.2, 2.3, and 2.4 are a bit "heavy" for a first reading in queueing systems, and it would be quite reasonable if the reader were to skip these sections at this point, proceeding directly to Section 2.5, in which the fundamental birth–death process is introduced and where we first encounter the use of z-transforms and Laplace transforms. Once these *preliminaries* in Part I are established one may proceed with the *elementary queueing theory* presented in Part II. We begin in Chapter 3 with the general equilibrium solution to birth–death processes and devote most of the chapter to providing simple yet important examples. Chapter 4 generalizes this treatment, and it is here where we discuss the method of stages and provide an introduction to networks of Markovian queues. Whereas Part II is devoted to algebraic and transform oriented calculations, Part III returns us once again to probabilistic (as well as transform) agruments. This discussion of *intermediate queueing theory* begins with the important M/G/1 queue (Chapter 5) and then proceeds to the dual G/M/1 queue and its natural generalization to the system G/M/m (Chapter 6). The material on collective marks in Chapter 7 develops the probabilistic interpretation of transforms. Finally, the *advanced material* in Part IV leads us to the queue G/G/1 in Chapter 8; this difficult system (whose mean wait cannot even be expressed simply in terms of the system parameters) is studied through the use of the spectral solution to Lindley's integral equation. An approximation to the precedence structure among chapters in these two volumes is given below. In this diagram we have represented chapters in Volume I as numbers enclosed in circles and have used small squares for Volume II. The shading for the Volume I nodes indicates an appropriate amount of material for a relatively leisurely first course in queueing systems that can easily be accomplished in one semester or can be comfortably handled in a one-quarter course. The shading of Chapter 2 is meant to indicate that Sections 2.2–2.4 may be omitted on a first reading, and the same applies to Sections 8.3 and 8.4. A more rapid one-semester pace and a highly accelerated one-quarter pace would include all of Volume I in a single course. We close Volume I with a summary of important equations, developed throughout the book, which are grouped together according to the class of queueing system involved; this list of results then serves as a "handbook" for later use by the reader in concisely summarizing the principal results of this text. The results

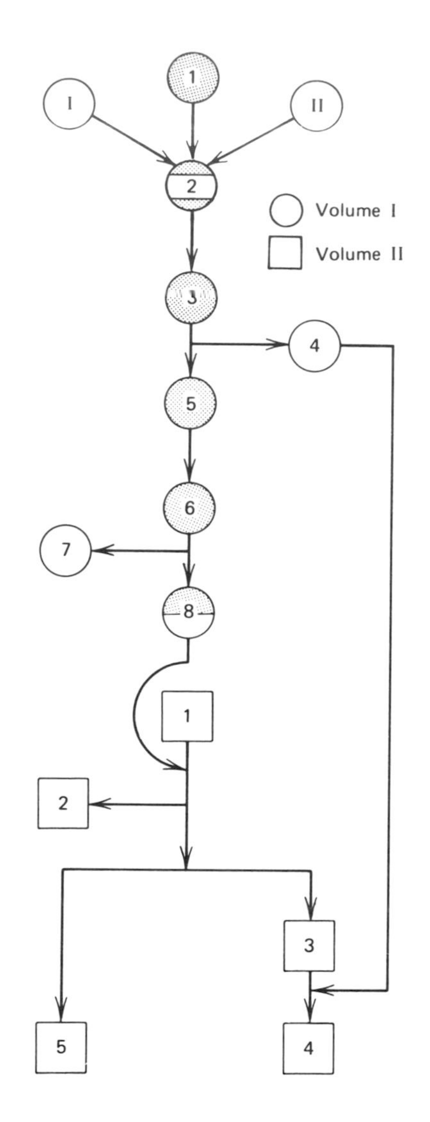

are keyed to the page where they appear in order to simplify the task of locating the explanatory material associated with each result.

Each chapter contains its own list of references keyed alphabetically to the author and year; for example, [KLEI 74] would reference this book. All equations of importance have been marked with the symbol ▬, and it is these which are included in the summary of important equations. Each chapter includes a set of exercises which, in some cases, extend the material in that chapter; the reader is urged to work them out.

Volume II represents material that comfortably fills a second course in queueing systems devoted to some extensions and principally to computer applications. The material in Chapter 1 is a review of the basic theory in Volume I. In Chapter 2 we study bounds, inequalities, and approximations in an attempt to capture the essential behavior of queueing systems, even including transient and nonstationary behavior; this is achieved by relaxing the requirement that we obtain exact results and thereby allows a much deeper penetration into meaningful results of importance in solving real-world problems. Chapter 3 lays the groundwork for our time-sharing studies by developing some of the basic notions of priority queueing systems. Chapter 4 studies a class of systems that may be described as highly preemptive priority queueing systems, but is more commonly known as the class of time-shared computer systems. We present single resource models as well as multiple resource models, which draw upon the material from Chapter 4 in Volume I. In this chapter we also make use of the diffusion approximation developed in Chapter 2. Finally, in Chapter 5 we introduce and study the very important class of computer-communication networks, using our previous queueing theoretic results, as well as some additional tools. This two-volume series begins by presenting the theory of queueing systems, which in itself is a worthwhile goal. The principal goal, however, and that which has motivated the author in his own work, is to develop these tools in order to apply them to real-world systems and, in this case, principally to computer systems.

The choice of material presented in Volume I (and especially in Volume II) represents my personal bias and preferences in the field. For example, I have chosen not to include a lengthy discussion of simulation methods or of methods for conducting a numerical analysis of queueing systems. These topics are certainly of importance in studying real-world queues, but do not lend themselves very well to textbook material; they are discussed briefly in the applications portions, but certainly are not emphasized.

It is perhaps worthwhile to devote a few words to the history and development of queueing theory. It is no surprise that the study of queueing systems began in the field of telephony when, during the first two decades of this century, A. K. Erlang developed the basic foundations of the theory long before probability theory was popularized or even well developed. This remarkable man established many of the principal results which we still use today. The 1920s were basically devoted to the application of his results, and it was not until the mid-1930s, when Feller introduced the concept of the birth–death process, that queueing was recognized by the world of mathematics as an object of serious interest. During and following World War II this theory played an important role in the development of the new field of operations research, which seemed to hold so much promise in the early post-war years. As the enchantment with operations research diminished in the face of the real world's complicated models, the mathematicians proceeded to advance the field of queueing theory rapidly and elegantly. The frontiers of this research proceeded into the far reaches of deep and complex mathematics. It was soon found that the really interesting models did not yield to solution and the field quieted down considerably. It was mainly with the advent of digital computers that once again the tools of queueing theory were brought to bear on a class of practical problems, but this time with great success. The fact is that at present, one of the few tools we have for analyzing the performance of computer systems is that of queueing theory, and this explains its popularity among engineers and scientists today. A wealth of new problems are being formulated in terms of this theory and new tools and methods are being developed to meet the challenge of these problems. Moreover, the application of digital computers in solving the equations of queueing theory has spawned new interest in the field. It is hoped that this two-volume series will provide the reader with an appreciation for and competence in the methods of analysis and application as we now see them.

I take great pleasure in closing this Preface by acknowledging those individuals and institutions that made it possible for me to bring this book into being. First, I would like to thank all those who participated in creating the stimulating environment of the Computer Science Department at UCLA, which encouraged and fostered my effort in this direction. Acknowledgment is due the Advanced Research Projects Agency of the Department of Defense, which enabled me to participate in some of the most exciting and advanced computer systems and networks ever developed. Furthermore, the John Simon Guggenheim Foundation provided me with a Fellowship for the academic year 1971–1972, during which time I was able to further pursue my investigations. Hundreds of students who have passed through my queueing-systems courses have in major and minor ways contributed to the creation of this book, and I am happy to acknowledge the special help offered by Arne Nilsson, Johnny Wong, Simon Lam, Fouad Tobagi, Farouk Kamoun, Robert Rice, and Thomas Sikes. My academic and professional colleagues have all been very supportive of this endeavour. To the typists I owe all. By far the largest portion of this book was typed by Charlotte LaRoche, and I will be forever in her debt. To Diana Skocypec and Cynthia Ellman I give my deepest thanks for carrying out the enormous task of proofreading and correction-making in a rapid, enthusiastic, and supportive fashion. Others who contributed in major ways are Barbara Warren, Jean Dubinsky, Jean D'Fucci, and Gloria Roy. I owe a great debt of thanks to my family (and especially to my wife, Stella) who have stood by me and supported me well beyond the call of duty or marriage contract. Lastly, I would certainly be remiss in omitting an acknowledgement to my ever-faithful dictating machine, which was constantly talking back to me.

LEONARD KLEINROCK

March, 1974

Contents

VOLUME I

PART I: PRELIMINARIES

PART II: ELEMENTARY QUEUEING THEORY

PART IV: ADVANCED MATERIAL

VOLUME II

QUEUEING SYSTEMS

VOLUME I: THEORY

PART I

PRELIMINARIES

It is difficult to see the forest for the trees (especially if one is in a mob rather than in a well-ordered queue). Likewise, it is often difficult to see the impact of a collection of mathematical results as you try to master them; it is only after one gains the understanding and appreciation for their application to real-world problems that one can say with confidence that he understands the use of a set of tools.

The two chapters contained in this preliminary part are each extreme in opposite directions. The first chapter gives a global picture of where queueing systems arise and why they are important. Entertaining examples are provided as we lure the reader on. In the second chapter, on random processes, we plunge deeply into mathematical definitions and techniques (quickly losing sight of our long-range goals); the reader is urged not to falter under this siege since it is perhaps the worst he will meet in passing through the text. Specifically, Chapter 2 begins with some very useful graphical means for displaying the dynamics of customer behavior in a queueing system. We then introduce stochastic processes through the study of customer arrival, behavior, and backlog in a very general queueing system and carefully lead the reader to one of the most significant results in queueing theory, namely, Little's result, using very simple arguments. Having thus introduced the concept of a stochastic process we then offer a rather compact treatment which compares many well-known (but not well-distinguished) processes and casts them in a common terminology and notation, leading finally to Figure 2.4 in which we see the basic relationships among these processes; the reader is quickly brought to realize the central role played by the Poisson process because of its position as the common intersection of all the stochastic processes considered in this chapter. We then give a treatment of Markov chains in discrete and continuous time; these sections are perhaps the toughest sledding for the novice, and it is perfectly acceptable if he passes over some of this material on a first reading. At the conclusion of Section 2.4 we find ourselves face to face with the important birth–death processes and it is here where things begin to take on a relationship to physical systems once again. In fact, it is not unreasonable for the reader to begin with Section 2.5 of this chapter since the treatment following is (almost) self-contained from there throughout the rest of the text. Only occasionally do we find a need for the more detailed material in Sections 2.3 and 2.4. If the reader perseveres through Chapter 2 he will have set the stage for the balance of the textbook.

I

Queueing Systems

One of life's more disagreeable activities, namely, waiting in line, is the delightful subject of this book. One might reasonably ask, "What does it profit a man to study such unpleasant phenomena?" The answer, of course, is that through understanding we gain compassion, and it is exactly this which we need since people will be waiting in longer and longer queues as civilization progresses, and we must find ways to tolerate these unpleasant situations. Think for a moment how much time is spent in one's daily activities waiting in some form of a queue: waiting for breakfast; stopped at a traffic light; slowed down on the highways and freeways; delayed at the entrance to one's parking facility; queued for access to an elevator; standing in line for the morning coffee; holding the telephone as it rings, and so on. The list is endless, and too often also are the queues.

The orderliness of queues varies from place to place around the world. For example, the English are terribly susceptible to formation of orderly queues, whereas some of the Mediterranean peoples consider the idea ludicrous (have you ever tried clearing the embarkation procedure at the Port of Brindisi?). A common slogan in the U.S. Army is, "Hurry up and wait." Such is the nature of the phenomena we wish to study.

1.1. SYSTEMS OF FLOW

Queueing systems represent an example of a much broader class of interesting dynamic systems, which, for convenience, we refer to as "systems of flow." A flow system is one in which some *commodity* flows, moves, or is transferred through one or more finite-capacity *channels* in order to go from one point to another. For example, consider the flow of automobile traffic through a road network, or the transfer of goods in a railway system, or the streaming of water through a dam, or the transmission of telephone or telegraph messages, or the passage of customers through a supermarket checkout counter, or the flow of computer programs through a time-sharing computer system. In these examples the commodities are the automobiles, the goods, the water, the telephone or telegraph messages, the customers, and the programs, respectively; the channel or channels are the road network, the railway network, the dam, the telephone or telegraph network, the supermarket checkout counter, and the computer processing system, respectively. The "finite capacity" refers to the fact that the channel can satisfy the demands (placed upon it by the commodity) at a finite rate only. It is clear that the analyses of many of these systems require analytic tools drawn from a variety of disciplines and, as we shall see, queueing theory is just one such discipline.

When one analyzes systems of flow, they naturally break into two classes: *steady* and *unsteady* flow. The first class consists of those systems in which the flow proceeds in a predictable fashion. That is, the quantity of flow is exactly known and is constant over the interval of interest; the time when that flow appears at the channel, and how much of a demand that flow places upon the channel is known and constant. These systems are trivial to analyze in the case of a *single channel*. For example, consider a pineapple factory in which empty tin cans are being transported along a conveyor belt to a point at which they must be filled with pineapple slices and must then proceed further down the conveyor belt for additional operations. In this case, assume that the cans arrive at a constant rate of one can per second and that the pineapple-filling operation takes nine-tenths of one second per can. These numbers are constant for all cans and all filling operations. Clearly this system will function in a reliable and smooth fashion as long as the assumptions stated above continue to exist. We may say that the *arrival rate* R is one can per second and the maximum service rate (or *capacity*) C is $1/0.9 = 1.11111 \cdots$ filling operations per second. The example above is for the case $R < C$. However, if we have the condition $R > C$, we all know what happens: cans and/or pineapple slices begin to inundate and overflow in the factory! Thus we see that *the mean capacity of the system must exceed the average flow requirements if chaotic congestion is to be avoided*; this is true for all systems of flow. This simple observation tells most of the story. Such systems are of little interest theoretically.

The more interesting case of steady flow is that of a *network* of channels. For stable flow, we obviously require that $R < C$ on each channel in the network. However we now run into some serious combinatorial problems. For example, let us consider a railway network in the fictitious land of Hatafla. See Figure 1.1. The scenario here is that figs grown in the city of Abra must be transported to the destination city of Cadabra, making use of the railway network shown. The numbers on each channel (section of railway) in Figure 1.1 refer to the maximum number of bushels of figs which that channel can handle per day. We are now confronted with the following fig flow problem: How many bushels of figs per day can be sent from Abra to Cadabra and in what fashion shall this flow of figs take place? The answer to such questions of maximal "traffic" flow in a variety of networks is nicely

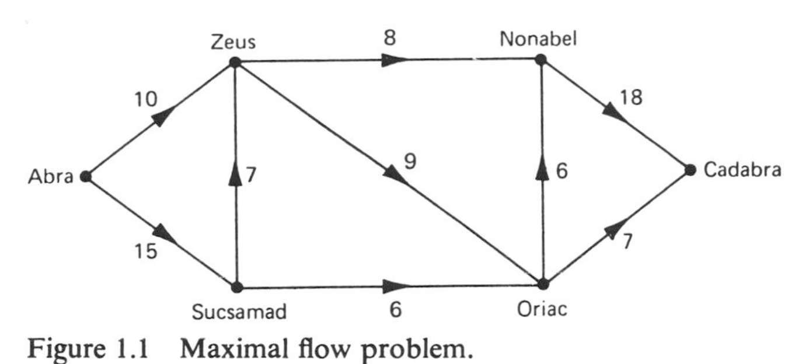

Figure 1.1 Maximal flow problem.

settled by a well-known result in network flow theory referred to as the *max-flow–min-cut* theorem. To state this theorem, we first define a *cut* as a set of channels which, once removed from the network, will separate all possible flow from the origin (Abra) to the destination (Cadabra). We define the *capacity* of such a cut to be the total fig flow that can travel across that cut in the direction from origin to destination. For example, one cut consists of the branches from Abra to Zeus, Sucsamad to Zeus, and Sucsamad to Oriac; the capacity of this cut is clearly 23 bushels of figs per day. The max-flow–min-cut theorem states that the maximum flow that can pass between an origin and a destination is the minimum capacity of all cuts. In our example it can be seen that the maximum flow is therefore 21 bushels of figs per day (work it out). In general, one must consider *all* cuts that separate a given origin and destination. This computation can be enormously time consuming. Fortunately, there exists an extremely powerful method for finding not only what is the maximum flow, but also which flow pattern achieves this maximum flow. This procedure is known as the *labeling algorithm* (due to Ford and Fulkerson [FORD 62]) and is efficient in that the computational requirement grows as a small power of the number of nodes; we present the algorithm in Volume II, Chapter 5.

In addition to maximal flow problems, one can pose numerous other interesting and worthwhile questions regarding flow in such networks. For example, one might inquire into the minimal cost network which will support a given flow if we assign costs to each of the channels. Also, one might ask the same questions in networks when more than one origin and destination exist. Complicating matters further, we might insist that a given network support flow of various kinds, for example, bushels of figs, cartons of cartridges and barrels of oil. This multicommodity flow problem is an extremely difficult one, and its solution typically requires considerable computational effort. These and numerous other significant problems in network flow theory are addressed in the comprehensive text by Frank and Frisch [FRAN 71] and we shall see them again in Volume II, Chapter 5. Network flow theory itself requires methods from graph theory, combinatorial mathematics, optimization theory, mathematical programming, and heuristic programming.

The *second* class into which systems of flow may be divided is the class of random or stochastic flow problems. By this we mean that the *times* at which demands for service (use of the channel) arrive are uncertain or unpredictable, and also that the *size* of the demands themselves that are placed upon the channel are unpredictable. The randomness, unpredictability, or unsteady nature of this flow lends considerable complexity to the solution and understanding of such problems. Furthermore, it is clear that most real-world systems fall into this category. Again, the simplest case is that of random flow through a *single* channel; whereas in the case of deterministic or steady flow discussed earlier in which the single-channel problems were trivial, we have now a case where these single-channel problems are extremely challenging and, in fact, techniques for solution to the single-channel or single-server problem comprise much of modern queueing theory.

For example, consider the case of a computer center in which computation requests are served making use of a batch service system. In such a system, requests for computation arrive at unpredictable times, and when they do arrive, they may well find the computer busy servicing other demands. If, in fact, the computer is idle, then typically a new demand will begin service and will be run until it is completed. On the other hand, if the system is busy, then this job will wait on a queue until it is selected for service from among those that are waiting. Until that job is carried to completion, it is usually the case that neither the computation center nor the individual who has submitted the program knows the extent of the demand in terms of computational effort that this program will place upon the system; in this sense the service requirement is indeed unpredictable.

A variety of natural questions present themselves to which we would like intelligent and complete answers. How long, for example, may a job expect to wait on queue before entering service? How many jobs will be serviced before the one just submitted? For what fraction of the day will the computation center be busy? How long will the intervals of continual busy work extend? Such questions require answers regarding the probability of certain periods and numbers or perhaps merely the average values for these quantities. Additional considerations, such as machine breakdown (a not uncommon condition), complicate the issue further; in this case it is clear that some preemptive event prevents the completion of the job currently in service. Other interesting effects can take place where jobs are not serviced according to their order of arrival. Time-shared computer systems, for example, employ rather complex scheduling and servicing algorithms, which, in fact, we explore in Volume II, Chapter 4.

The tools necessary for solving single-channel random-flow problems are contained and described within queueing theory, to which much of this text devotes itself. This requires a background in probability theory as well as an understanding of complex variables and some of the usual transform-calculus methods; this material is reviewed in Appendices I and II.

As in the case of deterministic flow, we may enlarge our scope of problems to that of *networks* of channels in which random flow is encountered. An example of such a system would be that of a computer network. Such a system consists of computers connected together by a set of communication lines where the capacity of these lines for carrying information is finite. Let us return to the fictitious land of Hatafla and assume that the railway network considered earlier is now in fact a computer network. Assume that users located at Abra require computational effort on the facility at Cadabra. The particular times at which these requests are made are themselves unpredictable, and the commands or instructions that describe these requests are also of unpredictable length. It is these commands which must be transmitted to Cadabra over our communication net as messages. When a message is inserted into the network at Abra, and after an appropriate decision rule (referred to as a routing procedure) is accessed, then the message proceeds through the network along some path. If a portion of this path is busy, and it may well be, then the message must queue up in front of the busy channel and wait for it to become free. Constant decisions must be made regarding the flow of messages and routing procedures. Hopefully, the message will eventually emerge at Cadabra, the computation will be performed, and the results will then be inserted into the network for delivery back at Abra.

It is clear that the problems exemplified by our computer network involve a variety of extremely complex queueing problems, as well as network flow and decision problems. In an earlier work [KLEI 64] the author addressed himself to certain aspects of these questions. We develop the analysis of these systems later in Volume II, Chapter 5.

Having thus classified* systems of flow, we hope that the reader understands where in the general scheme of things the field of queueing theory may be placed. The methods from this theory are central to analyzing most stochastic flow problems, and it is clear from an examination of the current literature that the field and in particular its applications are growing in a viable and purposeful fashion.

* The classification described above places queueing systems within the class of systems of flow. This approach identifies and emphasizes the fields of application for queueing theory. An alternative approach would have been to place queueing theory as belonging to the field of applied stochastic processes; this classification would have emphasized the mathematical structure of queueing theory rather than its applications. The point of view taken in this two-volume book is the former one, namely, with application of the theory as its major goal rather than extension of the mathematical formalism and results.

1.2. THE SPECIFICATION AND MEASURE OF QUEUEING SYSTEMS

In order to completely specify a queueing system, one must identify the stochastic processes that describe the arriving stream as well as the structure and discipline of the service facility. Generally, the arrival process is described in terms of the probability distribution of the *interarrival times* of customers and is denoted $A(t)$, where*

$$A(t) = P[\text{time between arrivals} \leq t] \tag{1.1}$$

The assumption in most of queueing theory is that these interarrival times are independent, identically distributed random variables (and, therefore, the stream of arrivals forms a stationary renewal process; see Chapter 2). Thus, only the distribution $A(t)$, which describes the time between arrivals, is usually of significance. The second statistical quantity that must be described is the amount of demand these arrivals place upon the channel; this is usually referred to as the *service time* whose probability distribution is denoted by $B(x)$, that is,

$$B(x) = P[\text{service time} \leq x] \tag{1.2}$$

Here service time refers to the length of time that a customer spends in the service facility.

Now regarding the structure and discipline of the service facility, one must specify a variety of additional quantities. One of these is the extent of *storage capacity* available to hold waiting customers and typically this quantity is described in terms of the variable K; often K is taken to be infinite. An additional specification involves the *number of service stations* available, and if more than one is available, then perhaps the distribution of service time will differ for each, in which case the distribution $B(x)$ will include a subscript to indicate that fact. On the other hand, it is sometimes the case that the arriving stream consists of more than one identifiable *class* of customers; in such a case the interarrival distribution $A(t)$ as well as the service distribution $B(x)$ may each be characteristic of each class and will be identified again by use of a subscript on these distributions. Another important structural description of a queueing system is that of the queueing *discipline*; this describes the order in which customers are taken from the queue and allowed into service. For example, some standard queueing disciplines are first-come–first-serve (FCFS), last-come–first-serve (LCFS), and random order of service. When the arriving customers are distinguishable according to groups, then we encounter the case of *priority* queueing disciplines in which priority

* The notation $P[A]$ denotes, as usual, the "probability of the event A."

among groups may be established. A further statement regarding the *availability* of the service facility is also necessary in case the service facility is occasionally required to pay attention to other tasks (as, for example, its own breakdown). Beyond this, queueing systems may enjoy customer behavior in the form of *defections* from the queue, *jockeying* among the many queues, *balking* before entering a queue, *bribing* for queue position, *cheating* for queue position, and a variety of other interesting and not-unexpected humanlike characteristics. We will encounter these as we move through the text in an orderly fashion (first-come–first-serve according to page number).

Now that we have indicated how one must specify a queueing system, it is appropriate that we identify the measures of performance and effectiveness that we shall obtain by analysis. Basically, we are interested in the *waiting time* for a customer, the *number of customers* in the system, the *length of a busy period* (the continuous interval during which the server is busy), the *length of an idle period*, and the current *work backlog* expressed in units of time. All these quantities are random variables and thus we seek their complete probabilistic description (i.e., their probability distribution function). Usually, however, to give the distribution function is to give more than one can easily make use of. Consequently, we often settle for the first few moments (mean, variance, etc.).

Happily, we shall begin with simple considerations and develop the tools in a straightforward fashion, paying attention to the essential details of analysis. In the following pages we will encounter a variety of simple queueing problems, simple at least in the sense of description and usually rather sophisticated in terms of solution. However, in order to do this properly, we first devote our efforts in the following chapter to describing some of the important random processes that make up the arrival and service processes in our queueing systems.

REFERENCES

FORD 62 Ford, L. R. and D. R. Fulkerson, *Flows in Networks*, Princeton University Press (Princeton, N.J.), 1962.

FRAN 71 Frank, H. and I. T. Frisch, *Communication, Transmission, and Transportation Networks*, Addison-Wesley (Reading, Mass.), 1971.

KLEI 64 Kleinrock, L., *Communication Nets; Stochastic Message Flow and Delay*, McGraw-Hill (New York), 1964, out of print. Reprinted by Dover (New York), 1972.

2

Some Important Random Processes*

We assume that the reader is familiar with the basic elementary notions, terminology, and concepts of probability theory. The particular aspects of that theory which we require are presented in summary fashion in Appendix II to serve as a review for those readers desiring a quick refresher and reminder; it is recommended that the material therein be reviewed, especially Section II.4 on transforms, generating functions, and characteristic functions.

Included in Appendix II are the following important definitions, concepts, and results:

- Sample space, events, and probability.
- Conditional probability, statistical independence, the law of total probability, and Bayes' theorem.
- A real random variable, its probability distribution function (PDF), its probability density function (pdf), and their simple properties.
- Events related to random variables and their probabilities.
- Joint distribution functions.
- Functions of a random variable and their density functions.
- Expectation.
- Laplace transforms, generating functions, and characteristic functions and their relationships and properties.†
- Inequalities and limit theorems.
- Definition of a stochastic process.

2.1. NOTATION AND STRUCTURE FOR BASIC QUEUEING SYSTEMS

Before we plunge headlong into a step-by-step development of queueing theory from its elementary notions to its intermediate and then finally to some advanced material, it is important first that we understand the basic

* Sections 2.2, 2.3, and 2.4 may be skipped on a first reading.

† Appendix I is a transform theory refresher. This material is also essential to the proper understanding of this text.

structure of queues. Also, we wish to provide the reader a glimpse as to where we are heading in this journey.

It is our purpose in this section to define some notation, both symbolic and graphic, and then to introduce one of the basic stochastic processes that we find in queueing systems. Further, we will derive a simple but significant result, which relates some first moments of importance in these systems. In so doing, we will be in a position to define the quantities and processes that we will spend many pages studying later in the text.

The system we consider is the very general queueing system G/G/m; recall (from the Preface) that this is a system whose interarrival time distribution $A(t)$ is completely arbitrary and whose service time distribution $B(x)$ is also completely arbitrary (all interarrival times and service times are assumed to be independent of each other). The system has m servers and order of service is also quite arbitrary (in particular, it need not be first-come–first-serve). We focus attention on the flow of customers as they arrive, pass through, and eventually leave this system; as such, we choose to number the customers with the subscript n and define C_n as follows:

$$C_n \text{ denotes the } n\text{th customer to enter the system} \tag{2.1}$$

Thus, we may portray our system as in Figure 2.1 in which the box represents the queueing system and the flow of customers both in and out of the system is shown. One can immediately define some random processes of interest. For example, we are interested in $N(t)$ where*

$$N(t) \triangleq \text{number of customers in the system at time } t \tag{2.2}$$

Another stochastic process of interest is the unfinished work $U(t)$ that exists in the system at time t, that is,

$$\begin{aligned} U(t) &\triangleq \text{the unfinished work in the system at time } t \\ &\triangleq \text{the remaining time required to empty the system of all} \\ &\qquad \text{customers present at time } t \end{aligned} \tag{2.3}$$

Whenever $U(t) > 0$, then the system is said to be busy, and only when $U(t) = 0$ is the system said to be idle. The duration and location of these busy and idle periods are also quantities of interest.

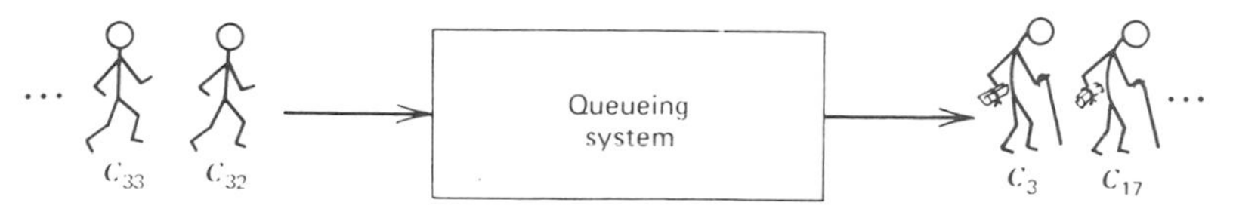

Figure 2.1 A general queueing system.

* The notation $\triangleq$ is to be read as "equals by definition."

The details of these stochastic processes may be observed first by defining the following variables and then by displaying these variables on an appropriate time diagram to be discussed below. We begin with the definitions. Recalling that the nth customer is denoted by C_n, we define his arrival time to the queueing system as

$$\tau_n \triangleq \text{arrival time for } C_n \tag{2.4}$$

and further define the interarrival time between C_{n-1} and C_n as

$$\begin{aligned} t_n &\triangleq \text{interarrival time between } C_{n-1} \text{ and } C_n \\ &= \tau_n - \tau_{n-1} \end{aligned} \tag{2.5}$$

Since we have assumed that all interarrival times are drawn from the distribution $A(t)$, we have that

$$P[t_n \le t] = A(t) \tag{2.6}$$

which is independent of n. Similarly, we define the service time for C_n as

$$x_n \triangleq \text{service time for } C_n \tag{2.7}$$

and from our assumptions we have

$$P[x_n \le x] = B(x) \tag{2.8}$$

The sequences $\{t_n\}$ and $\{x_n\}$ may be thought of as input variables for our queueing system; the way in which the system handles these customers gives rise to queues and waiting times that we must now define. Thus, we define the waiting time (time spent in the queue)* as

$$w_n \triangleq \text{waiting time (in queue) for } C_n \tag{2.9}$$

The total time spent in the system by C_n is the sum of his waiting time and service time, which we denote by

$$\begin{aligned} s_n &\triangleq \text{system time (queue plus service) for } C_n \\ &= w_n + x_n \end{aligned} \tag{2.10}$$

Thus we have defined for the nth customer his arrival time, "his" interarrival time, his service time, his waiting time, and his system time. We find it

* The terms "waiting time" and "queueing time" have conflicting definitions within the body of queueing-theory literature. The former sometimes refers to the total time spent in system, and the latter then refers to the total time spent on queue; however, these two definitions are occasionally reversed. We attempt to remove that confusion by defining waiting and queueing time to be the same quantity, namely, the time spent waiting on queue (but not being served); a more appropriate term perhaps would be "wasted time." The total time spent in the system will be referred to as "system time" (occasionally known as "flow time").

expedient at this point to elaborate somewhat further on notation. Let us consider the interarrival time t_n once again. We will have occasion to refer to the limiting random variable $\tilde{t}$ defined by

$$\tilde{t} \triangleq \lim_{n\to\infty} t_n \tag{2.11}$$

which we denote by $t_n \to \tilde{t}$. (We have already required that the interarrival times t_n have a distribution independent of n, but this will not necessarily be the case with many other random variables of interest.) The typical notation for the probability distribution function (PDF) will be

$$P[t_n \le t] = A_n(t) \tag{2.12}$$

and for the limiting PDF

$$P[\tilde{t} \le t] = A(t) \tag{2.13}$$

This we denote by $A_n(t) \to A(t)$; of course, for the interarrival time we have assumed that $A_n(t) = A(t)$, which gives rise to Eq. (2.6). Similarly, the probability density function (pdf) for t_n and $\tilde{t}$ will be $a_n(t)$ and $a(t)$, respectively, and will be denoted as $a_n(t) \to a(t)$. Finally, the Laplace transform (see Appendix II) of these pdf's will be denoted by $A_n^*(s)$ and $A^*(s)$, respectively, with the obvious notation $A_n^*(s) \to A^*(s)$. The use of the letter A (and a) is meant as a cue to remind the reader that they refer to the interarrival time. Of course, the moments of the interarrival time are of interest and they will be denoted as follows*:

$$E[t_n] \triangleq \bar{t}_n \tag{2.14}$$

According to our usual notation, the mean interarrival time for the limiting random variable will be given† by $\bar{t}$ in the sense that $\bar{t}_n \to \bar{t}$. As it turns out $\bar{t}$, which is the average interarrival time between customers, is used so frequently in our equations that a *special* notation has been adopted as follows:

$$\bar{t} \triangleq \frac{1}{\lambda} \tag{2.15}$$

Thus λ represents the *average arrival rate* of customers to our queueing system. Higher moments of the interarrival time are also of interest and so we define the kth moment by

$$E[\tilde{t}^k] \triangleq \overline{t^k} \triangleq a_k \qquad k = 0, 1, 2, \ldots \tag{2.16}$$

* The notation $E[\ ]$ denotes the expectation of the quantity within square brackets. As shown, we also adopt the overbar notation to denote expectation.

† Actually, we should use the notation t with a tilde *and* a bar, but this is excessive and will be simplified to $\bar{t}$. The same simplification will be applied to many of our other random variables.

In this last equation we have introduced the definition a_k to be the kth moment of the interarrival time $\tilde{t}$; this is fairly standard notation and we note immediately from the above that

$$\bar{t} = \frac{1}{\lambda} = a_1 \triangleq a \tag{2.17}$$

That is, three *special* notations exist for the mean interarrival time; in particular, the use of the symbol a is very common and various of these forms will be used throughout the text as appropriate. Summarizing the information with regard to the interarrival time we have the following shorthand glossary:

$$\begin{gathered} t_n = \text{interarrival time between } C_n \text{ and } C_{n-1} \\ t_n \to \tilde{t}, \quad A_n(t) \to A(t), \quad a_n(t) \to a(t), \quad A_n^*(s) \to A^*(s) \\ \bar{t}_n \to \bar{t} = \frac{1}{\lambda} = a_1 = a, \quad \overline{t_n^{\,k}} \to \overline{t^k} = a_k \end{gathered} \tag{2.18}$$

In a similar manner we identify the notation associated with x_n, w_n, and s_n as follows:

$$\begin{gathered} x_n = \text{service time for } C_n \\ x_n \to \tilde{x}, \quad B_n(x) \to B(x), \quad b_n(x) \to b(x), \quad B_n^*(s) \to B^*(s) \\ \bar{x}_n \to \bar{x} = \frac{1}{\mu} = b_1 = b, \quad \overline{x_n^{\,k}} \to \overline{x^k} = b_k \end{gathered} \tag{2.19}$$

$$\begin{gathered} w_n = \text{waiting time for } C_n \\ w_n \to \tilde{w}, \quad W_n(y) \to W(y), \quad w_n(y) \to w(y), \quad W_n^*(s) \to W^*(s) \\ \bar{w}_n \to \bar{w} = W, \quad \overline{w_n^{\,k}} \to \overline{w^k} \end{gathered} \tag{2.20}$$

$$\begin{gathered} s_n = \text{system time for } C_n \\ s_n \to \tilde{s}, \quad S_n(y) \to S(y), \quad s_n(y) \to s(y), \quad S_n^*(s) \to S^*(s) \\ \bar{s}_n \to \bar{s} = T, \quad \overline{s_n^{\,k}} \to \overline{s^k} \end{gathered} \tag{2.21}$$

All this notation is self-evident except perhaps for the occasional special symbols used for the first moment and occasionally the higher moments of the random variables involved (that is, the use of the symbols λ, a, μ, b, W, and T). The reader is, at this point, directed to the Glossary for a complete set of notation used in this book.

With the above notation we now suggest a *time-diagram notation* for queues, which permits a graphical view of the dynamics of our queueing system and also provides the details of the underlying stochastic processes. This diagram is shown in Figure 2.2. This particular figure is shown for a

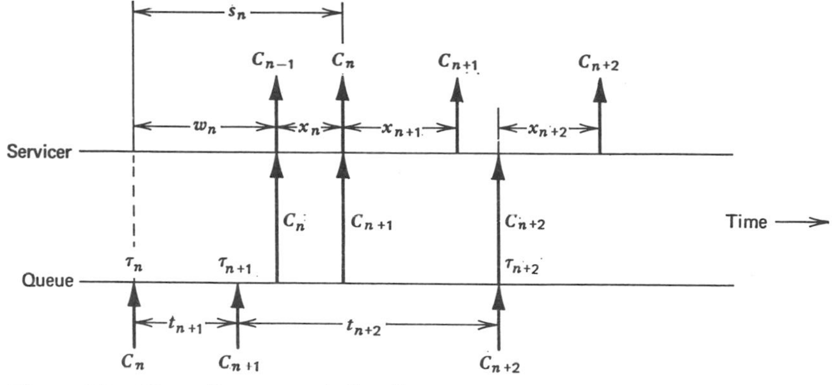

Figure 2.2 Time-diagram notation for queues.

first-come–first-serve order of service, but it is easy to see how the figure may also be made to represent any order of service. In this time diagram the lower horizontal time line represents the queue and the upper horizontal time line represents the service facility; moreover, the diagram shown is for the case of a single server, although this too is easily generalized. An arrow approaching the queue (or service) line from below indicates that an arrival has occurred to the queue (or service facility). Arrows emanating from the line indicate the departure of a customer from the queue (or service facility). In this figure we see that customer C_{n+1} arrives before customer C_n enters service; only when C_n departs from service may C_{n+1} enter service and, of course, these two events occur simultaneously. Notice that when C_{n+2} enters the system he finds it empty and so immediately proceeds through an empty queue directly into the service facility. In this diagram we have also shown the waiting time and the system time for C_n (note that $w_{n+2} = 0$). Thus, as time proceeds we can identify the number of customers in the system $N(t)$, the unfinished work $U(t)$, and also the idle and busy periods. We will find much use for this time-diagram notation in what follows.

In a general queueing system one expects that when the number of customers is large then so is the waiting time. One manifestation of this is a very simple relationship between the mean number in the queueing system, the mean arrival rate of customers to that system, and the mean system time for customers. It is our purpose next to derive that relationship and thereby familiarize ourselves a bit further with the underlying behavior of these systems. Referring back to Figure 2.1, let us position ourselves at the input of the queueing system and count how many customers enter as a function of time. We denote this by $\alpha(t)$ where

$$\alpha(t) \triangleq \text{number of arrivals in } (0, t) \tag{2.22}$$

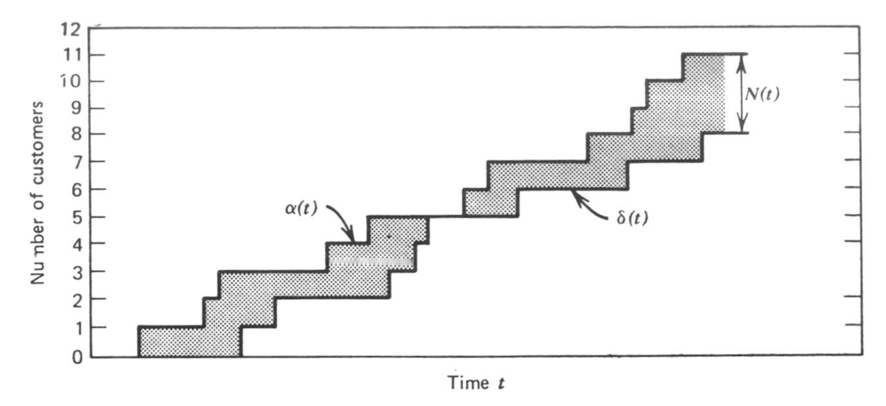

Figure 2.3 Arrivals and departures.

Alternatively, we may position ourselves at the output of the queueing system and count the number of departures that leave; this we denote by

$$\delta(t) \triangleq \text{number of departures in } (0, t) \tag{2.23}$$

Sample functions for these two stochastic processes are shown in Figure 2.3.

Clearly $N(t)$, the number in the system at time t, must be given by

$$N(t) = \alpha(t) - \delta(t)$$

On the other hand, the total area between these two curves up to some point, say t, represents the total time all customers have spent in the system (measured in units of customer-seconds) during the interval $(0, t)$; let us denote this cumulative area by $\gamma(t)$. Moreover, let λ_t be defined as the average arrival rate (customers per second) during the interval $(0, t)$; that is,

$$\lambda_t \triangleq \frac{\alpha(t)}{t} \tag{2.24}$$

We may define T_t as the system time per customer averaged over all customers in the interval $(0, t)$; since $\gamma(t)$ represents the accumulated customer-seconds up to time t, we may divide by the number of arrivals up to that point to obtain

$$T_t = \frac{\gamma(t)}{\alpha(t)}$$

Lastly, let us define $\bar{N}_t$ as the average number of customers in the queueing system during the interval $(0, t)$; this may be obtained by dividing the accumulated number of customer-seconds by the total interval length t

thusly

$$\bar{N}_t = \frac{\gamma(t)}{t}$$

From these last three equations we see

$$\bar{N}_t = \lambda_t T_t$$

Let us now assume that our queueing system is such that the following limits exist as $t \to \infty$:

$$\lambda = \lim_{t \to \infty} \lambda_t$$

$$T = \lim_{t \to \infty} T_t$$

Note that we are using our former definitions for λ and T representing the average customer arrival rate and the average system time, respectively. If these last two limits exist, then so will the limit for $\bar{N}_t$, which we denote by $\bar{N}$ now representing the average number of customers in the system; that is,

$$\bar{N} = \lambda T \qquad \blacksquare \; (2.25)$$

This last is the result we were seeking and is known as *Little's result*. It states that *the average number of customers in a queueing system is equal to the average arrival rate of customers to that system, times the average time spent in that system.** The above proof does not depend upon any specific assumptions regarding the arrival distribution $A(t)$ or the service time distribution $B(x)$; nor does it depend upon the number of servers in the system or upon the particular queueing discipline within the system. This result existed as a "folk theorem" for many years; the first to establish its validity in a formal way was J. D. C. Little [LITT 61] with some later simplifications by W. S. Jewell [JEWE 67], S. Eilon [EILO 69] and S. Stidham [STID 74]. It is important to note that we have not precisely defined the boundary around our queueing system. For example, the box in Figure 2.1 could apply to the entire system composed of queue and server, in which case $\bar{N}$ and T as defined refer to quantities for the entire system; on the other hand, we could have considered the boundary of the queueing system to contain only the queue itself, in which case the relationship would have been

$$\bar{N}_q = \lambda W \qquad \blacksquare \; (2.26)$$

where $\bar{N}_q$ represents the average number of customers in the queue and, as defined earlier, W refers to the average time spent waiting in the queue. As a third possible alternative the queueing system defined could have surrounded

* An intuitive proof of Little's result depends on the observation that an arriving customer should *find* the same average number, $\bar{N}$, in the system as he *leaves behind* upon his departure. This latter quantity is simply the arrival rate λ times his average time in system, T.

only the server (or servers) itself; in this case our equation would have reduced to

$$\bar{N}_s = \lambda \bar{x} \tag{2.27}$$

where $\bar{N}_s$ refers to the average number of customers in the service facility (or facilities) and $\bar{x}$, of course, refers to the average time spent in the service box. Note that it is always true that

$$T = \bar{x} + W \tag{2.28}$$

The queueing system could refer to a specific class of customers, perhaps based on priority or some other attribute of this class, in which case the same relationship would apply. In other words, the average arrival rate of customers to a "queueing system" times the average time spent by customers in that "system" is equal to the average number of customers in the "system," regardless of how we define that "system."

We now discuss a basic parameter ρ, which is commonly referred to as the *utilization factor*. The utilization factor is in a fundamental sense really the ratio R/C, which we introduced in Chapter 1. It is the ratio of the rate at which "work" enters the system to the maximum rate (capacity) at which the system can perform this work; the work an arriving customer brings into the system equals the number of seconds of service he requires. So, in the case of a single-server system, the definition for ρ becomes

$$\begin{aligned} \rho &\triangleq (\text{average arrival rate of customers}) \times (\text{average service time}) \\ &= \lambda \bar{x} \end{aligned} \tag{2.29}$$

This last is true since a single-server system has a maximum capacity for doing work, which equals 1 sec/sec and each arriving customer brings an amount of work equal to $\bar{x}$ sec; since, on the average, λ customers arrive per second, then $\lambda \bar{x}$ sec of work are brought in by customers each second that passes, on the average. In the case of multiple servers (say, m servers) the definition remains the same when one considers the ratio R/C, where now the work capacity of the system is m sec/sec; expressed in terms of system parameters we then have

$$\rho \triangleq \frac{\lambda \bar{x}}{m} \tag{2.30}$$

Equations (2.29) and (2.30) apply in the case when the maximum service rate is independent of the system state; if this is not the case, then a more careful definition must be provided. The rate at which work enters the system is sometimes referred to as the *traffic intensity* of the system and is usually expressed in *Erlangs*; in single-server systems, the utilization factor is equal to the traffic intensity whereas for (m) multiple servers, the traffic intensity equals $m\rho$. So long as $0 \leq \rho < 1$, then ρ may be interpreted as

$$\rho = E[\text{fraction of busy servers}] \tag{2.31}$$

[In the case of an infinite number of servers, the utilization factor ρ plays no important part, and instead we are interested in the *number* of busy servers (and its expectation).]

Indeed, for the system G/G/1 to be stable, it must be that $R < C$, that is, $0 \le \rho < 1$. Occasionally, we permit the case $\rho = 1$ within the range of stability (in particular for the system D/D/1). Stability here once again refers to the fact that limiting distributions for all random variables of interest exist, and that all customers are eventually served. In such a case we may carry out the following simple calculation. We let τ be an arbitrarily long time interval; during this interval we expect (by the law of large numbers) with probability 1 that the number of arrivals will be very nearly equal to $\lambda\tau$. Moreover, let us define p_0 as the probability that the server is idle at some randomly selected time. We may, therefore, say that during the interval τ, the server is busy for $\tau - \tau p_0$ sec, and so with probability 1, the number of customers served during the interval τ is very nearly $(\tau - \tau p_0)/\bar{x}$. We may now equate the number of arrivals to the number served during this interval, which gives, for large τ,

$$\lambda\tau \cong \frac{(\tau - \tau p_0)}{\bar{x}}$$

Thus, as $\tau \to \infty$ we have $\lambda\bar{x} = 1 - p_0$; using Definition (2.29) we finally have the important conclusion for G/G/1

$$\rho = 1 - p_0 \tag{2.32}$$

The interpretation here is that ρ is merely the fraction of time the server is busy; this supports the conclusion in Eq. (2.27) in which $\lambda\bar{x} = \rho$ was shown equal to the average number of customers in the service facility.

This, then, is a rapid look at an overall queueing system in which we have exposed some of the basic stochastic processes, as well as some of the important definitions and notation we will encounter. Moreover, we have established Little's result, which permits us to calculate the average number in the system once we have calculated the average time in the system (or vice versa). Now let us move on to a more careful study of the important stochastic processes in our queueing systems.

2.2*. DEFINITION AND CLASSIFICATION OF STOCHASTIC PROCESSES

At the end of Appendix II a definition is given for a stochastic process, which in essence states that it is a family of random variables $X(t)$ where the

* The reader may choose to skip Sections 2.2, 2.3, and 2.4 at this point and move directly to Section 2.5. He may then refer to this material only as he feels he needs to in the balance of the text.

random variables are "indexed" by the time parameter t. For example, the number of people sitting in a movie theater as a function of time is a stochastic process, as is also the atmospheric pressure in that movie theater as a function of time (at least those functions may be *modeled* as stochastic processes). Often we refer to a stochastic process as a random process. A random process may be thought of as describing the motion of a particle in some space. The classification of a random process depends upon three quantities: the *state space*; the *index (time) parameter*; and the *statistical dependencies* among the random variables $X(t)$ for different values of the index parameter t. Let us discuss each of these in order to provide the general framework for random processes.

First we consider the *state space*. The set of possible values (or states) that $X(t)$ may take on is called its state space. Referring to our analogy with regard to the motion of a particle, if the positions that particle may occupy are finite or countable, then we say we have a *discrete-state* process, often referred to as a *chain*. The state space for a chain is usually the set of integers $\{0, 1, 2, \ldots\}$. On the other hand, if the permitted positions of the particle are over a finite or infinite continuous interval (or set of such intervals), then we say that we have a *continuous-state* process.

Now for the *index (time) parameter*. If the permitted times at which changes in position may take place are finite or countable, then we say we have a *discrete-(time) parameter* process; if these changes in position may occur anywhere within (a set of) finite or infinite intervals on the time axis, then we say we have a *continuous-parameter* process. In the former case we often write X_n rather than $X(t)$. X_n is often referred to as a random or stochastic *sequence* whereas $X(t)$ is often referred to as a random or stochastic *process*.

The truly distinguishing feature of a stochastic process is the relationship of the random variables $X(t)$ or X_n to other members of the same family. As defined in Appendix II, one must specify the complete joint distribution function among the random variables (which we may think of as vectors denoted by the use of boldface) $\mathbf{X} = [X(t_1), X(t_2), \ldots]$, namely,

$$F_{\mathbf{X}}(\mathbf{x}; \mathbf{t}) \triangleq P[X(t_1) \le x_1, \ldots, X(t_n) \le x_n] \tag{2.33}$$

for all $\mathbf{x} = (x_1, x_2, \ldots, x_n)$, $\mathbf{t} = (t_1, t_2, \ldots, t_n)$, and n. As mentioned there, this is a formidable task; fortunately, many interesting stochastic processes permit a simpler description. In any case, it is the function $F_{\mathbf{X}}(\mathbf{x}; \mathbf{t})$ that really describes the dependencies among the random variables of the stochastic process. Below we describe some of the usual types of stochastic processes that are characterized by different kinds of dependency relations among their random variables. We provide this classification in order to give the reader a global view of this field so that he may better understand in which particular regions he is operating as we proceed with our study of queueing theory and its related stochastic processes.

(a) Stationary Processes. As we discuss at the very end of Appendix II, a stochastic process $X(t)$ is said to be stationary if $F_{\mathbf{X}}(\mathbf{x}; \mathbf{t})$ is invariant to shifts in time for all values of its arguments; that is, given any constant τ the following must hold:

$$F_{\mathbf{X}}(\mathbf{x}; \mathbf{t} + \tau) = F_{\mathbf{X}}(\mathbf{x}; \mathbf{t}) \tag{2.34}$$

where the notation $\mathbf{t} + \tau$ is defined as the vector $(t_1 + \tau, t_2 + \tau, \ldots, t_n + \tau)$.

An associated notion, that of *wide-sense stationarity*, is identified with the random process $X(t)$ if merely both the first and second moments are independent of the location on the time axis, that is, if $E[X(t)]$ is independent of t and if $E[X(t)X(t + \tau)]$ depends only upon τ and not upon t. Observe that all stationary processes are wide-sense stationary, but not conversely. The theory of stationary random processes is, as one might expect, simpler than that for nonstationary processes.

(b) Independent Processes. The simplest and most trivial stochastic process to consider is the random sequence in which $\{X_n\}$ forms a set of independent random variables, that is, the joint pdf defined for our stochastic process in Appendix II must factor into the product, thusly

$$\begin{aligned} f_{\mathbf{X}}(\mathbf{x}; \mathbf{t}) &\triangleq f_{X_1 \ldots X_n}(x_1, \ldots, x_n; t_1, \ldots, t_n) \\ &= f_{X_1}(x_1; t_1) \cdots f_{X_n}(x_n; t_n) \end{aligned} \tag{2.35}$$

In this case we are stretching things somewhat by calling such a sequence a random process since there is no structure or dependence among the random variables. In the case of a continuous random process, such an independent process may be defined, and it is commonly referred to as "white noise" (an example is the time derivative of Brownian motion).

(c) Markov Processes. In 1907 A. A. Markov published a paper [MARK 07] in which he defined and investigated the properties of what are now known as Markov processes. In fact, what he created was a simple and highly useful form of dependency among the random variables forming a stochastic process, which we now describe.

A Markov process with a discrete state space is referred to as a Markov chain. The discrete-time Markov chain is the easiest to conceptualize and understand. A set of random variables $\{X_n\}$ forms a Markov chain if the probability that the next value (state) is x_{n+1} depends only upon the current value (state) x_n and not upon any previous values. Thus we have a random sequence in which the dependency extends backwards one unit in time. That is, the way in which the entire past history affects the future of the process is completely summarized in the current value of the process.

In the case of a discrete-time Markov chain the instants when state changes may occur are preordained to be at the integers 0, 1, 2, . . . , n, ... In the case of the continuous-time Markov chain, however, the transitions between states may take place at any instant in time. Thus we are led to consider the random variable that describes how long the process remains in its current (discrete) state before making a transition to some other state. Because the Markov property insists that the past history be completely summarized in the specification of the current state, then we are not free to require that a specification also be given as to how long the process has been in its current state! This imposes a heavy constraint on the distribution of time that the process may remain in a given state. In fact, as we shall see in Eq. (2.85), this state time must be *exponentially* distributed. In a real sense, then, the exponential distribution is a continuous distribution which is "memoryless" (we will discuss this notion at considerable length later in this chapter). Similarly, in the discrete-time Markov chain, the process may remain in the given state for a time that must be *geometrically* distributed; this is the only *discrete* probability mass function that is memoryless. This memoryless property is required of all Markov chains and restricts the generality of the processes one would like to consider.

Expressed analytically the *Markov property* may be written as

$$P[X(t_{n+1}) = x_{n+1} \mid X(t_n) = x_n, X(t_{n-1}) = x_{n-1}, \ldots, X(t_1) = x_1]$$
$$= P[X(t_{n+1}) = x_{n+1} \mid X(t_n) = x_n] \qquad (2.36)$$

where $t_1 < t_2 < \cdots < t_n < t_{n+1}$ and x_i is included in some discrete state space.

The consideration of Markov processes is central to the study of queueing theory and much of this text is devoted to that study. Therefore, a good portion of this chapter deals with discrete-and continuous-time Markov chains.

(d) Birth–death Processes. A very important special class of Markov chains has come to be known as the birth–death process. These may be either discrete-or continuous-time processes in which the defining condition is that state transitions take place between *neighboring* states only. That is, one may choose the set of integers as the discrete state space (with no loss of generality) and then the birth–death process requires that if $X_n = i$, then $X_{n+1} = i - 1$, i, or $i + 1$ and no other. As we shall see, birth–death processes have played a significant role in the development of queueing theory. For the moment, however, let us proceed with our general view of stochastic processes to see how each fits into the general scheme of things.

(e) Semi-Markov Processes. We begin by discussing discrete-time semi-Markov processes. The discrete-time Markov chain had the property that at every unit interval on the time axis the process was required to make a transition from the current state to some other state (possibly back to the same state). The transition probabilities were completely arbitrary; however, the requirement that a transition be made at every unit time (which really came about because of the Markov property) leads to the fact that the time spent in a state is geometrically distributed [as we shall see in Eq. (2.66)]. As mentioned earlier, this imposes a strong restriction on the kinds of processes we may consider. If we wish to relax that restriction, namely, to permit an arbitrary distribution of time the process may remain in a state, then we are led directly into the notion of a discrete-time *semi-Markov process*; specifically, we now permit the times between state transitions to obey an *arbitrary* probability distribution. Note, however, that at the instants of state transitions, the process behaves just like an ordinary Markov chain and, in fact, at those instants we say we have an *imbedded* Markov chain.

Now the definition of a continuous-time semi-Markov process follows directly. Here we permit state transitions at any instant in time. However, as opposed to the Markov process which required an exponentially distributed time in state, we now permit an arbitrary distribution. This then affords us much greater generality, which we are happy to employ in our study of queueing systems. Here, again, the imbedded Markov process is defined at those instants of state transition. Certainly, the class of Markov processes is contained within the class of semi-Markov processes.

(f) Random Walks. In the study of random processes one often encounters a process referred to as a *random walk*. A random walk may be thought of as a particle moving among states in some (say, discrete) state space. What is of interest is to identify the *location* of the particle in that state space. The salient feature of a random walk is that the next position the process occupies is equal to the previous position plus a random variable whose value is drawn independently from an arbitrary distribution; this distribution, however, does not change with the state of the process.* That is, a sequence of random variables $\{S_n\}$ is referred to as a random walk (starting at the origin) if

$$S_n = X_1 + X_2 + \cdots + X_n \qquad n = 1, 2, \ldots \tag{2.37}$$

where $S_0 = 0$ and $X_1, X_2, \ldots$ is a sequence of independent random variables with a common distribution. The index n merely counts the number of state transitions the process goes through; of course, if the instants of these transitions are taken from a discrete set, then we have a discrete-time random

* Except perhaps at some boundary states.

walk, whereas if they are taken from a continuum, then we have a continuous-time random walk. In any case, we assume that the interval between these transitions is distributed in an arbitrary way and so a random walk is a special case of a semi-Markov process.* In the case when the common distribution for X_n is a discrete distribution, then we have a discrete-state random walk; in this case the transition probability p_{ij} of going from state i to state j will depend only upon the difference in indices $j - i$ (which we denote by q_{j-i}).

An example of a continuous-time random walk is that of Brownian motion; in the discrete-time case an example is the total number of heads observed in a sequence of independent coin tosses.

A random walk is occasionally referred to as a process with "independent increments."

(g) Renewal Processes. A renewal process is related† to a random walk. However, the interest is not in following a particle among many states but rather in *counting transitions* that take place as a function of time. That is, we consider the real time axis on which is laid out a sequence of points; the distribution of time between adjacent points is an arbitrary *common* distribution and each point corresponds to an instant of a state transition. We assume that the process begins in state 0 [i.e., $X(0) = 0$] and increases by unity at each transition epoch; that is, $X(t)$ equals the *number* of state transitions that have taken place by t. In this sense it is a special case of a random walk in which $q_1 = 1$ and $q_i = 0$ for $i \neq 1$. We may think of Eq. (2.37) as describing a renewal process in which S_n is the random variable denoting the *time* at which the nth transition takes place. As earlier, the sequence $\{X_n\}$ is a set of independent identically distributed random variables where X_n now represents the time between the $(n - 1)$th and nth transition. One should be careful to distinguish the interpretation of Eq. (2.37) when it applies to renewal processes as here and when it applies to a random walk as earlier. The difference is that here in the renewal process the equation describes the *time* of the nth renewal or transition, whereas in the random walk it describes the *state* of the process and the time between state transitions is some other random variable.

An important example of a renewal process is the set of arrival instants to the G/G/m queue. In this case, X_n is identified with the interarrival time.

* Usually, the distribution of time between intervals is of little concern in a random walk; emphasis is placed on the value (position) S_n after n transitions. Often, it is assumed that this distribution of interval time is memoryless, thereby making the random walk a special case of Markov processes; we are more generous in our definition here and permit an arbitrary distribution.

† It may be considered to be a special case of the random walk as defined in (f) above. A renewal process is occasionally referred to as a *recurrent* process.

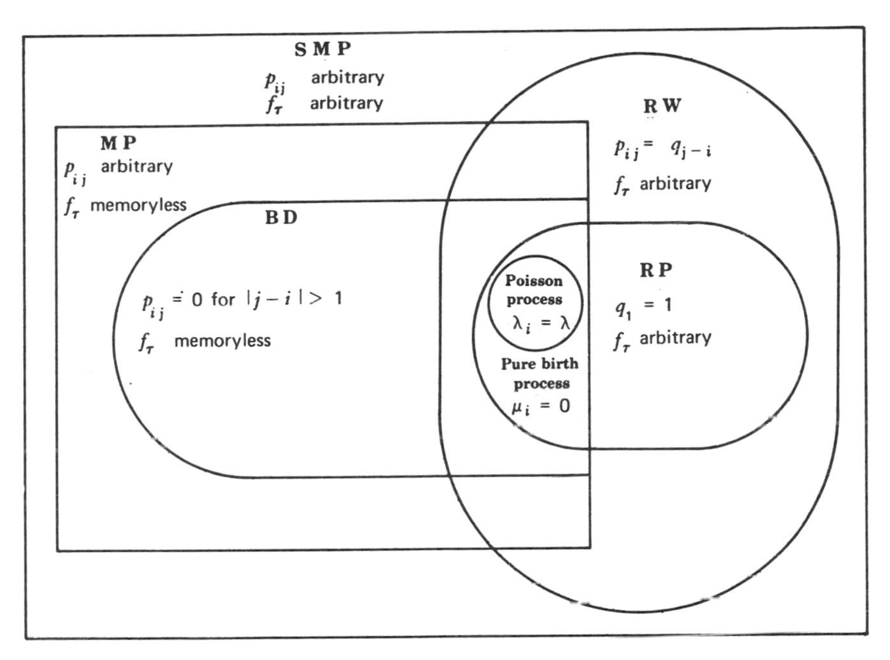

Figure 2.4 Relationships among the interesting random processes. **SMP**: Semi-Markov process; **MP**: Markov process; **RW**: Random walk; **RP**: Renewal process; **BD**: Birth-Death Process.

So there we have it—a self-consistent classification of some interesting stochastic processes. In order to aid the reader in understanding the relationship among Markov processes, semi-Markov processes, and their special cases, we have prepared the diagram of Figure 2.4, which shows this relationship for discrete-state systems. The figure is in the form of a Venn diagram. Moreover, the symbol p_{ij} denotes the probability of making a transition next to state j given that the process is currently in state i. Also, f_τ denotes the distribution of time between transitions; to say that "f_τ is memoryless" implies that if it is a discrete-time process, then f_τ is a geometric distribution, whereas if it is a continuous-time process, then f_τ is an exponential distribution. Furthermore, it is implied that f_τ may be a function both of the current and the next state for the process.

The figure shows that birth–death processes form a subset of Markov processes, which themselves form a subset of the class of semi-Markov processes. Similarly, renewal processes form a subset of random walk processes which also are a subset of semi-Markov processes. Moreover, there are some renewal processes that may also be classified as birth–death processes. Similarly, those Markov processes for which $p_{ij} = q_{j-i}$ (that is, where the transition probabilities depend only upon the difference of the indices) overlap those random walks where f_τ is memoryless. A random walk for which f_τ is memoryless and for which $q_{j-i} = 0$ when $|j - i| > 1$ overlaps the class of birth–death processes. If in addition to this last requirement our random walk has $q_1 = 1$, then we have a process that lies at the intersection of all five of the processes shown in the figure. This is referred to as a "pure birth" process; although f_τ must be memoryless, it may be a distribution which depends upon the state itself. If f_τ is independent of the state (thus giving a constant "birth rate") then we have a process that is figuratively and literally at the "center" of the study of stochastic processes and enjoys the nice properties of each! This very special case is referred to as the *Poisson process* and plays a major role in queueing theory. We shall develop its properties later in this chapter.

So much for the classification of stochastic processes at this point. Let us now elaborate upon the definition and properties of discrete-state Markov processes. This will lead us naturally into some of the elementary queueing systems. Some of the required theory behind the more sophisticated continuous-state Markov processes will be developed later in this work as the need arises. We begin with the simpler discrete-state, discrete-time Markov chains in the next section and follow that with a section on discrete-state, continuous-time Markov chains.

2.3. DISCRETE-TIME MARKOV CHAINS*

As we have said, Markov processes may be used to describe the motion of a particle in some space. We now consider discrete-time Markov chains, which permit the particle to occupy discrete positions and permit transitions between these positions to take place only at discrete times. We present the elements of the theory by carrying along the following contemporary example.

Consider the hippie who hitchhikes from city to city across the country. Let X_n denote the city in which we find our hippie at noon on day n. When he is in some particular city i, he will accept the first ride leaving in the evening from that city. We assume that the travel time between any two cities is negligible. Of course, it is possible that no ride comes along, in which case he will remain in city i until the next evening. Since vehicles heading for various neighboring cities come along in some unpredictable fashion, the hippie's position at some time in the future is clearly a random variable. It turns out that this random variable may properly be described through the use of a Markov chain.

* See footnote on p. 19.

We have the following definition

DEFINITION: The sequence of random variables $X_1, X_2, \ldots$ forms a discrete-time Markov chain if for all n $(n = 1, 2, \ldots)$ and all possible values of the random variables we have (for $i_1 < i_2 < \ldots < i_n$) that

$$P[X_n = j \mid X_1 = i_1, X_2 = i_2, \ldots, X_{n-1} = i_{n-1}] = P[X_n = j \mid X_{n-1} = i_{n-1}] \qquad ■ (2.38)$$

In terms of our example, this definition merely states that the city next to be visited by the hippie depends only upon the city in which he is currently located and not upon all the previous cities he has visited. In this sense the memory of the random process, or Markov chain, goes back only to the most recent position of the particle (hippie). When $X_n = j$ (the hippie is in city j on day n), then the system is said to be in state E_j at time n (or at the nth step). To get our hippie started on day 0 we begin with some initial probability distribution $P[X_0 = j]$. The expression on the right side of Eq. (2.38) is referred to as the (one-step) *transition probability* and gives the conditional probability of making a transition from state $E_{i_{n-1}}$ at step $n - 1$ to state E_j at the nth step in the process. It is clear that if we are given the initial state probability distribution and the transition probabilities, then we can uniquely find the probability of being in various states at time n [see Eqs. (2.55) and (2.56) below].

If it turns out that the transition probabilities are independent of n, then we have what is referred to as a *homogeneous* Markov chain and in that case we make the further definition

$$p_{ij} \triangleq P[X_n = j \mid X_{n-1} = i] \qquad (2.39)$$

which gives the probability of going to state E_j on the next step, given that we are currently at state i. What follows refers to homogeneous Markov chains only. These chains are such that their transition probabilities are stationary with time*; therefore, given the current city or state (pun) the probability of various states m steps into the future depends only upon m and not upon the current time; it is expedient to define the m-step transition probabilities as

$$p_{ij}^{(m)} \triangleq P[X_{n+m} = j \mid X_n = i] \qquad ■ (2.40)$$

From the Markov property given in Eq. (2.38) it is easy to establish the following recursive formula for calculating $p_{ij}^{(m)}$:

$$p_{ij}^{(m)} = \sum_k p_{ik}^{(m-1)} p_{kj} \qquad m = 2, 3, \ldots \qquad (2.41)$$

This equation merely says that if we are to travel from E_i to E_j in m steps,

* Note that although this is a Markov process with stationary transitions, it need *not* be a stationary random process.

then we must do so by first traveling from E_i to *some* state E_k in $m-1$ steps and then from E_k to E_j in one more step; the probability of these last two independent events (remember this is a Markov chain) is the product of the probability of each and if we sum this product over all possible intermediate states E_k, we arrive at $p_{ij}^{(m)}$.

We say that a Markov chain is *irreducible** if every state can be reached from every other state; that is, for each pair of states (E_i and E_j) there exists an integer m_0 (which may depend upon i and j) such that

$$p_{ij}^{(m_0)} > 0$$

Further, let A be the set of all states in a Markov chain. Then a subset of states A_1 is said to be *closed* if no one-step transition is possible from any state in A_1 to any state in A_1^c (the complement of the set A_1). If A_1 consists of a single state, say E_i, then it is called an *absorbing* state; a necessary and sufficient condition for E_i to be an absorbing state is $p_{ii} = 1$. If A is closed and does not contain any proper subset which is closed, then we have an *irreducible* Markov chain as defined above. On the other hand, if A contains proper subsets that are closed, then the chain is said to be *reducible*. If a closed subset of a reducible Markov chain contains no closed subsets of itself, then it is referred to as an *irreducible sub-Markov chain*; these subchains may be studied independently of the other states.

It may be that our hippie prefers not to return to a previously visited city. However, due to his mode of travel this may well happen, and it is important for us to define this quantity. Accordingly, let

$$f_j^{(n)} \triangleq P\,[\text{first return to } E_j \text{ occurs } n \text{ steps after leaving } E_j]$$

It is then clear that the probability of our hippie *ever* returning to city j is given by

$$f_j = \sum_{n=1}^{\infty} f_j^{(n)} = P[\text{ever returning to } E_j]$$

It is now possible to classify states of a Markov chain according to the value obtained for f_j. In particular, if $f_j = 1$ then state E_j is said to be *recurrent*; if on the other hand, $f_j < 1$, then state E_j is said to be *transient*. Furthermore, if the only possible steps at which our hippie can return to state E_j are $\gamma, 2\gamma, 3\gamma, \ldots$ (where $\gamma > 1$ and is the largest such integer), then state E_j is said to be *periodic* with period γ; if $\gamma = 1$, then E_j is *aperiodic*.

Considering states for which $f_j = 1$, we may then define the *mean recurrence time* of E_j as

$$M_j \triangleq \sum_{n=1}^{\infty} n f_j^{(n)} \tag{2.42}$$

* Many of the interesting Markov chains which one encounters in queueing theory are irreducible.

This is merely the average time to return to E_j. With this we may then classify states even further. In particular, if $M_j = \infty$, then E_j is said to be *recurrent null*, whereas if $M_j < \infty$, then E_j is said to be *recurrent nonnull*. Let us define $\pi_j^{(n)}$ to be the probability of finding the system in state E_j at the nth step, that is,

$$\pi_j^{(n)} \triangleq P[X_n = j] \tag{2.43}$$

We may now state (without proof) two important theorems. The first comments on the set of states for an irreducible Markov chain.

Theorem 1 *The states of an irreducible Markov chain are either all transient or all recurrent nonnull or all recurrent null. If periodic, then all states have the same period* γ.

Assuming that our hippie wanders forever, he will pass through the various cities of the nation many times, and we inquire as to whether or not there exists a *stationary* probability distribution $\{\pi_j\}$ describing his probability of being in city j at some time arbitrarily far into the future. [A probability distribution P_j is said to be a *stationary distribution* if when we choose it for our initial state distribution (that is, $\pi_j^{(0)} = P_j$) then for all n we will have $\pi_j^{(n)} = P_j$.] Solving for $\{\pi_j\}$ is a most important part of the analysis of Markov chains. Our second theorem addresses itself to this question.

Theorem 2 *In an irreducible and aperiodic homogeneous Markov chain the limiting probabilities*

$$\pi_j = \lim_{n\to\infty} \pi_j^{(n)} \tag{2.44}$$

always exist and are independent of the initial state probability distribution. Moreover, either

(*a*) *all states are transient or all states are recurrent null in which cases* $\pi_j = 0$ *for all* j *and there exists no* stationary *distribution, or*

(*b*) *all states are recurrent nonnull and then* $\pi_j > 0$ *for all* j, *in which case the set* $\{\pi_j\}$ *is a stationary probability distribution and*

$$\pi_j = \frac{1}{M_j} \tag{2.45}$$

In this case the quantities π_j *are* uniquely *determined through the following equations*

$$1 = \sum_i \pi_i \tag{2.46}$$

$$\pi_j = \sum_i \pi_i p_{ij} \tag{2.47}$$

We now introduce the notion of *ergodicity*. A state E_j is said to be ergodic if it is aperiodic, recurrent, and nonnull; that is, if $f_j = 1$, $M_j < \infty$, and $\gamma = 1$. If all states of a Markov chain are ergodic, then the Markov chain itself is said to be ergodic. Moreover, a Markov chain is said to be ergodic if the probability distribution $\{\pi_j^{(n)}\}$ as a function of n always converges to a limiting stationary distribution $\{\pi_j\}$, which is independent of the initial state distribution. It is easy to show that all states of a *finite** aperiodic irreducible Markov chain are ergodic. Moreover, among Foster's criteria [FELL 66] it can be shown that an irreducible and aperiodic Markov chain is ergodic if the set of linear equations given in Eq. (2.47) has a nonnull solution for which $\sum_j |\pi_j| < \infty$. The limiting probabilities $\{\pi_j\}$, of an ergodic Markov chain are often referred to as the *equilibrium* probabilities in the sense that the effect of the initial state distribution $\pi_j^{(0)}$ has disappeared.

By way of example, let's place the hippie in our fictitious land of Hatafla, and let us consider the network given in Figure 1.1 of Chapter 1. In order to simplify this example we will assume that the cities of Nonabel, Cadabra, and Oriac have been bombed out and that the resultant road network is as given in Figure 2.5. In this figure the ordered links represent permissible directions of road travel; the numbers on these links represent the probability (p_{ij}) that the hippie will be picked up by a car traveling over that road, given that he is hitchhiking from the city where the arrow emanates. Note that from the city of Sucsamad our hippie has probability 1/2 of remaining in that city until the next day. Such a diagram is referred to as a *state-transition* diagram. The parenthetical numbers following the cities will henceforth be used instead of the city names.

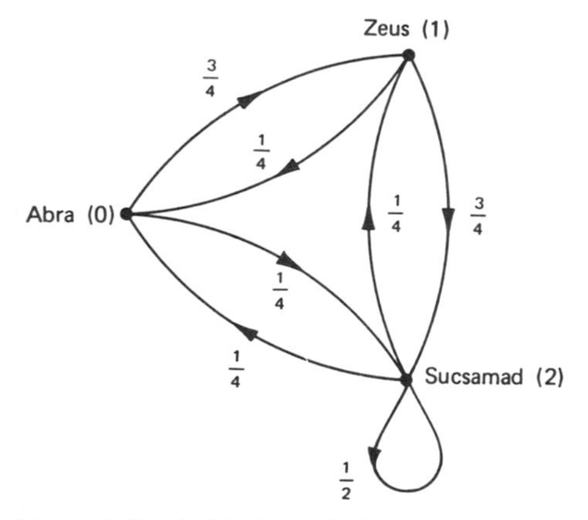

Figure 2.5 A Markov chain.

* A finite Markov chain is one with a finite number of states. If an irreducible Markov chain is of type (a) in Theorem 2 (i.e., recurrent null or transient) then it cannot be finite.

In order to continue our example we now define, in general, the *transition probability matrix* **P** as consisting of elements p_{ij}, that is,

$$\mathbf{P} = [p_{ij}] \tag{2.48}$$

If we further define the probability vector $\boldsymbol{\pi}$ as

$$\boldsymbol{\pi} = [\pi_0, \pi_1, \pi_2, \ldots] \tag{2.49}$$

then we may rewrite the set of relations in Eq. (2.47) as

$$\boldsymbol{\pi} = \boldsymbol{\pi}\mathbf{P} \tag{2.50}$$

For our example shown in Figure 2.5 we have

$$\mathbf{P} = \begin{bmatrix} 0 & \frac{3}{4} & \frac{1}{4} \\ \frac{1}{4} & 0 & \frac{3}{4} \\ \frac{1}{4} & \frac{1}{4} & \frac{1}{2} \end{bmatrix}$$

and so we may solve Eq. (2.50) by considering the three equations derivable from it, that is,

$$\begin{aligned} \pi_0 &= 0\pi_0 + \frac{1}{4}\pi_1 + \frac{1}{4}\pi_2 \\ \pi_1 &= \frac{3}{4}\pi_0 + 0\pi_1 + \frac{1}{4}\pi_2 \\ \pi_2 &= \frac{1}{4}\pi_0 + \frac{3}{4}\pi_1 + \frac{1}{2}\pi_2 \end{aligned} \tag{2.51}$$

Note from Eq. (2.51) that the first of these three equations equals the negative sum of the second and third, indicating that there is a linear dependence among them. It *always* will be the case that one of the equations will be linearly dependent on the others, and it is therefore necessary to introduce the additional conservation relationship as given in Eq. (2.46) in order to solve the system. In our example we then require

$$1 = \pi_0 + \pi_1 + \pi_2 \tag{2.52}$$

Thus the solution is obtained by simultaneously solving any two of the equations given by Eq. (2.51) along with Eq. (2.52). Solving we obtain

$$\pi_0 = \frac{1}{5} = 0.20$$
$$\pi_1 = \frac{7}{25} = 0.28 \qquad (2.53)$$
$$\pi_2 = \frac{13}{25} = 0.52$$

This gives us the equilibrium (stationary) state probabilities. It is clear that this is an ergodic Markov chain (it is finite and irreducible).

Often we are interested in the transient behavior of the system. The transient behavior involves solving for $\pi_j^{(n)}$, the probability of finding our hippie in city j at time n. We also define the probability vector at time n as

$$\boldsymbol{\pi}^{(n)} \triangleq [\pi_0^{(n)}, \pi_1^{(n)}, \pi_2^{(n)}, \ldots] \qquad \blacksquare\,(2.54)$$

Now using the definition of transition probability and making use of Definition (2.48) we have a method for calculating $\boldsymbol{\pi}^{(1)}$ expressible in terms of $\mathbf{P}$ and the initial state distribution $\boldsymbol{\pi}^{(0)}$. That is,

$$\boldsymbol{\pi}^{(1)} = \boldsymbol{\pi}^{(0)}\mathbf{P}$$

Similarly, we may calculate the state probabilities at the second step by

$$\begin{aligned} \boldsymbol{\pi}^{(2)} &= \boldsymbol{\pi}^{(1)}\mathbf{P} \\ &= [\boldsymbol{\pi}^{(0)}\mathbf{P}]\mathbf{P} \\ &= \boldsymbol{\pi}^{(0)}\mathbf{P}^2 \end{aligned}$$

From this last we can then generalize to the result

$$\boldsymbol{\pi}^{(n)} = \boldsymbol{\pi}^{(n-1)}\mathbf{P} \qquad n = 1, 2, \ldots \qquad \blacksquare\,(2.55)$$

which may be solved recursively to obtain

$$\boldsymbol{\pi}^{(n)} = \boldsymbol{\pi}^{(0)}\mathbf{P}^n \qquad n = 1, 2, \ldots \qquad \blacksquare\,(2.56)$$

Equation (2.55) gives the general method for calculating the state probabilities n steps into a process, given a transition probability matrix $\mathbf{P}$ and an initial state vector $\boldsymbol{\pi}^{(0)}$. From our earlier definitions, we have the stationary probability vector

$$\boldsymbol{\pi} = \lim_{n\to\infty} \boldsymbol{\pi}^{(n)}$$

assuming the limit exists. (From Theorem 2, we know that this will be the case if we have an irreducible aperiodic homogeneous Markov chain.)

Then, from Eq. (2.55) we find

$$\lim_{n\to\infty} \boldsymbol{\pi}^{(n)} = \lim_{n\to\infty} \pi^{(n-1)}\mathbf{P}$$

and so

$$\boldsymbol{\pi} = \boldsymbol{\pi}\mathbf{P}$$

which is Eq. (2.50) again. Note that the solution for $\boldsymbol{\pi}$ is independent of the initial state vector. Applying this to our example, let us assume that our hippie begins in the city of Abra at time 0 with probability 1, that is

$$\boldsymbol{\pi}^{(0)} = [1, 0, 0] \tag{2.57}$$

From this we may calculate the sequence of values $\boldsymbol{\pi}^{(n)}$ and these are given in the chart below. The limiting value $\boldsymbol{\pi}$ as given in Eq. (2.53) is also entered in this chart.

n	0	1	2	3	4	$\cdots$	∞
$\pi_0^{(n)}$	1	0	0.250	0.187	0.203		0.20
$\pi_1^{(n)}$	0	0.75	0.062	0.359	0.254		0.28
$\pi_2^{(n)}$	0	0.25	0.688	0.454	0.543		0.52

We may alternatively have chosen to assume that the hippie begins in the city of Zeus with probability 1, which would give rise to the initial state vector

$$\boldsymbol{\pi}^{(0)} = [0, 1, 0] \tag{2.58}$$

and which results in the following table:

n	0	1	2	3	4	$\cdots$	∞
$\pi_0^{(n)}$	0	0.25	0.187	0.203	0.199		0.20
$\pi_1^{(n)}$	1	0	0.375	0.250	0.289		0.28
$\pi_2^{(n)}$	0	0.75	0.438	0.547	0.512		0.52

Similarly, beginning in the city of Sucsamad we find

$$\boldsymbol{\pi}^{(0)} = [0, 0, 1] \tag{2.59}$$

n	0	1	2	3	4	$\cdots$	∞
$\pi_0^{(n)}$	0	0.25	0.187	0.203	0.199		0.20
$\pi_1^{(n)}$	0	0.25	0.313	0.266	0.285		0.28
$\pi_2^{(n)}$	1	0.50	0.500	0.531	0.516		0.52

From these calculations we may make a number of observations. First, we see that after only four steps the quantities $\pi_i^{(n)}$ for a given value of i are almost identical regardless of the city in which we began. The rapidity with which these quantities converge, as we shall soon see, depends upon the eigenvalues of **P**. In all cases, however, we observe that the limiting values at infinity are rapidly approached and, as stated earlier, are independent of the initial position of the particle.

In order to get a better physical feel for what is occurring, it is instructive to follow the probabilities for the various states of the Markov chain as time evolves. To this end we introduce the notion of *baricentric coordinates*, which are extremely useful in portraying probability vectors. Consider a probability vector with N components (i.e., a Markov process with N states in our case) and a tetrahedron in $N - 1$ dimensions. In our example $N = 3$ and so our tetrahedron becomes an equilateral triangle in two dimensions. In general, we let the height of this tetrahedron be unity. Any probability vector $\boldsymbol{\pi}^{(n)}$ may be represented as a point in this $N - 1$ space by identifying each component of that probability vector with a distance from one face of the tetrahedron. That is, we measure from face j a distance equal to the probability associated with that component $\pi_j^{(n)}$; if we do this for each face and therefore for each component, we will specify one point within the tetrahedron and that point correctly identifies our probability vector. Each unique probability vector will map into a unique point in this space, and it is easy to determine the probability measure from its location in that space. In our example we may plot the three initial state vectors as given in Eqs. (2.57)–(2.59) as shown in Figure 2.6. The numbers in parentheses represent which probability components are to be measured from the face associated with those numbers. The initial state vector corresponding to Eq. (2.59), for

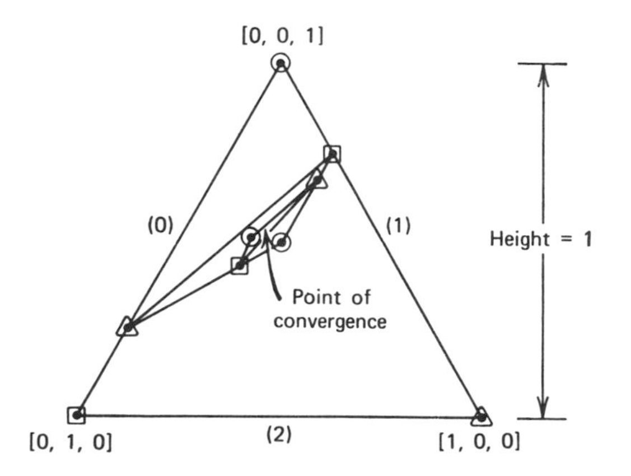

Figure 2.6 Representation of the convergence of a Markov chain.

example, will appear at the apex of the triangle and is indicated as such. In our earlier calculations we followed the progress of our probability vectors beginning with three initial state probability vectors. Let us now follow these paths simultaneously and observe, for example, that the vector [0, 0, 1], following Eq. (2.59), moves to the position [0.25, 0.25, 0.5]; the vector [0, 1, 0] moves to the position [0.25, 0, 0.75], and the vector [1, 0, 0] moves to the position [0, 0.75, 0.25]. Now it is clear that had we started with an initial state vector anywhere within the original equilateral triangle, that point would have been mapped into the *interior* of the smaller triangle, which now joins the three points just referred to and which represent possible positions of the original state vectors. We note from the figure that this new triangle is a shrunken version of the original triangle. If we now continue to map these three points into the second step of the process as given by the three charts above, we find an even smaller triangle interior to both the first and the second triangles, and this region represents the possible locations of any original state vector after *two* steps into the process. Clearly, this shrinking will continue until we reach a convergent point. This convergent point will in the limit be exactly that given by Eq. (2.53)! Thus we can see the way in which the possible positions of our probability vectors move around in our space.

The calculation of the transient response $\boldsymbol{\pi}^{(n)}$ from Eqs. (2.55) or (2.56) is extremely tedious if we desire more than just the first few terms. In order to obtain the general solution, we often resort to transform methods. Below we demonstrate this method in general and then apply it to our hippie hitchhiking example. This will give us an opportunity to apply the z-transform calculations that we have introduced in Appendix I.* Our point of departure is Eq. (2.55). That equation is a difference equation among vectors. The fact that it is a difference equation suggests the use of z-transforms as in Appendix I, and so we naturally define the following *vector* transform (the vectors in no way interfere with our transform approach except that we must be careful when taking inverses):

$$\boldsymbol{\Pi}(z) \triangleq \sum_{n=0}^{\infty} \boldsymbol{\pi}^{(n)} z^n \tag{2.60}$$

This transform will certainly exist in the unit disk, that is, $|z| \leq 1$. We now apply the transform method to Eq. (2.55) over its range of application ($n = 1, 2, \ldots,$); this we do by first multiplying that equation by z^n and then summing from 1 to infinity, thus

$$\sum_{n=1}^{\infty} \boldsymbol{\pi}^{(n)} z^n = \sum_{n=1}^{\infty} \boldsymbol{\pi}^{(n-1)} \mathbf{P} z^n$$

* The steps involved in applying this method are summarized on pp. 74–5 of this chapter.

We have now reduced our infinite set of difference equations to a single algebraic equation. Following through with our method we must now try to identify our vector transform $\mathbf{\Pi}(z)$. Our left-hand side contains all but the initial term of this transform and so we have

$$\mathbf{\Pi}(z) - \boldsymbol{\pi}^{(0)} = z\left(\sum_{n=1}^{\infty} \boldsymbol{\pi}^{(n-1)} z^{n-1}\right)\mathbf{P}$$

The parenthetical term on the right-hand side of this last equation is recognized as $\mathbf{\Pi}(z)$ simply by changing the index of summation. Thus we find

$$\mathbf{\Pi}(z) - \boldsymbol{\pi}^{(0)} = z\mathbf{\Pi}(z)\mathbf{P}$$

z is merely a scalar in this vector equation and may be moved freely across vectors and matrices. Solving this matrix equation we immediately come up with a general solution for our vector transform:

$$\mathbf{\Pi}(z) = \boldsymbol{\pi}^{(0)}[\mathbf{I} - z\mathbf{P}]^{-1} \tag{2.61}$$

where $\mathbf{I}$ is the identity matrix and the (-1) notation implies the matrix inverse. If we can invert this equation, we will have, by the uniqueness of transforms, the transient solution; that is, using the double-headed, double-barred arrow notation as in Appendix I to denote transform pairs, we have

$$\mathbf{\Pi}(z) \Leftrightarrow \boldsymbol{\pi}^{(n)} = \boldsymbol{\pi}^{(0)}\mathbf{P}^n \tag{2.62}$$

In this last we have taken advantage of Eq. (2.56). Comparing Eqs. (2.61) and (2.62) we have the obvious transform pair

$$[\mathbf{I} - z\mathbf{P}]^{-1} \Leftrightarrow \mathbf{P}^n \tag{2.63}$$

Of course $\mathbf{P}^n$ is precisely what we are looking for in order to obtain our transient solution since this will directly give us $\boldsymbol{\pi}^{(n)}$ from Eq. (2.56). All that is required, therefore, is that we form the matrix inverse indicated in Eq. (2.63). In general this becomes a rather complex task when the number of states in our Markov chain is at all large. Nevertheless, this is one formal procedure for carrying out the transient analysis.

Let us apply these techniques to our hippie hitchhiking example. Recall that the transition probability matrix $\mathbf{P}$ was given by

$$\mathbf{P} = \begin{bmatrix} 0 & \frac{3}{4} & \frac{1}{4} \\ \frac{1}{4} & 0 & \frac{3}{4} \\ \frac{1}{4} & \frac{1}{4} & \frac{1}{2} \end{bmatrix}$$

First we must form

$$\mathbf{I} - z\mathbf{P} = \begin{bmatrix} 1 & -\frac{3}{4}z & -\frac{1}{4}z \\ -\frac{1}{4}z & 1 & -\frac{3}{4}z \\ -\frac{1}{4}z & -\frac{1}{4}z & 1 - \frac{1}{2}z \end{bmatrix}$$

Next, in order to find the inverse of this matrix we must form its determinant thus:

$$\det(\mathbf{I} - z\mathbf{P}) = 1 - \frac{1}{2}z - \frac{7}{16}z^2 - \frac{1}{16}z^3$$

which factors nicely into

$$\det(\mathbf{I} - z\mathbf{P}) = (1 - z)\left(1 + \frac{1}{4}z\right)^2$$

It is easy to show that $z = 1$ is always a root of the determinant for an irreducible Markov chain (and, as we shall see, gives rise to our equilibrium solution). We now proceed with the calculation of the matrix inverse using the usual methods to arrive at

$$[\mathbf{I} - z\mathbf{P}]^{-1} = \frac{1}{(1 - z)[1 + (1/4)z]^2} \times \begin{bmatrix} 1 - \frac{1}{2}z - \frac{3}{16}z^2 & \frac{3}{4}z - \frac{5}{16}z^2 & \frac{1}{4}z + \frac{9}{16}z^2 \\ \frac{1}{4}z + \frac{1}{16}z^2 & 1 - \frac{1}{2}z - \frac{1}{16}z^2 & \frac{3}{4}z + \frac{1}{16}z^2 \\ \frac{1}{4}z + \frac{1}{16}z^2 & \frac{1}{4}z + \frac{3}{16}z^2 & 1 - \frac{3}{16}z^2 \end{bmatrix}$$

Having found the matrix inverse, we are now faced with finding the inverse transform of this matrix which will yield $\mathbf{P}^n$. This we do as usual by carrying out a partial fraction expansion (see Appendix I). The fact that we have a matrix presents no problem; we merely note that each element in the matrix is itself a rational function of z which must be expanded in partial fractions term by term. (This task is simplified if the matrix is written as the sum of three matrices: a constant matrix; a constant matrix times z; and a constant matrix times z^2.) Since we have three roots in the denominator of our rational functions we expect three terms in our partial fraction expansion. Carrying out this expansion and separating the three terms we find

$$[\mathbf{I} - z\mathbf{P}]^{-1} = \frac{1/25}{1-z}\begin{bmatrix} 5 & 7 & 13 \\ 5 & 7 & 13 \\ 5 & 7 & 13 \end{bmatrix} + \frac{1/5}{(1+z/4)^2}\begin{bmatrix} 0 & -8 & 8 \\ 0 & 2 & -2 \\ 0 & 2 & -2 \end{bmatrix}$$

$$+ \frac{1/25}{1+z/4}\begin{bmatrix} 20 & 33 & -53 \\ -5 & 8 & -3 \\ -5 & -17 & 22 \end{bmatrix} \qquad (2.64)$$

We observe immediately from this expansion that the matrix associated with the root $(1 - z)$ gives precisely the equilibrium solution we found by direct methods [see Eq. (2.53)]; the fact that each row of this matrix is identical reflects the fact that the equilibrium solution is independent of the initial state. The other matrices associated with roots greater than unity in absolute value will always be what are known as differential matrices (each of whose rows must sum to zero). Inverting on z we finally obtain (by our tables in Appendix I)

$$\mathbf{P}^n = \frac{1}{25}\begin{bmatrix} 5 & 7 & 13 \\ 5 & 7 & 13 \\ 5 & 7 & 13 \end{bmatrix} + \frac{1}{5}(n+1)\left(-\frac{1}{4}\right)^n\begin{bmatrix} 0 & -8 & 8 \\ 0 & 2 & -2 \\ 0 & 2 & -2 \end{bmatrix}$$

$$+ \frac{1}{25}\left(-\frac{1}{4}\right)^n\begin{bmatrix} 20 & 33 & -53 \\ -5 & 8 & -3 \\ -5 & -17 & 22 \end{bmatrix} \qquad n = 0, 1, 2, \ldots \quad (2.65)$$

This is then the complete solution since application of Eq. (2.56) directly gives $\boldsymbol{\pi}^{(n)}$, which is the transient solution we were seeking. Note that for $n = 0$ we obtain the identity matrix whereas for $n = 1$ we must, of course, obtain the transition probability matrix $\mathbf{P}$. Furthermore, we see that in this case we have two transient matrices, which decay in the limit leaving only the constant matrix representing our equilibrium solution. When we think about the decay of the transient, we are reminded of the shrinking triangles in Figure 2.6. Since the transients decay at a rate related to the characteristic values (one over the zeros of the determinant) we therefore expect the permitted positions in Figure 2.6 to decay with n in a similar fashion. In fact, it can be shown that these triangles shrink by a constant factor each time n increases by 1. This shrinkage factor for any Markov process can be shown to be equal to the absolute value of the product of the characteristic values of its transition probability matrix; in our example we have characteristic values equal to 1, 1/4, 1/4. Their product is 1/16 and this indeed is the factor by which the area of our triangles decreases each time n is increased.

This method of transform analysis is extended in two excellent volumes by Howard [HOWA 71] in which he treats such problems and discusses additional approaches such as the flow-graph method of analysis.

Throughout this discussion of discrete-time Markov chains we have not explicitly addressed ourselves to the memoryless property* of the time that the system spends in a given state. Let us now prove that the number of time units that the system spends in the same state is *geometrically* distributed; the geometric distribution is the unique discrete memoryless distribution. Let us assume the system has just entered state E_i. It will remain in this state at the next step with probability p_{ii}; similarly, it will leave this state at the next step with probability $1 - p_{ii}$. If indeed it does remain in this state at the next step, then the probability of its remaining for an additional step is again p_{ii}, and similarly the conditional probability of its leaving at this second step is given by $1 - p_{ii}$. And so it goes. Furthermore, due to the Markov property the fact that it has remained in a given state for a known number of steps in no way affects the probability that it leaves at the next step. Since these probabilities are independent, we may then write

$$P[\text{system remains in } E_i \text{ for exactly } m \text{ additional steps given that it has just entered } E_i] = (1 - p_{ii})p_{ii}^{\ m} \quad (2.66)$$

This, of course, is the geometric distribution as we claimed. A similar argument will be given later for the continuous-time Markov chain.

So far we have concerned ourselves principally with homogeneous Markov processes. Recall that a homogeneous Markov chain is one for which the transition probabilities are independent of time. Among the quantities we were able to calculate was the m-step transition probability $p_{ij}^{(m)}$, which gave the probability of passing from state E_i to state E_j in m steps; the recursive formula for this calculation was given in Eq. (2.41). We now wish to take a more general point of view and permit the transition probabilities to depend upon time. We intend to derive a relationship not unlike Eq. (2.41), which will form our point of departure for many further developments in the application of Markov processes to queueing problems. For the time being we continue to restrict ourselves to discrete-time, discrete-state Markov chains.

Generalizing the homogeneous definition for the multistep transition probabilities given in Eq. (2.40) we now define

$$p_{ij}(m, n) \triangleq P[X_n = j \mid X_m = i] \quad \blacksquare\ (2.67)$$

which gives the probability that the system will be in state E_j at step n, given

* The memoryless property is discussed in some detail later.

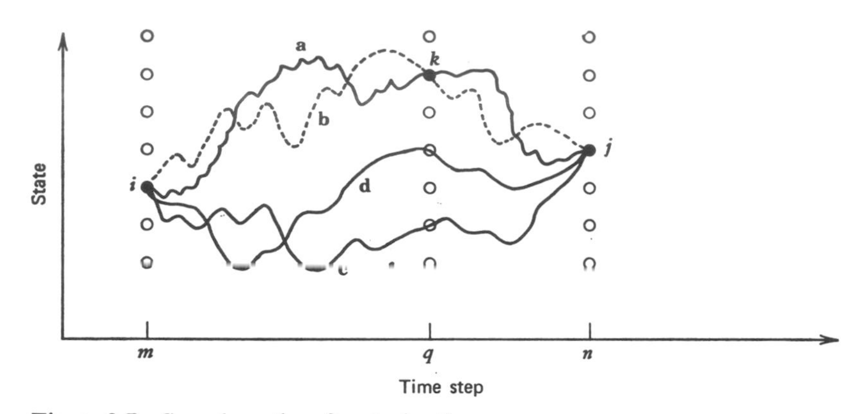

Figure 2.7 Sample paths of a stochastic process.

that it was in state E_i at step m, where $n \geq m$. As discussed in the homogeneous case, it certainly must be true that if our process goes from state E_i at time m to state E_j at time n, then at some intermediate time q it must have passed through some state E_k. This is depicted in Figure 2.7. In this figure we have shown four sample paths of a stochastic process as it moves from state E_i at time m to state E_j at time n. We have plotted the state of the process vertically and the discrete time steps horizontally. (We take the liberty of drawing continuous curves rather than a sequence of points for convenience.) Note that sample paths a and b both pass through state E_k at time q, whereas sample paths c and d pass through other intermediate states at time q. We are certain of one thing only, namely, that we must pass through *some* intermediate state at time q. We may then express $p_{ij}(m, n)$ as the sum of probabilities for all of these (mutually exclusive) intermediate states, that is,

$$p_{ij}(m, n) = \sum_k P[X_n = j, X_q = k \mid X_m = i] \tag{2.68}$$

for $m \leq q \leq n$. This last equation must hold for any stochastic process (not necessarily Markovian) since we are considering all mutually exclusive and exhaustive possibilities. From the definition of conditional probability we may rewrite this last equation as

$$p_{ij}(m, n) = \sum_k P[X_q = k \mid X_m = i]P[X_n = j \mid X_m = i, X_q = k] \tag{2.69}$$

Now we invoke the Markov property and observe that

$$P[X_n = j \mid X_m = i, X_q = k] = P[X_n = j \mid X_q = k]$$

Applying this to Eq. (2.69) and making use of our definition in Eq. (2.67) we finally arrive at

$$p_{ij}(m, n) = \sum_k p_{ik}(m, q)p_{kj}(q, n) \quad ■ \tag{2.70}$$

for $m \leq q \leq n$. Equation (2.70) is known as the *Chapman–Kolmogorov* equation for discrete-time Markov processes. Were this a homogeneous Markov chain then from the definition in Eq. (2.40) we would have the relationship $p_{ij}(m,n) = p_{ij}^{(n-m)}$ and in the case when $n = q + 1$ our Chapman–Kolmogorov equation would reduce to our earlier Eq. (2.41). The Chapman–Kolmogorov equation states that we can partition any $n - m$ step transition probability into the sum of products of a $q - m$ and an $n - q$ step transition probability to and from the intermediate states that might have been occupied at some time q within the interval. Indeed we are permitted to choose any partitioning we wish, and we will take advantage of this shortly.

It is convenient at this point to write the Chapman–Kolmogorov equation in matrix form. We have in the past defined $\mathbf{P}$ as the matrix containing the elements p_{ij} in the case of a homogeneous Markov chain. Since these quantities may now depend upon time, we define $\mathbf{P}(n)$ to be the one-step transition probability matrix at time n, that is,

$$\mathbf{P}(n) \triangleq [p_{ij}(n, n + 1)] \qquad (2.71)$$

Of course, $\mathbf{P}(n) = \mathbf{P}$ if the chain is homogeneous. Also, for the homogeneous case we found that the n-step transition probability matrix was equal to $\mathbf{P}^n$. In the nonhomogeneous case we must make a new definition and for this purpose we use the symbol $\mathbf{H}(m, n)$ to denote the following multistep transition probability matrix:

$$\mathbf{H}(m, n) \triangleq [p_{ij}(m, n)] \qquad (2.72)$$

Note that $\mathbf{H}(n, n + 1) = \mathbf{P}(n)$ and that in the homogeneous case $\mathbf{H}(m, m + n) = \mathbf{P}^n$. With these definitions we may then rewrite the Chapman-Kolmogorov equation in matrix form as

$$\mathbf{H}(m, n) = \mathbf{H}(m, q)\mathbf{H}(q, n) \qquad (2.73)$$

for $m \leq q \leq n$. To complete the definition we require that $\mathbf{H}(n, n) = \mathbf{I}$, where $\mathbf{I}$ is the identity matrix. All of the matrices we are considering are square matrices with dimensionality equal to the number of states of the Markov chain. A solution to Eq. (2.73) will consist of expressing $\mathbf{H}(m, n)$ in terms of the given matrices $\mathbf{P}(n)$.

As mentioned above, we are free to choose q to lie anywhere in the interval between m and n. Let us begin by choosing $q = n - 1$. In this case Eq. (2.70) becomes

$$p_{ij}(m, n) = \sum_k p_{ik}(m, n - 1)p_{kj}(n - 1, n) \qquad (2.74)$$

which in matrix form may be written as

$$\mathbf{H}(m, n) = \mathbf{H}(m, n - 1)\mathbf{P}(n - 1) \qquad (2.75)$$

Equations (2.74) and (2.75) are known as the *forward* Chapman–Kolmogorov equations for discrete-time Markov chains since they are written at the forward (most recent time) end of the interval. On the other hand, we could have chosen $q = m + 1$, in which case we obtain

$$p_{ij}(m, n) = \sum_k p_{ik}(m, m + 1)p_{kj}(m + 1, n) \qquad (2.76)$$

whose matrix form is

$$\mathbf{H}(m, n) = \mathbf{P}(m)\mathbf{H}(m + 1, n) \qquad \blacksquare\ (2.77)$$

These last two are referred to as the *backward* Chapman–Kolmogorov equations since they occur at the backward (oldest time) end of the interval.

Since the forward and backward equations both describe the same discrete-time Markov chain, we would expect their solutions to be the same, and indeed this is the case. The general form of the solution is

$$\mathbf{H}(m, n) = \mathbf{P}(m)\mathbf{P}(m + 1) \cdots \mathbf{P}(n - 1) \qquad m \le n - 1 \qquad \blacksquare\ (2.78)$$

That this solves Eqs. (2.75) and (2.77) may be established by direct substitution. We observe in the homogeneous case that this yields $\mathbf{H}(m, n) = \mathbf{P}^{n-m}$ as we have seen earlier. By similar arguments we find that the time-dependent probabilities $\{\pi_j^{(n)}\}$ defined earlier may now be obtained through the following equation:

$$\boldsymbol{\pi}^{(n+1)} = \boldsymbol{\pi}^{(n)}\mathbf{P}(n) \qquad \blacksquare$$

whose solution is

$$\boldsymbol{\pi}^{(n+1)} = \boldsymbol{\pi}^{(0)}\mathbf{P}(0)\mathbf{P}(1) \cdots \mathbf{P}(n) \qquad \blacksquare\ (2.79)$$

These last two equations correspond to Eqs. (2.55) and (2.56), respectively, for the homogeneous case. The Chapman–Kolmogorov equations give us a means for describing the time-dependent probabilities of many interesting queueing systems that we develop in later chapters.*

Before leaving discrete-time Markov chains, we wish to introduce the special case of discrete time *birth–death processes*. A birth–death process is an example of a Markov process that may be thought of as modeling changes in the size of a population. In what follows we say that the system is in state E_k when the population consists of k members. We further assume that changes in population size occur by at most one; that is, a "birth" will change the population's size to one greater, whereas a "death" will lower the population size to one less. In considering birth–death processes we do not permit multiple births or bulk disasters; such possibilities will be considered

* It is clear from this development that all Markov processes must satisfy the Chapman–Kolmogorov equations. Let us note, however, that all processes that satisfy the Chapman–Kolmogorov equation are not necessarily Markov processes; see, for example, p. 203 of [PARZ 62].

later in the text and correspond to random walks. We will consider the Markov chain to be homogeneous in that the transition probabilities p_{ij} do not change with time; however, certainly they will be a function of the state of the system. Thus we have that for our discrete-time birth–death process

$$p_{ij} = \begin{cases} d_i & j = i - 1 \\ 1 - b_i - d_i & j = i \\ b_i & j = i + 1 \\ 0 & \text{otherwise} \end{cases} \tag{2.80}$$

Here d_i is the probability that at the next time step a single death will occur, driving the population size down to $i - 1$, given that the population size now is i. Similarly, b_i is the probability that a single birth will occur, given that the current size is i, thereby driving the population size to $i + 1$ at the next time step. $1 - b_i - d_i$ is the probability that neither of these events will occur and that at the next time step the population size will not change. Only these three possibilities are permitted. Clearly $d_0 = 0$, since we can have no deaths when there is no one in the population to die. However, contrary to intuition we do permit $b_0 > 0$; this corresponds to a birth when there are no members in the population. Whereas this may seem to be spontaneous generation, or perhaps divine creation, it does provide a meaningful model in terms of queueing theory. The model is as follows: The population corresponds to the customers in the queueing system; a death corresponds to a customer departure from that system; and a birth corresponds to a customer arrival to that system. Thus we see it is perfectly feasible to have an arrival (a birth) to an empty system! The stationary probability transition matrix for the general birth–death process then appears as follows:

$$\mathbf{P} = \begin{bmatrix} 1-b_0 & b_0 & 0 & 0 & \cdots & 0 & 0 & 0 & 0 & \cdots \\ d_1 & 1-b_1-d_1 & b_1 & 0 & \cdots & 0 & 0 & 0 & 0 & \\ 0 & d_2 & 1-b_2-d_2 & b_2 & \cdots & 0 & 0 & 0 & 0 & \\ \cdot & & \cdot & & & & & & & \\ \cdot & & & \cdot & & & & & & \\ \cdot & & & & \cdot & & & & & \\ 0 & & & & & d_i & 1-b_i-d_i & b_i & 0 & \cdots \\ \cdot & & & & & & \cdot & & & \\ \cdot & & & & & & & \cdot & & \\ \cdot & & & & & & & \cdot & & \end{bmatrix}$$

If we are dealing with a finite chain, then the last row of this matrix would be $[0\ 0 \cdots 0\ d_N\ 1 - d_N]$, which illustrates the fact that no births are permitted when the population has reached its maximum size N. We see that the $\mathbf{P}$

matrix has nonzero terms only along the main diagonal and along the diagonals directly above and below it. This is a highly specialized form for the transition probability matrix, and as such we might expect that it can be solved. To solve the birth–death process means to find the solution for the state probabilities $\boldsymbol{\pi}^{(n)}$. As we have seen, the general form of solution for these probabilities is given in Eqs. (2.55) and (2.56) and the equation that describes the limiting solution (as $n \to \infty$) is given in Eq. (2.50). We also demonstrated earlier the z-transform method for finding the solution. Of course, due to this special structure of the birth–death transition matrix, we might expect a more explicit solution. We defer discussion of the solution to the material on continuous-time Markov chains, which we now investigate.

2.4. CONTINUOUS-TIME MARKOV CHAINS*

If we allow our particle in motion to occupy positions (take-on values) from a discrete set, but permit it to change positions or states at *any* point in time, then we say we have a continuous-time Markov chain. We may continue to use our example of the hippie hitchhiking from city to city, where now his transitions between cities may occur at any time of day or night. We let $X(t)$ denote the city in which we find our hippie at time t. $X(t)$ will take on values from a discrete set, which we will choose to be the ordered integers and which will be in one-to-one correspondence with the cities which our hippie may visit.

In the case of a continuous-time Markov chain, we have the following definition:

DEFINITION: The random process $X(t)$ forms a continuous-time Markov chain if for all integers n and for any sequence $t_1, t_2, \ldots, t_{n+1}$ such that $t_1 < t_2 < \cdots < t_{n+1}$ we have

$$P[X(t_{n+1}) = j \mid X(t_1) = i_1, X(t_2) = i_2, \ldots, X(t_n) = i_n] = P[X(t_{n+1}) = j \mid X(t_n) = i_n] \qquad (2.81)$$

This definition† is the continuous-time version of that given in Eq. (2.38). The interpretation here is also the same, namely, that the future of our hippie's travels depends upon the past only through the current city in which we find him. The development of the theory for continuous time parallels that for discrete time quite directly as one might expect and, therefore, our explanations will be a bit more concise. Moreover, we will not overly concern

* See footnote on p. 19.

† An alternate definition for a discrete-state continuous-time Markov process is that the following relation must hold:

$$P[X(t) = j \mid X(\tau) \quad \text{for} \quad \tau_1 \le \tau \le \tau_2 < t] = P[X(t) = j \mid X(\tau_2)]$$

ourselves with some of the deeper questions of convergence of limits in passing from discrete to continuous time; for a careful treatment the reader is referred to [PARZ 62, FELL 66].

Earlier we stated for any Markov process that the time which the process spends in any state must be "memoryless"; this implies that the discrete-time Markov chains must have geometrically distributed state times [which we have already proved in Eq. (2.66)] and that continuous-time Markov chains must have exponentially distributed state times. Let us now prove this last statement. For this purpose let τ_i be a random variable that represents the time which the process spends in state E_i. Recall the Markov property which states that the way in which the past trajectory of the process influences the future development is completely specified by giving the current state of the process. In particular, we need not specify how *long* the process has been in its current state. This means that the remaining time in E_i must have a distribution that depends only upon i and not upon how long the process has been in E_i. We may write this in the following form:

$$P[\tau_i > s + t \mid \tau_i > s] = h(t)$$

where $h(t)$ is a function only of the additional time t (and not of the expended time s)*. We may rewrite this conditional probability as follows:

$$\begin{aligned} P[\tau_i > s + t \mid \tau_i > s] &= \frac{P[\tau_i > s + t, \tau_i > s]}{P[\tau_i > s]} \\ &= \frac{P[\tau_i > s + t]}{P[\tau_i > s]} \end{aligned}$$

This last step follows since the event $\tau_i > s + t$ implies the event $\tau_i > s$. Rewriting this last equation and introducing $h(t)$ once again we find

$$P[\tau_i > s + t] = P[\tau_i > s]h(t) \tag{2.82}$$

Setting $s = 0$ and observing that $P[\tau_i > 0] = 1$ we have immediately that

$$P[\tau_i > t] = h(t)$$

Using this last equation in Eq. (2.82) we then obtain

$$P[\tau_i > s + t] = P[\tau_i > s]P[\tau_i > t] \tag{2.83}$$

for $s, t \geq 0$. (Setting $s = t = 0$ we again require $P[\tau_i > 0] = 1$.) We now show that the only continuous distribution satisfying Eq. (2.83) is the

* The symbol s is used as a time variable in this section only and should not be confused with its use as a transform variable elsewhere.

exponential distribution. First we have, by definition, the following general relationship:

$$\frac{d}{dt}(P[\tau_i > t]) = \frac{d}{dt}(1 - P[\tau_i \le t])$$
$$= -f_{\tau_i}(t) \tag{2.84}$$

where we use the notation $f_{\tau_i}(t)$ to denote the pdf for τ_i. Now let us differentiate Eq. (2.83) with respect to s, yielding

$$\frac{dP[\tau_i > s + t]}{ds} = -f_{\tau_i}(s)P[\tau_i > t]$$

where we have taken advantage of Eq. (2.84). Dividing both sides by $P[\tau_i > t]$ and setting $s = 0$ we have

$$\frac{dP[\tau_i > t]}{P[\tau_i > t]} = -f_{\tau_i}(0)\, ds$$

If we integrate this last from 0 to t we obtain

$$\log_e P[\tau_i > t] = -f_{\tau_i}(0)t$$

or

$$P[\tau_i > t] = e^{-f_{\tau_i}(0)t}$$

Now we use Eq. (2.84) again to obtain the pdf for τ_i as

$$f_{\tau_i}(t) = f_{\tau_i}(0)e^{-f_{\tau_i}(0)t} \tag{2.85}$$

which holds for $t \ge 0$. There we have it: the pdf for the time the process spends in state E_i is exponentially distributed with the parameter $f_{\tau_i}(0)$, which may depend upon the state E_i. We will have much more to say about this exponential distribution and its importance in Markov processes shortly.

In the case of a discrete-time homogeneous Markov chain we defined the transition probabilities as $p_{ij} = P[X_n = j \mid X_{n-1} = i]$ and also the m-step transition probabilities as $p_{ij}^{(m)} = P[X_{n+m} = j \mid X_n = i]$; these quantities were independent of n due to the homogeneity of the Markov chain. In the case of the nonhomogeneous Markov chain we found it necessary to identify points along the time axis in an absolute fashion and were led to the important transition probability definition $p_{ij}(m, n) = P[X_n = j \mid X_m = i]$. In a completely analogous way we must now define for our continuous-time Markov chains the following time-dependent transition probability:

$$p_{ij}(s, t) \triangleq P[X(t) = j \mid X(s) = i] \tag{2.86}$$

where $X(t)$ is the position of the particle at time $t \ge s$. Just as we considered three successive time instants $m \le q \le n$ for the discrete case, we may consider the following three successive time instants for our continuous time chain $s \le u \le t$. We may then refer back to Figure 2.7 and identify some sample paths for what we will now consider to be a continuous-time Markov chain; the critical observation once again is that in passing from state E_i at time s to state E_j at time t, the process must pass through some intermediate state E_k at the intermediate time u. We then proceed exactly as we did in deriving Eq. (2.70) and arrive at the following Chapman–Kolmogorov equation for continuous-time Markov chains:

$$p_{ij}(s, t) = \sum_k p_{ik}(s, u) p_{kj}(u, t) \tag{2.87}$$

where $i, j = 0, 1, 2, \ldots$. We may put this equation into matrix form if we first define the matrix consisting of elements $p_{ij}(s, t)$ as

$$\mathbf{H}(s, t) \triangleq [p_{ij}(s, t)] \tag{2.88}$$

Then the Chapman–Kolmogorov equation becomes

$$\mathbf{H}(s, t) = \mathbf{H}(s, u)\mathbf{H}(u, t) \qquad s \le u \le t \tag{2.89}$$

[We define $\mathbf{H}(t, t) = \mathbf{I}$, the identity matrix.]

In the case of a homogeneous discrete-time Markov chain we found that the matrix equation $\boldsymbol{\pi} = \boldsymbol{\pi}\mathbf{P}$ had to be investigated in order to determine if the chain was ergodic, and so on; also, the transient solution in the nonhomogeneous case could be determined from $\boldsymbol{\pi}^{(n+1)} = \boldsymbol{\pi}^{(0)}\mathbf{P}(0)\mathbf{P}(1) \cdots \mathbf{P}(n)$, which was given in terms of the time-dependent transition probabilities $p_{ij}(m, n)$. For the continuous-time Markov chain the one-step transition probabilities are replaced by the infinitesimal rates to be defined below; as we shall see they are given in terms of the time derivative of $p_{ij}(s, t)$ as $t \to s$.

What we wish now to do is to form the continuous-time analog of the forward and backward equations. So far we have reached Eq. (2.89), which is analogous to Eq. (2.73) in the discrete-time case. We wish to extract the analog for Eqs. (2.74)–(2.77), which show both the term-by-term and matrix form of the forward and backward equations, respectively. We choose to do this in the case of the forward equation, for example, by starting with Eq. (2.75), namely, $\mathbf{H}(m, n) = \mathbf{H}(m, n-1)\mathbf{P}(n-1)$, and allowing the unit time interval to shrink toward zero. To this end we use this last equation and form the following difference:

$$\begin{aligned} \mathbf{H}(m, n) - \mathbf{H}(m, n-1) &= \mathbf{H}(m, n-1)\mathbf{P}(n-1) - \mathbf{H}(m, n-1) \\ &= \mathbf{H}(m, n-1)[\mathbf{P}(n-1) - \mathbf{I}] \end{aligned} \tag{2.90}$$

We must now consider some limits. Just as in the discrete case we defined $\mathbf{P}(n) = \mathbf{H}(n, n+1)$, we find it convenient in this continuous-time case to define the following matrix:

$$\mathbf{P}(t) \triangleq [p_{ij}(t, t + \Delta t)] \tag{2.91}$$

Furthermore we identify the matrix $\mathbf{H}(s, t)$ as the limit of $\mathbf{H}(m, n)$ as our time interval shrinks; similarly we see that the limit of $\mathbf{P}(n)$ will be $\mathbf{P}(t)$. Returning to Eq. (2.90) we now divide both sides by the time step, which we denote by Δt, and take the limit as $\Delta t \to 0$. Clearly then the left-hand side limits to the derivative, resulting in

$$\frac{\partial \mathbf{H}(s, t)}{\partial t} = \mathbf{H}(s, t)\mathbf{Q}(t) \qquad s \leq t \tag{2.92}$$

where we have defined the matrix $\mathbf{Q}(t)$ as the following limit:

$$\mathbf{Q}(t) = \lim_{\Delta t \to 0} \frac{\mathbf{P}(t) - \mathbf{I}}{\Delta t} \tag{2.93}$$

This matrix $\mathbf{Q}(t)$ is known as the *infinitesimal generator* of the transition matrix function $\mathbf{H}(s, t)$. Another more descriptive name for $\mathbf{Q}(t)$ is the *transition rate* matrix; we will use both names interchangeably. The elements of $\mathbf{Q}(t)$, which we denote by $q_{ij}(t)$, are the rates that we referred to earlier. They are defined as follows:

$$q_{ii}(t) = \lim_{\Delta t \to 0} \frac{p_{ii}(t, t + \Delta t) - 1}{\Delta t} \tag{2.94}$$

$$q_{ij}(t) = \lim_{\Delta t \to 0} \frac{p_{ij}(t, t + \Delta t)}{\Delta t} \qquad i \neq j \tag{2.95}$$

These limits have the following interpretation. If the system at time t is in state E_i then the probability that a transition occurs (to any state other than E_i) during the interval $(t, t + \Delta t)$ is given by $-q_{ii}(t)\,\Delta t + o(\Delta t)$.* Thus we may say that $-q_{ii}(t)$ is the *rate* at which the process departs from state E_i when it is in that state. Similarly, given that the system is in state E_i at time t, the conditional probability that it will make a transition from this state to state E_j in the time interval $(t, t + \Delta t)$ is given by $q_{ij}(t)\,\Delta t + o(\Delta t)$. Thus

* As usual, the notation $o(\Delta t)$ denotes *any* function that goes to zero with Δt faster than Δt itself, that is,

$$\lim_{\Delta t \to 0} \frac{o(\Delta t)}{\Delta t} = 0$$

More generally, one states that the function $g(t)$ is $o(\gamma(t))$ as $t \to t_1$ if

$$\lim_{t \to t_1} \left| \frac{g(t)}{\gamma(t)} \right| = 0$$

See also Chapter 8, p. 284 for a definition of $O(\cdot)$.

$q_{ij}(t)$ is the *rate* at which the process moves from E_i to E_j, given that the system is currently in the state E_i. Since it is always true that $\sum_j p_{ij}(s, t) = 1$ then we see that Eqs. (2.94) and (2.95) imply that

$$\sum_j q_{ij}(t) = 0 \qquad \text{for all } i \tag{2.96}$$

Thus we have interpreted the terms in Eq. (2.92); this is nothing more than the *forward* Chapman–Kolmogorov equation for the continuous-time Markov chain.

In a similar fashion, beginning with Eq. (2.77) we may derive the *backward* Chapman–Kolmogorov equation

$$\frac{\partial \mathbf{H}(s, t)}{\partial s} = -\mathbf{Q}(s)\mathbf{H}(s, t) \qquad s \le t \qquad \blacksquare \tag{2.97}$$

The forward and backward matrix equations just derived may be expressed through their individual terms as follows. The forward equation gives us [with the additional condition that the passage to the limit in Eq. (2.95) is uniform in i for fixed j]

$$\frac{\partial p_{ij}(s, t)}{\partial t} = q_{jj}(t)p_{ij}(s, t) + \sum_{k \ne j} q_{kj}(t)p_{ik}(s, t) \tag{2.98}$$

The initial state E_i at the initial time s affects the solution of this set of differential equations only through the initial conditions

$$p_{ij}(s, s) = \begin{cases} 1 & \text{if } j = i \\ 0 & \text{if } j \ne i \end{cases}$$

From the backward matrix equation we obtain

$$\frac{\partial p_{ij}(s, t)}{\partial s} = -q_{ii}(s)p_{ij}(s, t) - \sum_{k \ne i} q_{ik}(s)p_{kj}(s, t) \tag{2.99}$$

The "initial" conditions for this equation are

$$p_{ij}(t, t) = \begin{cases} 1 & \text{if } i = j \\ 0 & \text{if } i \ne j \end{cases}$$

These equations [(2.98) and (2.99)] uniquely determine the transition probabilities $p_{ij}(s, t)$ and must, of course, also satisfy Eq. (2.87) as well as the initial conditions.

In matrix notation we may exhibit the solution to the forward and backward Eqs. (2.92) and (2.97), respectively, in a straightforward manner; the result is*

$$\mathbf{H}(s, t) = \exp\left[\int_s^t \mathbf{Q}(u)\, du\right] \qquad \blacksquare\ (2.100)$$

We observe that this solution also satisfies Eq. (2.89) and is a continuous-time analog to the discrete-time solution given in Eq. (2.78).

Now for the state probabilities themselves: In analogy with $\pi_j^{(n)}$ we now define

$$\pi_j(t) \triangleq P[X(t) = j] \qquad \blacksquare\ (2.101)$$

as well as the vector of these probabilities

$$\boldsymbol{\pi}(t) \triangleq [\pi_0(t), \pi_1(t), \pi_2(t), \ldots] \qquad (2.102)$$

If we are given the initial state distribution $\boldsymbol{\pi}(0)$ then we can solve for the time-dependent state probabilities from

$$\boldsymbol{\pi}(t) = \boldsymbol{\pi}(0)\mathbf{H}(0, t) \qquad (2.103)$$

where a general solution may be seen from Eq. (2.100) to be

$$\boldsymbol{\pi}(t) = \boldsymbol{\pi}(0) \exp\left[\int_0^t \mathbf{Q}(u)\, du\right] \qquad \blacksquare\ (2.104)$$

This corresponds to the discrete-time solution given in Eq. (2.79). The matrix differential equation corresponding to Eq. (2.103) is easily seen to be

$$\frac{d\boldsymbol{\pi}(t)}{dt} = \boldsymbol{\pi}(t)\mathbf{Q}(t)$$

This last is similar in form to Eq. (2.92) and may be expressed in terms of its elements as

$$\frac{d\pi_j(t)}{dt} = q_{jj}(t)\pi_j(t) + \sum_{k \neq j} q_{kj}(t)\pi_k(t) \qquad (2.105)$$

The similarity between Eqs. (2.105) and (2.98) is not accidental. The latter describes the probability that the process is in state E_j at time t given that it was in state E_i at time s. The former merely gives the probability that the system is in state E_j at time t; information as to where the process began is given in the initial state probability vector $\boldsymbol{\pi}(0)$. If indeed $\pi_k(0) = 1$ for $k = i$ and $\pi_k(0) = 0$ for $k \neq i$, then we are stating for sure that the system was in state E_i at time 0. In this case $\pi_j(t)$ will be identically equal to $p_{ij}(0, t)$. Both forms for this probability are often used; the form $p_{ij}(s, t)$ is used when

* The expression $e^{\mathbf{P}t}$ where $\mathbf{P}$ is a square matrix is defined as the following matrix power series:

$$e^{\mathbf{P}t} = \mathbf{I} + \mathbf{P}t + \mathbf{P}^2\frac{t^2}{2!} + \mathbf{P}^3\frac{t^3}{3!} + \cdots$$

we want to specifically show the initial state; the form $\pi_j(t)$ is used when we choose to neglect or imply the initial state.

We now consider the case where our continuous-time Markov chain is *homogeneous*. In this case we drop the dependence upon time and adopt the following notation:

$$p_{ij}(t) \triangleq p_{ij}(s, s+t) \tag{2.106}$$

$$q_{ij} \triangleq q_{ij}(t) \quad i, j = 1, 2, \ldots \tag{2.107}$$

$$\mathbf{H}(t) \triangleq \mathbf{H}(s, s+t) = [p_{ij}(t)] \quad \blacksquare \tag{2.108}$$

$$\mathbf{Q} \triangleq \mathbf{Q}(t) = [q_{ij}] \tag{2.109}$$

In this case we may list in rapid order the corresponding results. First, the Chapman–Kolmogorov equations become

$$p_{ij}(s+t) = \sum_k p_{ik}(s) p_{kj}(t) \quad \blacksquare$$

and in matrix form*

$$\mathbf{H}(s+t) = \mathbf{H}(s)\mathbf{H}(t) \quad \blacksquare$$

The forward and backward equations become, respectively,

$$\frac{dp_{ij}(t)}{dt} = q_{jj} p_{ij}(t) + \sum_{k \neq j} q_{kj} p_{ik}(t) \tag{2.110}$$

and

$$-\frac{dp_{ij}(t)}{dt} = -q_{ii} p_{ij}(t) - \sum_{k \neq i} q_{ik} p_{kj}(t) \tag{2.111}$$

and in matrix form these become, respectively,

$$\frac{d\mathbf{H}(t)}{dt} = \mathbf{H}(t)\mathbf{Q} \quad \blacksquare \tag{2.112}$$

and

$$\frac{d\mathbf{H}(t)}{dt} = \mathbf{Q}\mathbf{H}(t) \quad \blacksquare \tag{2.113}$$

with the common initial condition $\mathbf{H}(0) = \mathbf{I}$. The solution for this matrix is given by

$$\mathbf{H}(t) = e^{\mathbf{Q}t} \quad \blacksquare$$

Now for the state probabilities themselves we have the differential equation

$$\frac{d\pi_j(t)}{dt} = q_{jj}\pi_j(t) + \sum_{k \neq j} q_{kj}\pi_k(t) \tag{2.114}$$

which in matrix form is

$$\frac{d\boldsymbol{\pi}(t)}{dt} = \boldsymbol{\pi}(t)\mathbf{Q} \quad \blacksquare$$

* The corresponding discrete-time result is simply $\mathbf{P}^{m+n} = \mathbf{P}^m\mathbf{P}^n$.

For an irreducible homogeneous Markov chain it can be shown that the following limits always exist and are independent of the initial state of the chain, namely,

$$\lim_{t\to\infty} p_{ij}(t) = \pi_j$$

This set $\{\pi_j\}$ will form the limiting state probability distribution. For an ergodic Markov chain we will have the further limit, which will be independent of the initial distribution, namely,

$$\lim_{t\to\infty} \pi_j(t) = \pi_j$$

This limiting distribution is given uniquely as the solution of the following system of linear equations:

$$q_{jj}\pi_j + \sum_{k\neq j} q_{kj}\pi_k = 0 \tag{2.115}$$

In matrix form this last equation may be expressed as

$$\boldsymbol{\pi}\mathbf{Q} = 0 \quad \blacksquare \tag{2.116}$$

where we have used the obvious notation $\boldsymbol{\pi} = [\pi_0, \pi_1, \pi_2, \ldots]$. This last equation coupled with the probability conservation relation, namely,

$$\sum_j \pi_j = 1 \tag{2.117}$$

uniquely gives us our limiting state probabilities. We compare the Eq. (2.116) with our earlier equation for discrete-time Markov chains, namely, $\boldsymbol{\pi} = \boldsymbol{\pi}\mathbf{P}$; here $\mathbf{P}$ was the matrix of transition *probabilities*, whereas the infinitesimal generator $\mathbf{Q}$ is a matrix of transition *rates*.

This completes our discussion of discrete-state Markov chains. In the table on pp. 402–403, we summarize the major results for the four cases considered here. For a further discussion, the reader is referred to [BHAR 60].

Having discussed discrete-state Markov chains (both in discrete and continuous time) it would seem natural that we next consider continuous-state Markov processes. This we will not do, but rather we postpone consideration of such material until we require it [viz., in Chapter 5 we consider Takács' integrodifferential equation for M/G/1, and in Chapter 2 (Volume II) we develop the Fokker–Planck equation for use in the diffusion approximation for queues]. One would further expect that following the study of Markov processes, we would then investigate renewal processes, random walks, and finally, semi-Markov processes. Here too, we choose to postpone such discussions until they are needed later in the text (e.g., the discussion in Chapter 5 of Markov chains imbedded in semi-Markov processes).

Indeed it is fair to say that much of the balance of this textbook depends upon additional material from the theory of stochastic processes and will be developed as needed. For the time being we choose to specialize the results we have obtained from the continuous-time Markov chains to the class of birth–death processes, which, as we have forewarned, play a major role in queueing systems analysis. This will lead us directly to the important Poisson process.

2.5. BIRTH–DEATH PROCESSES

Earlier in this chapter we said that a birth–death process is the special case of a Markov process in which transitions from state E_k are permitted only to neighboring states E_{k+1}, E_k, and E_{k-1}. This restriction permits us to carry the solution much further in many cases. These processes turn out to be excellent models for all of the material we will study under elementary queueing theory in Chapter 3, and as such forms our point of departure for the study of queueing systems. The discrete-time birth–death process is of less interest to us than the continuous-time case, and, therefore, discrete-time birth–death processes are not considered explicitly in the following development; needless to say, an almost parallel treatment exists for that case. Moreover, transitions of the form from state E_i back to E_i are of direct interest only in the discrete-time Markov chains; in the continuous-time Markov chains, the rate at which the process returns to the state that it currently occupies is infinite, and the astute reader should have observed that we very carefully subtracted this term out of our definition for $q_{ii}(t)$ in Eq. (2.94). Therefore, our main interest will focus on continuous-time birth–death processes with discrete state space in which transitions only to neighboring states E_{k+1} or E_{k-1} from state E_k are permitted.*

Earlier we described a birth–death process as one that is appropriate for modeling changes in the size of a population. Indeed, when the process is said to be in state E_k we will let this denote the fact that the population at that time is of size k. Moreover, a transition from E_k to E_{k+1} will signify a "birth" within the population, whereas a transition from E_k to E_{k-1} will denote a "death" in the population.

Thus we consider changes in size of a population where transitions from state E_k take place to *nearest neighbors* only. Regarding the nature of births and deaths, we introduce the notion of a *birth rate* λ_k, which describes the

* This is true in the one-dimensional case. Later, in Chapter 4, we consider multidimensional systems for which the states are described by discrete vectors, and then each state has two neighbors in *each* dimension. For example, in the two-dimensional case, the state descriptor is a couplet (k_1, k_2) denoted by E_{k_1,k_2} whose four neighbors are E_{k_1-1,k_2}, E_{k_1,k_2-1}, E_{k_1+1,k_2}, and E_{k_1,k_2+1}.

rate at which births occur when the population is of size k. Similarly, we define a *death rate* μ_k, which is the rate at which deaths occur when the population is of size k. Note that these birth and death rates are independent of time and depend only on E_k; thus we have a continuous-time homogeneous Markov chain of the birth–death type. We adopt this special notation since it leads us directly into the queueing system notation; note that, in terms of our earlier definitions, we have

$$\lambda_k = q_{k,k+1}$$

and

$$\mu_k = q_{k,k-1}$$

The nearest-neighbor condition requires that $q_{kj} = 0$ for $|k - j| > 1$. Moreover, since we have previously shown in Eq. (2.96) that $\sum_j q_{kj} = 0$, then we require

$$q_{kk} = -(\mu_k + \lambda_k) \tag{2.118}$$

Thus our infinitesimal generator for the general homogeneous birth–death process takes the form

$$\mathbf{Q} = \begin{bmatrix} -\lambda_0 & \lambda_0 & 0 & 0 & 0 & \\ \mu_1 & -(\lambda_1 + \mu_1) & \lambda_1 & 0 & 0 & \\ 0 & \mu_2 & -(\lambda_2 + \mu_2) & \lambda_2 & 0 & \cdots \\ 0 & 0 & \mu_3 & -(\lambda_3 + \mu_3) & \lambda_3 & \\ & & \cdot & & \cdot & \\ & & \cdot & & \cdot & \\ & & \cdot & & \cdot & \end{bmatrix}$$

Note that except for the main, upper, and lower diagonals, all terms are zero.

To be more explicit, the assumptions we need for the birth–death process are that it is a homogeneous Markov chain $X(t)$ on the states $0, 1, 2, \ldots,$ that births and deaths are independent (this follows directly from the Markov property), *and*

B_1: $P[\text{exactly 1 birth in } (t, t + \Delta t) \mid k \text{ in population}]$
$$= \lambda_k \Delta t + o(\Delta t)$$

D_1: $P[\text{exactly 1 death in } (t, t + \Delta t) \mid k \text{ in population}]$
$$= \mu_k \Delta t + o(\Delta t)$$

B_2: $P[\text{exactly 0 births in } (t, t + \Delta t) \mid k \text{ in population}]$
$$= 1 - \lambda_k \Delta t + o(\Delta t)$$

D_2: $P[\text{exactly 0 deaths in } (t, t + \Delta t) \mid k \text{ in population}]$
$$= 1 - \mu_k \Delta t + o(\Delta t)$$

From these assumptions we see that multiple births, multiple deaths, or in fact, both a birth and a death in a small time interval are prohibited in the sense that each such multiple event is of order $o(\Delta t)$.

What we wish to solve for is the probability that the population size is k at some time t; this we denote by*

$$P_k(t) \triangleq P[X(t) = k] \tag{2.119}$$

This calculation could be carried out *directly* by using our result in Eq. (2.114) for $\pi_j(t)$ and our specific values for q_{ij}. However, since the derivation of these equations for the birth–death process is so straightforward and follows from first principles, we choose *not* to use the heavy machinery we developed in the previous section, which tends to camouflage the simplicity of the basic approach, but rather to rederive them below. The reader is encouraged to identify the parallel steps in this development and compare them to the more general steps taken earlier. Note in terms of our previous definition that $P_k(t) = \pi_k(t)$. Moreover, we are "suppressing" the initial conditions temporarily, and will introduce them only when required.

We begin by expressing the Chapman–Kolmogorov dynamics, which are quite trivial in this case. In particular, we focus on the possible motions of our particle (that is, the number of members in our population) during an interval $(t, t + \Delta t)$. We will find ourselves in state E_k at time $t + \Delta t$ if one of the three following (mutually exclusive and exhaustive) eventualities occurred:

1. that we had k in the population at time t and no state changes occurred;
2. that we had $k - 1$ in the population at time t and we had a birth during the interval $(t, t + \Delta t)$;
3. that we had $k + 1$ members in the population at time t and we had one death during the interval $(t, t + \Delta t)$.

These three cases are portrayed in Figure (2.8). The probability for the first of these possibilities is merely the probability $P_k(t)$ that we were in state E_k at time t times the probability $p_{kk}(\Delta t)$ that we moved from state E_k to state E_k (i.e., had neither a birth nor a death) during the next Δt seconds; this is represented by the first term on the right-hand side of Eq. (2.120) below. The second and third terms on the right-hand side of that equation correspond, respectively, to the second and third cases listed above. We need not concern ourselves specifically with transitions from states other than nearest neighbors to state E_k since we have assumed that such transitions in an interval of

* We use $X(t)$ here to denote the number in system at time t to be consistent with the use of $X(t)$ for our general stochastic process. Certainly we could have used $N(t)$ as defined earlier; we use $N(t)$ outside of this chapter.

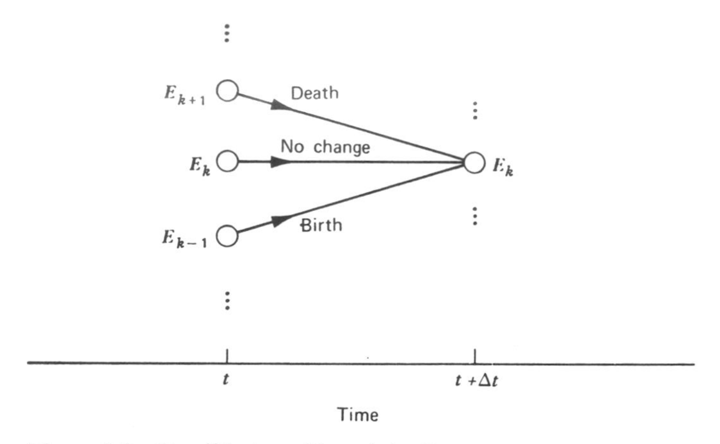

Figure 2.8 Possible transitions into E_k.

duration Δt are of order $o(\Delta t)$. Thus we may write

$$\begin{aligned} P_k(t + \Delta t) = P_k(t)p_{k,k}(\Delta t) \\ + P_{k-1}(t)p_{k-1,k}(\Delta t) \\ + P_{k+1}(t)p_{k+1,k}(\Delta t) \\ + o(\Delta t) \qquad k \geq 1 \end{aligned} \tag{2.120}$$

We may add the three probabilities above since these events are clearly mutually exclusive. Of course, Eq. (2.120) only makes sense in the case for $k \geq 1$, since clearly we could not have had -1 members in the population. For the case $k = 0$ we need the special boundary equation given by

$$\begin{aligned} P_0(t + \Delta t) = P_0(t)p_{00}(\Delta t) \\ + P_1(t)p_{10}(\Delta t) \\ + o(\Delta t) \qquad k = 0 \end{aligned} \tag{2.121}$$

Furthermore, it is also clear for all values of t that we must conserve our probability, and this is expressed in the following equation:

$$\sum_{k=0}^{\infty} P_k(t) = 1 \tag{2.122}$$

To solve the system represented by Eqs. (2.120)–(2.122) we must make use of our assumptions B_1, D_1, B_2, and D_2, in order to evaluate the coefficients in these equations. Carrying out this operation our equations convert to

$$
\begin{aligned}
P_k(t+\Delta t) = P_k(t)[1 - \lambda_k\,\Delta t + o(\Delta t)][1 - \mu_k\,\Delta t + o(\Delta t)] \\
+ P_{k-1}(t)[\lambda_{k-1}\,\Delta t + o(\Delta t)] \\
+ P_{k+1}(t)[\mu_{k+1}\,\Delta t + o(\Delta t)] \\
+ o(\Delta t) \qquad k \geq 1 \quad (2.123)
\end{aligned}
$$

$$
\begin{aligned}
P_0(t+\Delta t) = P_0(t)[1 - \lambda_0\,\Delta t + o(\Delta t)] \\
+ P_1(t)[\mu_1\,\Delta t + o(\Delta t)] \\
+ o(\Delta t) \qquad k = 0 \quad (2.124)
\end{aligned}
$$

In Eq. (2.124) we have used the assumption that it is impossible to have a death when the population is of size 0 (i.e., $\mu_0 = 0$) and the assumption that one indeed can have a birth when the population size is 0 ($\lambda_0 \geq 0$). Expanding the right-hand side of Eqs. (2.123) and (2.124) we have

$$
\begin{aligned}
P_k(t+\Delta t) &= P_k(t) - (\lambda_k + \mu_k)\,\Delta t P_k(t) + \lambda_{k-1}\,\Delta t P_{k-1}(t) \\
&\quad + \mu_{k+1}\,\Delta t P_{k+1}(t) + o(\Delta t) \qquad k \geq 1 \\
P_0(t+\Delta t) &= P_0(t) - \lambda_0\,\Delta t P_0(t) + \mu_1\,\Delta t P_1(t) + o(\Delta t) \qquad k = 0
\end{aligned}
$$

If we now subtract $P_k(t)$ from both sides of each equation and divide by Δt, we have the following:

$$
\frac{P_k(t+\Delta t) - P_k(t)}{\Delta t} = -(\lambda_k + \mu_k)P_k(t) + \lambda_{k-1}P_{k-1}(t) + \mu_{k+1}P_{k+1}(t) + \frac{o(\Delta t)}{\Delta t} \qquad k \geq 1 \quad (2.125)
$$

$$
\frac{P_0(t+\Delta t) - P_0(t)}{\Delta t} = -\lambda_0 P_0(t) + \mu_1 P_1(t) + \frac{o(\Delta t)}{\Delta t} \qquad k = 0 \quad (2.126)
$$

Taking the limit as Δt approaches 0 we see that the left-hand sides of Eqs. (2.125) and (2.126) represent the formal derivative of $P_k(t)$ with respect to t and also that the term $o(\Delta t)/\Delta t$ goes to 0. Consequently, we have the resulting equations:

$$
\begin{aligned}
\frac{dP_k(t)}{dt} &= -(\lambda_k + \mu_k)P_k(t) + \lambda_{k-1}P_{k-1}(t) + \mu_{k+1}P_{k+1}(t) \qquad k \geq 1 \\
\frac{dP_0(t)}{dt} &= -\lambda_0 P_0(t) + \mu_1 P_1(t) \qquad k = 0
\end{aligned}
\qquad \blacksquare\ (2.127)
$$

The set of equations given by (2.127) is clearly a set of *differential-difference* equations and represents the dynamics of our probability system; we recognize them as Eq. (2.114) and their solution will give the behavior of $P_k(t)$. It remains for us to solve them. (Note that this set was obtained by essentially using the Chapman–Kolmogorov equations.)

In order to solve Eqs. (2.127) for the time-dependent behavior $P_k(t)$ we now require our initial conditions; that is, we must specify $P_k(0)$ for $k = 0, 1, 2, \ldots$. In addition, we further require that Eq. (2.122) be satisfied.

Let us pause temporarily to describe a simple *inspection* technique for finding the differential-difference equations given above. We begin by observing that an alternate way for displaying the information contained in the **Q** matrix is by means of the *state-transition-rate diagram*. In such a diagram the state E_k is represented by an oval surrounding the number k. Each nonzero infinitesimal rate q_{ij} (the elements of the **Q** matrix) is represented in the state-transition-rate diagram by a directed branch pointing from E_i to E_j and labeled with the value q_{ij}. Furthermore, since it is clear that the terms along the main diagonal of **Q** contain no new information [see Eqs. (2.96) and (2.118)] we do not include the "self"-loop from E_i back to E_i. Thus the state-transition-rate diagram for the general birth–death process is as shown in Figure 2.9.

In viewing this figure we may truly think of a particle in motion moving among these states; the branches identify the permitted transitions and the branch labels give the infinitesimal rates at which these transitions take place. We emphasize that the labels on the ordered links refer to birth and death *rates* and *not to probabilities*. If one wishes to convert these labels to probabilities, one must multiply each by the quantity dt to obtain the probability of such a transition occurring in the next interval of time whose duration is dt. In that case it is also necessary to put self-loops on each state indicating the probability that in the next interval of time dt the system remains in the given state. Note that the state-transition-rate diagram contains exactly the same information as does the transition-rate matrix **Q**.

Concentrating on state E_k we observe that one may enter it only from state E_{k-1} or from state E_{k+1} and similarly one leaves state E_k only by entering state E_{k-1} or state E_{k+1}. From this picture we see why such processes are referred to as "nearest-neighbor" birth–death processes.

Since we are considering a dynamic situation it is clear that the difference between the rate at which the system enters E_k and the rate at which the system leaves E_k must be equal to the rate of change of "flow" into that state. This

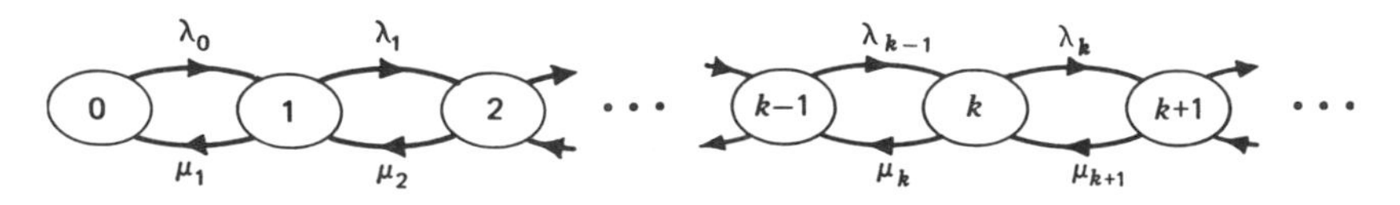

Figure 2.9 State-transition-rate diagram for the birth–death process.

notion is *crucial* and provides for us a simple intuitive means for writing down the equations of motion for the probabilities $P_k(t)$. Specifically, if we focus upon state E_k we observe that the rate at which *probability* "flows" into this state at time t is given by

$$\text{Flow rate into } E_k = \lambda_{k-1}P_{k-1}(t) + \mu_{k+1}P_{k+1}(t)$$

whereas the flow rate out of that state at time t is given by

$$\text{Flow rate out of } E_k = (\lambda_k + \mu_k)P_k(t)$$

Clearly the difference between these two is the effective probability flow rate into this state, that is,

$$\frac{dP_k(t)}{dt} = \lambda_{k-1}P_{k-1}(t) + \mu_{k+1}P_{k+1}(t) - (\lambda_k + \mu_k)P_k(t) \qquad \blacksquare\ (2.128)$$

But this is exactly Eq. (2.127)! Of course, we have not attended to the details for the boundary state E_0 but it is easy to see that the rate argument just given leads to the correct equation for $k = 0$. Observe that each term in Eq. (2.128) is of the form: probability of being in a particular state at time t multiplied by the infinitesimal rate of leaving that state. It is clear that what we have done is to draw an imaginary boundary surrounding state E_k and have calculated the probability flow rates crossing that boundary, where we place opposite signs on flows entering as opposed to leaving; this total computation is then set equal to the time derivative of the probability flow rate into that state.

Actually there is no reason for selecting a single state as the "system" for which the flow equations must hold. In fact one may enclose any number of states within a contour and then write a flow equation for all flow crossing that boundary. The only danger in dealing with such a conglomerate set is that one may write down a *dependent* set of equations rather than an independent set; on the other hand, if one systematically encloses each state singly and writes down a conservation law for each, then one is guaranteed to have an independent set of equations for the system with the qualification that the conservation of probability given by Eq. (2.122) must also be applied.* Thus we have a simple inspection technique for arriving at the equations of motion for the birth–death process. As we shall see later this approach is perfectly suitable for other Markov processes (including semi-Markov processes) and will be used extensively; these observations also lead us to the notion of global and local balance equations (see Chapter 4).

At this point it is important for the reader to recognize and accept the fact that the birth–death process described above is capable of providing the

* When the number of states is finite (say, K states) then *any* set of $K - 1$ single-node state equations will be independent. The additional equation needed is Eq. (2.122).

framework for discussing a large number of important and interesting problems in queueing theory. The direct solution for appropriate special cases of Eq. (2.127) provides for us the transient behavior of these queueing systems and is of less interest to this book than the equilibrium or steady-state behavior of queues.* However, for purposes of illustration and to elaborate further upon these equations, we now consider some important examples.

The simplest system to consider is a *pure birth* system in which we assume $\mu_k = 0$ for all k (note that we have now entered the next-to-innermost circle in Figure 2.4!). Moreover, to simplify the problem we will assume that $\lambda_k = \lambda$ for all $k = 0, 1, 2, \ldots$. (Now we have entered the innermost circle! We therefore expect some marvelous properties to emerge.) Substituting this into our Eqs. (2.127) we have

$$\begin{aligned} \frac{dP_k(t)}{dt} &= -\lambda P_k(t) + \lambda P_{k-1}(t) \qquad k \geq 1 \\ \frac{dP_0(t)}{dt} &= -\lambda P_0(t) \qquad k = 0 \end{aligned} \tag{2.129}$$

For simplicity we assume that the system begins at time 0 with 0 members, that is,

$$P_k(0) = \begin{cases} 1 & k = 0 \\ 0 & k \neq 0 \end{cases} \tag{2.130}$$

Solving for $P_0(t)$ we have immediately

$$P_0(t) = e^{-\lambda t}$$

Inserting this last into Eq. (2.129) for $k = 1$ results in

$$\frac{dP_1(t)}{dt} = -\lambda P_1(t) + \lambda e^{-\lambda t}$$

The solution to this differential equation is clearly

$$P_1(t) = \lambda t e^{-\lambda t}$$

Continuing by induction, then, we finally have as a solution to Eq. (2.129)

$$P_k(t) = \frac{(\lambda t)^k}{k!} e^{-\lambda t} \qquad k \geq 0, t \geq 0 \qquad \blacksquare \tag{2.131}$$

This is the celebrated *Poisson* distribution. It is a pure birth process with constant birth rate λ and gives rise to a sequence of birth epochs which are

* Transient behavior is discussed elsewhere in this text, notably in Chapter 2 (Vol. II). For an excellent treatment the reader is referred to [COHE 69].

said to constitute a *Poisson process*. Let us study the Poisson process more carefully and show its relationship to the exponential distribution.

The Poisson process is central to much of elementary and intermediate queueing theory and is widely used in their development. The special position of this process comes about for two reasons. First, as we have seen, it is the "innermost circle" in Figure 2.4 and, therefore, enjoys a number of marvelous and simplifying analytical and probabilistic properties; this will become undeniably apparent in our subsequent development. The second reason for its great importance is that, in fact, numerous natural physical and organic processes exhibit behavior that is probably meaningfully modeled by Poisson processes. For example, as Fry [FRY 28] so graphically points out, one of the first observations of the Poisson process was that it properly represented the number of army soldiers killed due to being kicked (in the head?) by their horses. Other examples include the sequence of gamma rays emitting from a radioactive particle, and the sequence of times at which telephone calls are originated in the telephone network. In fact, it was shown by Palm [PALM 43] and Khinchin [KHIN 60] that in many cases the sum of a large number of independent stationary renewal processes (each with an arbitrary distribution of renewal time) will tend to a Poisson process. This is an important limit theorem and explains why Poisson processes appear so often in nature where the aggregate effect of a large number of individuals or particles is under observation.

Since this development is intended for our use in the study of queueing systems, let us immediately adopt queueing notation and also condition ourselves to discussing a Poisson process as the *arrival of customers* to some queueing facility rather than as the birth of new members in a population. Thus λ is the average rate at which these customers arrive. With the initial condition in Eq. (2.130), $P_k(t)$ gives the probability that k arrivals occur during the time interval $(0, t)$. It is intuitively clear, since the average arrival rate is λ per second, that the average number of arrivals in an interval of length t must be λt. Let us carry out the calculation of this last intuitive statement. Defining K as the number of arrivals in this interval of length t [previously we used $\alpha(t)$] we have

$$\begin{aligned} E[K] &= \sum_{k=0}^{\infty} kP_k(t) \\ &= e^{-\lambda t}\sum_{k=0}^{\infty} k\frac{(\lambda t)^k}{k!} \\ &= e^{-\lambda t}\sum_{k=1}^{\infty} \frac{(\lambda t)^k}{(k-1)!} \\ &= e^{-\lambda t}\lambda t\sum_{k=0}^{\infty} \frac{(\lambda t)^k}{k!} \end{aligned}$$

By definition, we know that $e^x = 1 + x + x^2/2! + \cdots$ and so we get

$$E[K] = \lambda t \qquad \blacksquare \; (2.132)$$

Thus clearly the expected number of arrivals in $(0, t)$ is equal to λt.

We now proceed to calculate the variance of the number of arrivals. In order to do this we find it convenient to first calculate the following moment

$$\begin{aligned} E[K(K-1)] &= \sum_{k=0}^{\infty} k(k-1)P_k(t) \\ &= e^{-\lambda t}\sum_{k=0}^{\infty} k(k-1)\frac{(\lambda t)^k}{k!} \\ &= e^{-\lambda t}(\lambda t)^2 \sum_{k=2}^{\infty} \frac{(\lambda t)^{k-2}}{(k-2)!} \\ &= e^{-\lambda t}(\lambda t)^2 \sum_{k=0}^{\infty} \frac{(\lambda t)^k}{k!} \\ &= (\lambda t)^2 \end{aligned}$$

Now forming the variance in terms of this last quantity and in terms of $E[K]$, we have

$$\begin{aligned} \sigma_K^2 &= E[K(K-1)] + E[K] - (E[K])^2 \\ &= (\lambda t)^2 + \lambda t - (\lambda t)^2 \end{aligned}$$

and so

$$\sigma_K^2 = \lambda t \qquad \blacksquare \; (2.133)$$

Thus we see that the mean and variance of the Poisson process are identical and each equal to λt.

In Figure 2.10 we plot the family of curves $P_k(t)$ as a function of k and as a function of λt (a convenient normalizing form for t).

Recollect from Eq. (II.27) in Appendix II that the z-transform (probability generating function) for the probability mass distribution of a discrete random variable K where

$$g_k = P[K = k]$$

is given by

$$\begin{aligned} G(z) &= E[z^K] \\ &= \sum_k z^k g_k \end{aligned}$$

for $|z| \le 1$. Applying this to the Poisson distribution derived above we have

$$\begin{aligned} E[z^K] &= \sum_{k=0}^{\infty} z^k P_k(t) \\ &= \sum_{k=0}^{\infty} e^{-\lambda t}\frac{(\lambda t z)^k}{k!} \\ &= e^{-\lambda t + \lambda t z} \end{aligned}$$

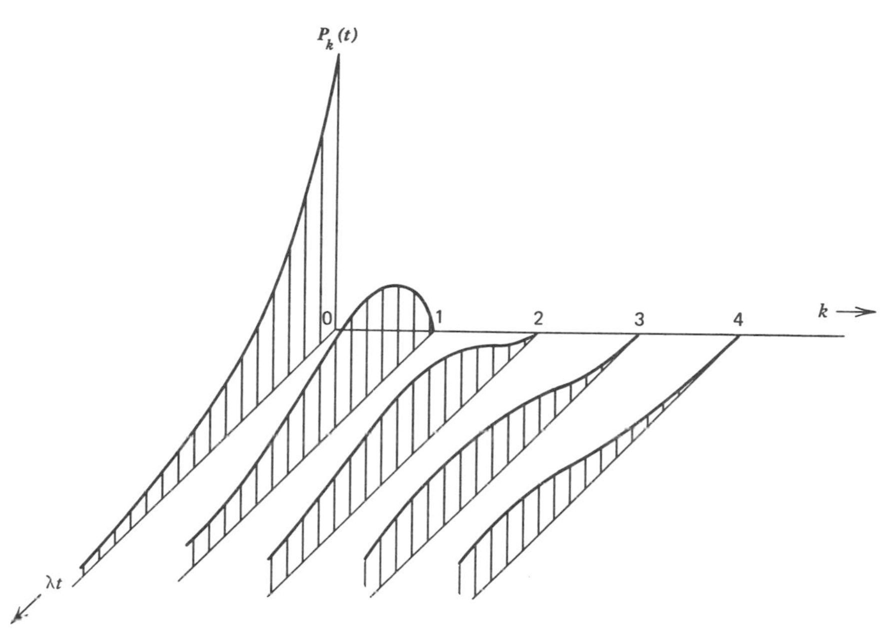

Figure 2.10 The Poisson distribution.

and so

$$G(z) = E[z^K] = e^{\lambda t(z-1)} \qquad (2.134)$$

We shall make considerable use of this result for the z-transform of a Poisson distribution. For example, we may now easily calculate the mean and variance as given in Eqs. (2.132) and (2.133) by taking advantage of the special properties of the z-transform (see Appendix II) as follows*:

$$G^{(1)}(1) = \frac{\partial}{\partial z} E[z^K] \bigg|_{z=1} = E[K]$$

Applying this to the Poisson distribution, we get

$$\begin{aligned} E[K] &= \lambda t e^{\lambda t(z-1)} \big|_{z=1} \\ &= \lambda t \end{aligned}$$

Also

$$\sigma_K^{\ 2} = G^{(2)}(1) + G^{(1)}(1) - [G^{(1)}(1)]^2$$

Thus, for the Poisson distribution,

$$\begin{aligned} \sigma_K^{\ 2} &= (\lambda t)^2 e^{\lambda t(z-1)} \big|_{z=1} + \lambda t - (\lambda t)^2 \\ &= \lambda t \end{aligned}$$

This confirms our earlier calculations.

* The shorthand notation for derivatives given in Eq. (II.25) should be reviewed.

Figure 2.11 Partitioning of the interval $(0, t)$.

We have introduced the Poisson process here as a pure birth process and we have found an expression for $P_k(t)$, the probability distribution for the number of arrivals during a given interval of length t. Now let us consider the joint distribution of the arrival instants when it is known beforehand that exactly k arrivals have occurred during that interval. We break the interval $(0, t)$ into $2k + 1$ intervals as shown in Figure 2.11. We are interested in A_k, which is defined to be the event that exactly one arrival occurs in each of the intervals $\{\beta_i\}$ and that no arrival occurs in any of the intervals $\{\alpha_i\}$. We wish to calculate the probability that the event A_k occurs given that exactly k arrivals have occurred in the interval $(0, t)$; from the definition of conditional probability we thus have

$$P[A_k \mid \text{exactly } k \text{ arrivals in } (0, t)] = \frac{P[A_k \text{ and exactly } k \text{ arrivals in } (0, t)]}{P[\text{exactly } k \text{ arrivals in } (0, t)]} \tag{2.135}$$

When we consider Poisson arrivals in nonoverlapping intervals, we are considering independent events whose joint probability may be calculated as the product of the individual probabilities (i.e., the Poisson process has independent increments). We note from Eq. (2.131), therefore, that

$$P[\text{one arrival in interval of length } \beta_i] = \lambda\beta_i e^{-\lambda\beta_i}$$

and

$$P[\text{no arrival in interval of length } \alpha_i] = e^{-\lambda\alpha_i}$$

Using this in Eq. (2.135) we have directly

$$\begin{aligned} &P[A_k \mid \text{exactly } k \text{ arrivals in } (0, t)] \\ &\quad= \frac{(\lambda\beta_1\lambda\beta_2 \cdots \lambda\beta_k e^{-\lambda\beta_1}e^{-\lambda\beta_2} \cdots e^{-\lambda\beta_k})(e^{-\lambda\alpha_1}e^{-\lambda\alpha_2} \cdots e^{-\lambda\alpha_{k+1}})}{[(\lambda t)^k/k!]e^{-\lambda t}} \\ &\quad= \frac{\beta_1\beta_2 \cdots \beta_k}{t^k} k! \end{aligned} \tag{2.136}$$

On the other hand, let us consider a new process that selects k points in the interval $(0, t)$ independently where each point is uniformly distributed over this interval. Let us now make the same calculation that we did for the Poisson process, namely,

$$P[A_k \mid \text{exactly } k \text{ arrivals in } (0, t)] = \left(\frac{\beta_1}{t}\right)\left(\frac{\beta_2}{t}\right) \cdots \left(\frac{\beta_k}{t}\right) k! \tag{2.137}$$

where the term $k!$ comes about since we do not distinguish among the permutations of the k points among the k chosen intervals. We observe that the two conditional probabilities given in Eqs. (2.136) and (2.137) are the same and, therefore, conclude that if an interval of length t contains exactly k arrivals from a Poisson process, then the joint distribution of the instants when these arrivals occurred is the same as the distribution of k points uniformly distributed over the same interval.

Furthermore, it is easy to show from the properties of our birth process that the Poisson process is one with independent increments; that is, defining $X(s, s + t)$ as the number of arrivals in the interval $(s, s + t)$ then the following is true:

$$P[X(s, s + t) = k] = \frac{(\lambda t)^k e^{-\lambda t}}{k!}$$

regardless of the location of this interval.

We would now like to investigate the intimate relationship between the Poisson process and the exponential distribution. This distribution also plays a central role in queueing theory. We consider the random variable $\tilde{t}$, which we recall is the *time between adjacent arrivals* in a queueing system, and whose PDF and pdf are given by $A(t)$ and $a(t)$, respectively, as already agreed for the interarrival times. From its definition, then, $a(t)\,\Delta t + o(\Delta t)$ is the probability that the next arrival occurs at least t sec and at most $(t + \Delta t)$ sec from the time of the last arrival.

Since the definition of $A(t)$ is merely the probability that the time between arrivals is $\leq t$, it must clearly be given by

$$A(t) = 1 - P[\tilde{t} > t]$$

But $P[\tilde{t} > t]$ is just the probability that *no* arrivals occur in $(0, t)$, that is, $P_0(t)$. Therefore, we have

$$A(t) = 1 - P_0(t)$$

and so from Eq. (2.131), we obtain the PDF (in the Poisson case)

$$A(t) = 1 - e^{-\lambda t} \qquad t \geq 0 \tag{2.138}$$

Differentiating, we obtain the pdf

$$a(t) = \lambda e^{-\lambda t} \qquad t \geq 0 \tag{2.139}$$

This is the well-known *exponential* distribution; its pdf and PDF are shown in Figure 2.12.

What we have shown by Eqs. (2.138) and (2.139) is that for a Poisson arrival process, the time between arrivals is exponentially distributed; thus we say that the Poisson arrival process has exponential interarrival times.

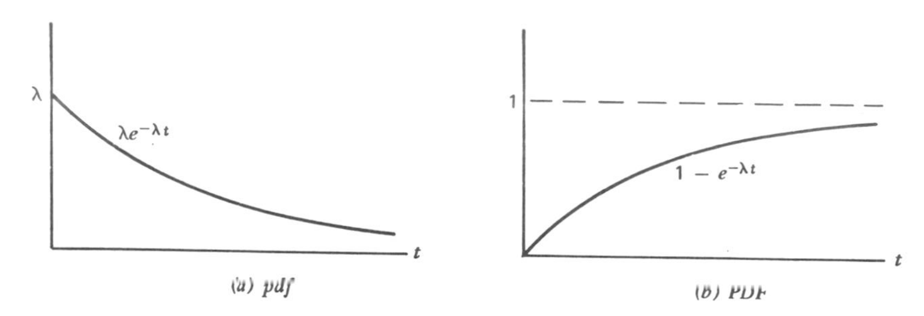

Figure 2.12 Exponential distribution.

The most amazing characteristic of the exponential distribution is that it has the remarkable *memoryless* property, which we introduced in our discussion of Markov processes. As the name indicates, the past history of a random variable that is distributed exponentially plays no role in predicting its future; precisely, we mean the following. Consider that an arrival has just occurred at time 0. If we inquire as to what our feeling is regarding the distribution of time until the next arrival, we clearly respond with the pdf given in Eq. (2.139). Now let some time pass, say, t_0 sec, during which no arrival occurs. We may at this point in time again ask, "What is the probability that the next arrival occurs t sec from *now*?" This question is the same question we asked at time 0 except we now know that the time between arrivals is at least t_0 sec. To answer the second question, we carry out the following calculations:

$$P[\tilde{t} \leq t + t_0 \mid \tilde{t} > t_0] = \frac{P[t_0 < \tilde{t} \leq t + t_0]}{P[\tilde{t} > t_0]}$$

$$= \frac{P[\tilde{t} \leq t + t_0] - P[\tilde{t} \leq t_0]}{P[\tilde{t} > t_0]}$$

Due to Eq. (2.138) we then have

$$P[\tilde{t} \leq t + t_0 \mid \tilde{t} > t_0] = \frac{1 - e^{-\lambda(t+t_0)} - (1 - e^{-\lambda t_0})}{1 - (1 - e^{-\lambda t_0})}$$

and so

$$P[\tilde{t} \leq t + t_0 \mid \tilde{t} > t_0] = 1 - e^{-\lambda t} \tag{2.140}$$

This result shows that the distribution of remaining time until the next arrival, given that t_0 sec has elapsed since the last arrival, is identically equal to the unconditional distribution of interarrival time. The impact of this statement is that our probabilistic feeling regarding the time until a future arrival occurs is independent of how long it has been since the last arrival occurred. That is, the future of an exponentially distributed random variable

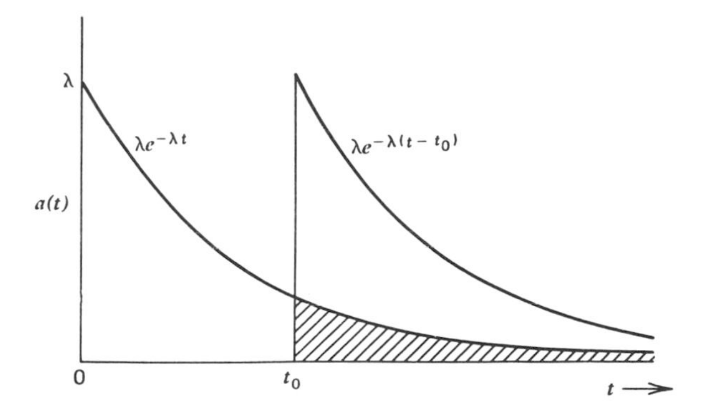

Figure 2.13 The memoryless property of the exponential distribution.

is independent of the past history of that variable and this distribution remains constant in time. The exponential distribution is the *only* continuous distribution with this property. (In the case of a discrete random variable we have seen that the geometric distribution is the only discrete distribution with that same property.) We may further appreciate the nature of this memoryless property by considering Figure 2.13. In this figure we show the exponential density $\lambda e^{-\lambda t}$. Now given that t_0 sec has elapsed, in order to calculate the density function for the time until the next arrival, what one must do is to take that portion of the density function lying to the right of the point t_0 (shown shaded) and recognize that this region represents our probabilistic feeling regarding the future; the portion of the density function in the interval from 0 to t_0 is past history and involves no more uncertainty. In order to make the shaded region into a bona fide density function, we must magnify it in order to increase its total area to unity; the appropriate magnification takes place by dividing the function representing the tail of this distribution by the area of the shaded region (which must, of course, be $P[\tilde{t} > t]$). This operation is identical to the operation of creating a conditional distribution by dividing a joint distribution by the probability of the condition. Thus the shaded region magnifies into the second indicated curve in Figure 2.13. This new function is an *exact replica* of the original density function as shown from time 0, except that it is shifted t_0 sec to the right. No other density function has the property that its tail everywhere possesses the exact same shape as the entire density function.

We now use the memoryless property of the exponential distribution in order to close the circle regarding the relationship between the Poisson and exponential distributions. Equation (2.140) gives an expression for the PDF of the interarrival time conditioned on the fact that it is at least as large as t_0. Let us position ourselves at time t_0 and ask for the probability that the next arrival occurs within the next Δt sec. From Eq. (2.140) we have

$$\begin{aligned} P[\tilde{t} \le t_0 + \Delta t \mid \tilde{t} > t_0] &= 1 - e^{-\lambda \Delta t} \\ &= 1 - \left[1 - \lambda \Delta t + \frac{(\lambda \Delta t)^2}{2!} - \cdots\right] \\ &= \lambda \Delta t + o(\Delta t) \end{aligned} \tag{2.141}$$

Equation (2.141) tells us, given that an arrival has not yet occurred, that the probability of it occurring in the next interval of length Δt sec is $\lambda \Delta t + o(\Delta t)$. But this is exactly assumption B_1 from the opening paragraphs of this section. Furthermore, the probability of no arrival in the interval $(t_0, t_0 + \Delta t)$ is calculated as

$$\begin{aligned} P[\tilde{t} > t_0 + \Delta t \mid \tilde{t} > t_0] &= 1 - P[\tilde{t} \le t_0 + \Delta t \mid \tilde{t} > t_0] \\ &= 1 - (1 - e^{-\lambda \Delta t}) \\ &= e^{-\lambda \Delta t} \\ &= 1 - \lambda \Delta t + \frac{(\lambda \Delta t)^2}{2!} - \cdots \\ &= 1 - \lambda \Delta t + o(\Delta t) \end{aligned}$$

This corroborates assumption B_2. Furthermore,

$$\begin{aligned} &P[2 \text{ or more arrivals in } (t_0, t_0 + \Delta t)] \\ &\qquad = 1 - P[\text{none in } (t_0, t_0 + \Delta t)] - P[\text{one in } (t_0, t_0 + \Delta t)] \\ &\qquad = 1 - [1 - \lambda \Delta t + o(\Delta t)] - [\lambda \Delta t + o(\Delta t)] \\ &\qquad = o(\Delta t) \end{aligned}$$

This corroborates the "multiple-birth" assumption. Our conclusion, then, is that the assumption of exponentially distributed interarrival times (which are independent one from the other) implies that we have a Poisson process, which implies we have a constant birth rate. The converse implications are also true. This relationship is shown graphically in Figure 2.14 in which the symbol $\leftrightarrow$ here denotes implication in both directions.

Let us now calculate the mean and variance for the exponential distribution as we did for the Poisson process. We shall proceed using two methods (the direct method and the transform method). We have

$$\begin{aligned} E[\tilde{t}] \triangleq \bar{t} &= \int_0^\infty t a(t)\, dt \\ &= \int_0^\infty t \lambda e^{-\lambda t}\, dt \end{aligned}$$

We use a trick here to evaluate the (simple) integral by recognizing that the integrand is no more than the partial derivative of the following integral,

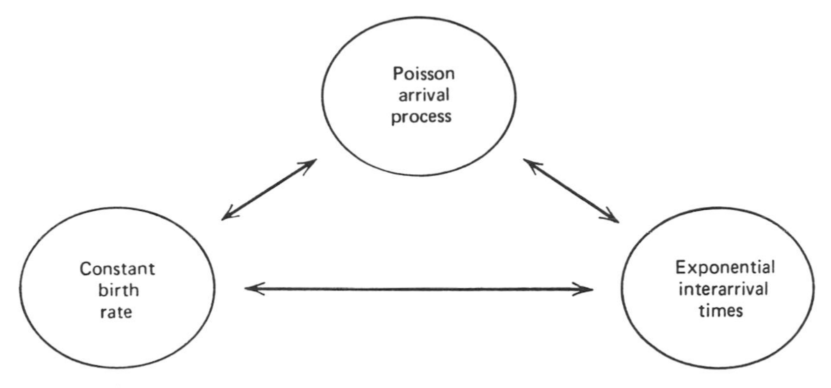

Figure 2.14 The memoryless triangle.

which may be evaluated by inspection:

$$\int_0^\infty t\lambda e^{-\lambda t}\,dt = -\lambda \frac{\partial}{\partial\lambda}\int_0^\infty e^{-\lambda t}\,dt$$
$$= -\lambda \frac{\partial}{\partial\lambda}\left(\frac{1}{\lambda}\right) = -\lambda\left(-\frac{1}{\lambda^2}\right)$$

and so

$$\bar{t} = \frac{1}{\lambda} \tag{2.142}$$

Thus we have that the average interarrival time for an exponential distribution is given by $1/\lambda$. This result is intuitively pleasing if we examine Eq. (2.141) and observe that the probability of an arrival in an interval of length Δt is given by $\lambda\,\Delta t$ [+ $o(\Delta t)$] and thus λ itself must be the average rate of arrivals; thus the average time between arrivals must be $1/\lambda$. In order to evaluate the variance, we first calculate the second moment for the interarrival time as follows:

$$E[(\tilde{t})^2] = \int_0^\infty t^2 a(t)\,dt$$
$$= \int_0^\infty t^2\lambda e^{-\lambda t}\,dt$$
$$= \lambda\frac{\partial^2}{\partial\lambda^2}\int_0^\infty e^{-\lambda t}\,dt$$
$$= \lambda\frac{\partial^2}{\partial\lambda^2}\left(\frac{1}{\lambda}\right)$$
$$= \frac{2}{\lambda^2}$$

Thus the variance is given by

$$\sigma_{\tilde{t}}^2 = E[(\tilde{t})^2] - (\bar{t})^2$$
$$= \frac{2}{\lambda^2} - \left(\frac{1}{\lambda}\right)^2$$

and so

$$\sigma_{\tilde{t}}^2 = \frac{1}{\lambda^2} \tag{2.143}$$

As usual, these two moments could more easily have been calculated by first considering the Laplace transform of the probability density function for this random variable. The notation for the Laplace transform of the interarrival pdf is $A^*(s)$. In this special case of the exponential distribution we then have the following:

$$A^*(s) \triangleq \int_0^\infty e^{-st} a(t)\, dt$$
$$= \int_0^\infty e^{-st} \lambda e^{-\lambda t}\, dt$$

and so

$$A^*(s) = \frac{\lambda}{s + \lambda} \tag{2.144}$$

Equation (2.144) thus gives the Laplace transform for the exponential density function. From Appendix II we recognize that the mean of this density function is given by

$$\bar{t} = -\left.\frac{dA^*(s)}{ds}\right|_{s=0}$$
$$= \left.\frac{\lambda}{(s+\lambda)^2}\right|_{s=0}$$
$$= \frac{1}{\lambda}$$

The second moment is also calculated in a similar fashion:

$$E[(\tilde{t})^2] = \left.\frac{d^2A^*(s)}{ds^2}\right|_{s=0}$$
$$= \left.\frac{2\lambda}{(s+\lambda)^3}\right|_{s=0}$$
$$= \frac{2}{\lambda^2}$$

and so

$$\sigma_{\tilde{t}}^2 = E[(\tilde{t})^2] - (\bar{t})^2$$
$$= \frac{1}{\lambda^2}$$

Thus we see the ease with which moments can be calculated by making use of transforms.

Note also, that the coefficient of variation [see Eq. (II.23)] for the exponential is

$$C_a \triangleq \frac{\sigma_{\tilde{t}}}{\bar{t}} = 1 \tag{2.145}$$

It will be of further interest to us later in the text to be able to calculate the pdf for the time interval X required in order to collect k arrivals from a Poisson process. Let us define this random variable in terms of the random variables t_n where t_n = time between nth and $(n-1)$th arrival (where the "zeroth" arrival is assumed to occur at time 0). Thus

$$X = \sum_{n=1}^{k} t_n$$

We define $f_X(x)$ to be the pdf for this random variable. From Appendix II we should immediately recognize that the density of X is given by the convolution of the densities on each of the t_n's, since they are independently distributed. Of course, this convolution operation is a bit lengthy to carry out, so let us use our further result in Appendix II, which tells us that the Laplace transform of the pdf for the sum of independent random variables is equal to the product of the Laplace transforms of the density for each. In our case each t_n has a common exponential distribution and therefore the Laplace transform for the pdf of X will merely be the kth power of $A^*(s)$ where $A^*(s)$ is given by Eq. (2.144); that is, defining

$$X^*(s) = \int_0^\infty e^{-sx} f_X(x)\, dx$$

for the Laplace transform of the pdf of our sum, we have

$$X^*(s) = [A^*(s)]^k$$

thus

$$X^*(s) = \left(\frac{\lambda}{s+\lambda}\right)^k \tag{2.146}$$

We must now invert this transform. Fortunately, we identify the needed transform pair as entry 10 in Table I.4 of Appendix I. Thus the density function we are looking for, which describes the time required to observe k arrivals, is given by

$$f_X(x) = \frac{\lambda(\lambda x)^{k-1}}{(k-1)!} e^{-\lambda x} \qquad x \geq 0 \tag{2.147}$$

This family of density functions (one for each value of k) is referred to as the family of *Erlang* distributions. We will have considerable use for this family later when we discuss the method of stages, in Chapter 4.

So much for the Poisson arrival process and its relation to the exponential distribution. Let us now return to the birth–death equations and consider a more general pure birth process in which we permit state-dependent birth rates λ_k (for the Poisson process, we had $\lambda_k = \lambda$). We once again insist that the death rates $\mu_k = 0$. From Eq. (2.127) this yields the set of equations

$$\begin{aligned} \frac{dP_k(t)}{dt} &= -\lambda_k P_k(t) + \lambda_{k-1} P_{k-1}(t) \qquad & k \geq 1 \\ \frac{dP_0(t)}{dt} &= -\lambda_0 P_0(t) & k = 0 \end{aligned} \tag{2.148}$$

Again, let us assume the initial distribution as given in Eq. (2.130), which states that (with probability one) the population begins with 0 members at time 0. Solving for $P_0(t)$ we have

$$P_0(t) = e^{-\lambda_0 t}$$

The general solution* for $P_k(t)$ is given below with an explicit expression for the first two values of k:

$$P_k(t) = e^{-\lambda_k t}\left[\lambda_{k-1}\int_0^t P_{k-1}(x) e^{\lambda_k x}\, dx + P_k(0)\right] \qquad k = 0, 1, 2, \ldots \tag{2.149}$$

$$P_1(t) = \frac{\lambda_0(e^{-\lambda_0 t} - e^{-\lambda_1 t})}{\lambda_1 - \lambda_0}$$

$$P_2(t) = \frac{\lambda_0 \lambda_1}{\lambda_1 - \lambda_0}\left[\frac{e^{-\lambda_2 t} - e^{-\lambda_1 t}}{\lambda_2 - \lambda_1} - \frac{e^{-\lambda_2 t} - e^{-\lambda_0 t}}{\lambda_2 - \lambda_0}\right]$$

As a third example of the time-dependent solution to the birth–death equations let us consider a *pure death* process in which a population is initiated with, say, N members and all that can happen to this population is that members die; none are born. Thus $\lambda_k = 0$ for all k, and $\mu_k = \mu \geq 0$

* The validity of this solution is easily verified by substituting Eq. (2.149) into Eq. (2.148).

for $k = 1, 2, \ldots, N$. For this constant death rate process we have

$$\frac{dP_k(t)}{dt} = -\mu P_k(t) + \mu P_{k+1}(t) \qquad 0 < k < N$$

$$\frac{dP_N(t)}{dt} = -\mu P_N(t) \qquad k = N$$

$$\frac{dP_0(t)}{dt} = \mu P_1(t) \qquad k = 0$$

Proceeding as earlier and using induction we obtain the solution

$$P_k(t) = \frac{(\mu t)^{N-k}}{(N-k)!} e^{-\mu t} \qquad 0 < k \le N \tag{2.150}$$
$$\frac{dP_0(t)}{dt} = \frac{\mu(\mu t)^{N-1}}{(N-1)!} e^{-\mu t} \qquad k = 0$$

Note the similarity of this last result to the Erlang distribution.

The last case we consider is a birth–death process in which all birth coefficients are equal to λ for $k \ge 0$ and all death coefficients are equal to μ for $k \ge 1$. This birth–death process with constant coefficients is of primary importance and forms perhaps the simplest interesting model of a queueing system. It is the celebrated M/M/1 queue; recall that the notation denotes a single-server queue with a Poisson arrival process and an exponential distribution for service time (from our earlier discussion we recognize that this is the memoryless system). Thus we may say

$$\begin{array}{c} \text{M/M/1} \\ \swarrow\!\!\nearrow \qquad \nwarrow\!\!\searrow \\ \left.\begin{array}{l}\lambda_k = \lambda \\ \mu_k = \mu\end{array}\right\} \longleftrightarrow \left\{\begin{array}{l}A(t) = 1 - e^{-\lambda t} \\ B(x) = 1 - e^{-\mu x}\end{array}\right. \end{array} \tag{2.151}$$

It should be clear why $A(t)$ is of exponential form from our earlier discussion relating the exponential interarrival distribution with the Poisson arrival process. In a similar fashion, since the death rate is constant ($\mu_k = \mu$, $k = 1, 2, \ldots$) then the same reasoning leads to the observation that the time between deaths is also exponentially distributed (in this case with a parameter μ). However, deaths correspond in the queueing system to service completions and, therefore, the service-time distribution $B(x)$ must be of exponential form. The interpretation of the condition $\mu_0 = 0$, which says that the death rate is zero when the population size is zero, corresponds in our queueing system to the condition that no service may take place when no customers are present. The behavior of the system M/M/1 will be studied throughout this text as we introduce new methods and new measures of performance; we will constantly check our sophisticated advanced techniques against this example since it affords one of the simplest applications of many of these advanced methods. Moreover, much of the behavior manifested in this system is characteristic of more complex queueing system behavior, and so a careful study here will serve to familiarize the reader with some important queueing phenomena.

Now for our first exposure to the M/M/1 system behavior. From the general equation for $P_k(t)$ given in Eq. (2.127) we find for this case that the corresponding differential-difference equations are

$$\begin{aligned} \frac{dP_k(t)}{dt} &= -(\lambda + \mu)P_k(t) + \lambda P_{k-1}(t) + \mu P_{k+1}(t) \qquad & k \geq 1 \\ \frac{dP_0(t)}{dt} &= -\lambda P_0(t) + \mu P_1(t) & k = 0 \end{aligned} \tag{2.152}$$

Many methods are available for solving this set of equations. Here, we choose to use the method of z-transforms developed in Appendix I. We have already seen one application of this method earlier in this chapter [when we defined the transform in Eq. (2.60) and applied it to the system of equations (2.55) to obtain the algebraic equation (2.61)]. Recall that the steps involved in applying the method of z-transforms to the solution of a set of difference equations may be summarized as follows:

1. Multiply the kth equation by z^k.
2. Sum all those equations that have the same form (typically true for $k = K, K + 1, \ldots$).
3. In this single equation, attempt to identify the z-transform for the unknown function. If all but a finite set of terms for the transform are present, then add the missing terms to get the function and then explicitly subtract them out in the equation.
4. Make use of the K "boundary" equations (namely, those that were omitted in step 2 above for $k = 0, 1, \ldots, K - 1$) to eliminate unknowns in the transformed equation.
5. Solve for the desired transform in the resulting algebraic, matrix or

differential* equation. Use the conservation relationship, Eq. (2.122), to eliminate the last unknown term.†

6. Invert the solution to get an explicit solution in terms of k.
7. If step 6 cannot be carried out, then moments may be obtained by differentiating with respect to z and setting $z = 1$.

Let us apply this method to Eq. (2.152). First we define the time-dependent transform

$$P(z, t) \triangleq \sum_{k=0}^{\infty} P_k(t) z^k \tag{2.153}$$

Next we multiply the kth differential equation by z^k (step 1) and then sum over all permitted k ($k = 1, 2, \ldots$) (step 2) to yield a single differential equation for the z-transform of $P_k(t)$:

$$\sum_{k=1}^{\infty} \frac{dP_k(t)}{dt} z^k = -(\lambda + \mu) \sum_{k=1}^{\infty} P_k(t) z^k + \lambda \sum_{k=1}^{\infty} P_{k-1}(t) z^k + \mu \sum_{k=1}^{\infty} P_{k+1}(t) z^k$$

Property 14 from Table I.1 in Appendix I permits us to move the differentiation operator outside the summation sign in this last equation. This summation then appears very much like $P(z, t)$ as defined above, except that it is missing the term for $k = 0$; the same is true of the first summation on the right-hand side of this last equation. In these two cases we need merely add and subtract the term $P_0(t)z^0$, which permits us to form the transform we are seeking. The second summation on the right-hand side is clearly $\lambda z P(z, t)$ since it contains an extra factor of z, but no missing terms. The last summation is missing a factor of z as well as the first two terms of this sum. We have now

* We sometimes obtain a differential equation at this stage if our original set of difference equations was, in fact, a set of differential-difference equations. When this occurs, we are effectively back to step 1 of this procedure as far as the differential variable (usually time) is concerned. We then proceed through steps 1–5 a second time using the Laplace transform for this new variable; our transform multiplier becomes e^{-st}, our sums become integrals, and our "tricks" become the properties associated with Laplace transforms (see Appendix I). Similar "returns to step 1" occur whenever a function of more than one variable is transformed; for each discrete variable, we require a z-transform and, for each continuous variable, we require a Laplace transform.

† When additional unknowns remain, we must appeal to the analyticity of the transform and observe that in its region of analyticity the transform must have a zero to cancel each pole (singularity) if the transform is to remain bounded. These additional conditions completely remove any remaining unknowns. This procedure will often be used and explained in the next few chapters.

carried out step 3, which yields

$$\frac{\partial}{\partial t}[P(z, t) - P_0(t)]$$

$$= -(\lambda + \mu)[P(z, t) - P_0(t)] + \lambda z P(z, t) + \frac{\mu}{z}[P(z, t) - P_0(t) - P_1(t)z] \tag{2.154}$$

The equation for $k = 0$ has so far not been used and we now apply it as described in step 4 ($K = 1$), which permits us to eliminate certain terms in Eq. (2.154):

$$\frac{\partial}{\partial t} P(z, t) = -\lambda P(z, t) - \mu[P(z, t) - P_0(t)] + \lambda z P(z, t) + \frac{\mu}{z}[P(z, t) - P_0(t)]$$

Rearranging this last equation we obtain the following linear, first-order (partial) differential equation for $P(z, t)$:

$$z \frac{\partial}{\partial t} P(z, t) = (1 - z)[(\mu - \lambda z)P(z, t) - \mu P_0(t)] \tag{2.155}$$

This differential equation requires further transforming, as mentioned in the first footnote to step 5. We must therefore define the Laplace transform for our function $P(z, t)$ as follows*:

$$P^*(z, s) \triangleq \int_{0^+}^{\infty} e^{-st} P(z, t)\, dt \tag{2.156}$$

Returning to step 1, applying this transform to Eq. (2.155), and taking advantage of property 11 in Table I.3 in Appendix I, we obtain

$$z[sP^*(z, s) - P(z, 0^+)] = (1 - z)[(\mu - \lambda z)P^*(z, s) - \mu P_0^*(s)] \tag{2.157}$$

where we have defined $P_0^*(s)$ to be the Laplace transform of $P_0(t)$, that is,

$$P_0^*(s) \triangleq \int_0^{\infty} e^{-st} P_0(t)\, dt \tag{2.158}$$

We have now transformed the set of differential-difference equations for $P_k(t)$ both on the discrete variable k and on the continuous variable t. This has led us to Eq. (2.157), which is a simple algebraic equation in our twice-transformed function $P^*(z, s)$, and this we may write as

$$P^*(z, s) = \frac{zP(z, 0^+) - \mu(1 - z)P_0^*(s)}{sz - (1 - z)(\mu - \lambda z)} \tag{2.159}$$

* For convenience we take the lower limit of integration to be 0^+ rather than our usual convention of using 0^- with the nonnegative random variables we often deal with. As a consequence, we must include the initial condition $P(z, 0^+)$ in Eq. (2.157).

Let us carry this argument just a bit further. From the definition in Eq. (2.153) we see that

$$P(z, 0^+) = \sum_{k=0}^{\infty} P_k(0^+)z^k \tag{2.160}$$

Of course, $P_k(0^+)$ is just our initial condition; whereas earlier we took the simple point of view that the system was empty at time 0 [that is, $P_0(0^+) = 1$ and all other terms $P_k(0^+) = 0$ for $k \neq 0$], we now generalize and permit i customers to be present at time 0, that is,

$$P_k(0^+) = \begin{cases} 1 & k = i \\ 0 & k \neq i \end{cases} \tag{2.161}$$

When $i = 0$ we have our original initial condition. Substituting Eq. (2.161) into Eq. (2.160) we see immediately that

$$P(z, 0^+) = z^i$$

which we may place into Eq. (2.159) to obtain

$$P^*(z, s) = \frac{z^{i+1} - \mu(1 - z)P_0^*(s)}{sz - (1 - z)(\mu - \lambda z)} \tag{2.162}$$

We are almost finished with step 5 except for the fact that the unknown function $P_0^*(s)$ appears in our equation. The second footnote to step 5 tells us how to proceed. From here on the analysis becomes a bit complex and it is beyond our desire at this point to continue the calculation; instead we relegate the excruciating details to the exercises below (see Exercise 2.20). It suffices to say that $P_0^*(s)$ is determined through the denominator roots of Eq. (2.162), which then leaves us with an explicit expression for our double transform. We are now at step 6 and must attempt to invert on both the transform variables; the exercises require the reader to show that the result of this inversion yields the final solution for our transient analysis, namely,

$$P_k(t) = e^{-(\lambda+\mu)t}\left[\rho^{(k-i)/2}I_{k-i}(at) + \rho^{(k-i-1)/2}I_{k+i+1}(at) + (1 - \rho)\rho^k \sum_{j=k+i+2}^{\infty} \rho^{-j/2}I_j(at)\right] \quad \blacksquare \tag{2.163}$$

where

$$\rho = \frac{\lambda}{\mu} \tag{2.164}$$

$$a = 2\mu\rho^{1/2} \tag{2.165}$$

and

$$I_k(x) \triangleq \sum_{m=0}^{\infty} \frac{(x/2)^{k+2m}}{(k + m)!\, m!} \qquad k \geq -1 \tag{2.166}$$

where $I_k(x)$ is the modified Bessel function of the first kind of order k. This last expression is most disheartening. What it has to say is that an appropriate model for the *simplest interesting* queueing system (discussed further in the next chapter) leads to an ugly expression for the time-dependent behavior of its state probabilities. As a consequence, we can only hope for greater complexity and obscurity in attempting to find time-dependent behavior of more general queueing systems.

More will be said about time-dependent results later in the text. Our main purpose now is to focus upon the *equilibrium* behavior of queueing systems rather than upon their transient behavior (which is far more difficult). In the next chapter the equilibrium behavior for birth–death queueing systems will be studied and in Chapter 4 more general Markovian queues in equilibrium will be considered. Only when we reach Chapter 5, Chapter 8, and then Chapter 2 (Volume II) will the time-dependent behavior be considered again. Let us now proceed to the simplest equilibrium behavior.

REFERENCES

BHAR 60 Bharucha-Reid, A. T., *Elements of the Theory of Markov Processes and Their Applications*, McGraw-Hill (New York) 1960.

COHE 69 Cohen, J., *The Single Server Queue*, North Holland (Amsterdam), 1969.

EILO 69 Eilon, S., "A Simpler Proof of $L = \lambda W$," *Operations Research*, **17,** 915–916 (1969).

FELL 66 Feller, W., *An Introduction to Probability Theory and Its Applications*, Vol. II, Wiley (New York), 1966.

FRY 28 Fry, T. C., *Probability and Its Engineering Uses*, Van Nostrand, (New York), 1928.

HOWA 71 Howard, R. A., *Dynamic Probabilistic Systems*, Vol. I (Markov Models) and Vol. II (Semi-Markov and Decision Processes), Wiley (New York), 1971.

JEWE 67 Jewell, W. S., "A Simple Proof of $L = \lambda W$," *Operations Research*, **15,** 1109–1116 (1967).

KHIN 60 Khinchin, A. J., *Mathematical Methods in the Theory of Queueing*, Griffin (London), 1960.

LITT 61 Little, J. D. C., "A Proof of the Queueing Formula $L = \lambda W$," *Operations Research*, **9,** 383–387 (1961).

MARK 07 Markov, A. A., "Extension of the Limit Theorems of Probability Theory to a Sum of Variables Connected in a Chain," The Notes of the Imperial Academy of Sciences of St. Petersburg VIII Series, Physio-Mathematical College, Vol. XXII, No. 9, December 5, 1907.

PALM 43 Palm, C., "Intensitätsschwankungen im Fernsprechverkehr," *Ericsson Technics*, **44,** 1–89 (1943).

PARZ 62 Parzen, E., *Stochastic Processes*, Holden Day (San Francisco), 1962.

STID 74 Stidham, S., Jr., "A Last Word on $L = \lambda W$," *Operations Research*, **22,** 417–421 (1974).

EXERCISES

2.1. Consider K independent sources of customers where the interarrival time between customers for each source is exponentially distributed with parameter λ_k (i.e., each source is a Poisson process). Now consider the arrival stream, which is formed by merging the input from each of the K sources defined above. Prove that this merged stream is also Poisson with parameter $\lambda = \lambda_1 + \lambda_2 + \cdots + \lambda_K$.

2.2. Referring back to the previous problem, consider this merged Poisson stream and now assume that we wish to break it up into several branches. Let p_i be the probability that a customer from the merged stream is assigned to the substream i. If the overall rate is λ customers per second, and if the substream probabilities p_i are chosen for each customer independently, then show that each of these substreams is a Poisson process with rate λp_i.

2.3. Let $\{X_j\}$ be a sequence of identically distributed mutually independent Bernoulli random variables (with $P[X_j = 1] = p$, and $P[X_j = 0] = 1 - p$). Let $S_N = X_1 + \cdots + X_N$ be the sum of a random number N of the random variables X_j, where N has a Poisson distribution with mean λ. Prove that S_N has a Poisson distribution with mean λp. (In general, the distribution of the sum of a random number of independent random variables is called a compound distribution.)

2.4. Find the pdf for the smallest of K independent random variables, each of which is exponentially distributed with parameter λ.

2.5. Consider the homogeneous Markov chain whose state diagram is

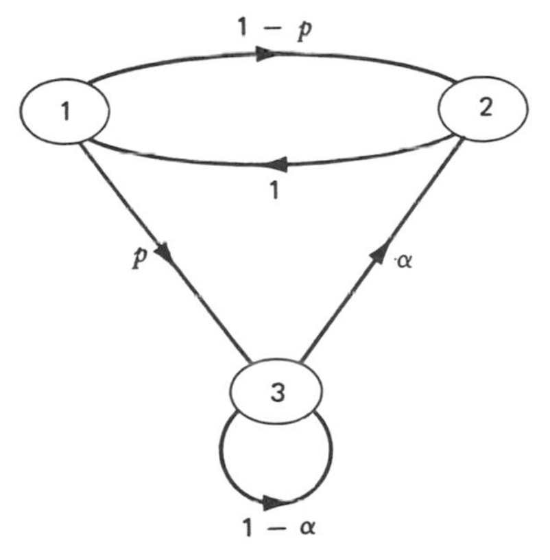

(a) Find $\mathbf{P}$, the probability transition matrix.
(b) Under what conditions (if any) will the chain be irreducible and aperiodic?

(c) Solve for the equilibrium probability vector $\boldsymbol{\pi}$.
(d) What is the mean recurrence time for state E_2?
(e) For which values of α and p will we have $\pi_1 = \pi_2 = \pi_3$? (Give a physical interpretation of this case.)

2.6. Consider the discrete-state, discrete-time Markov chain whose transition probability matrix is given by

$$P = \begin{bmatrix} \frac{1}{2} & \frac{1}{2} \\ \frac{3}{4} & \frac{1}{4} \end{bmatrix}$$

(a) Find the stationary state probability vector $\boldsymbol{\pi}$.
(b) Find $[\mathbf{I} - z\mathbf{P}]^{-1}$.
(c) Find the general form for $\mathbf{P}^n$.

2.7. Consider a Markov chain with states $E_0, E_1, E_2, \ldots$ and with transition probabilities

$$p_{ij} = e^{-\lambda} \sum_{n=0}^{j} \binom{i}{n} p^n q^{i-n} \frac{\lambda^{j-n}}{(j-n)!}$$

where $p + q = 1$ $(0 < p < 1)$.
(a) Is this chain irreducible? Periodic? Explain.
(b) We wish to find

$$\pi_i = \text{equilibrium probability of } E_i$$

Write π_i in terms of p_{ij} and π_j for $j = 0, 1, 2, \ldots$.
(c) From (b) find an expression relating $P(z)$ to $P[1 + p(z - 1)]$, where

$$P(z) = \sum_{i=0}^{\infty} \pi_i z^i$$

(d) Recursively (i.e., repeatedly) apply the result in (c) to itself and show that the nth recursion gives

$$P(z) = e^{\lambda(z-1)(1+p+p^2+\cdots+p^{n-1})} P[1 + p^n(z - 1)]$$

(e) From (d) find $P(z)$ and then recognize π_i.

2.8. Show that any point in or on the equilateral triangle of unit height shown in Figure 2.6 represents a three-component probability vector in the sense that the sum of the distances from any such point to each of the three sides must always equal unity.

2.9. Consider a pure birth process with constant birth rate λ. Let us consider an interval of length T, which we divide up into m segments each of length T/m. Define $\Delta t = T/m$.

(a) For Δt small, find the probability that a single arrival occurs in each of exactly k of the m intervals and that no arrivals occur in the remaining $m - k$ intervals.

(b) Consider the limit as $\Delta t \to 0$, that is, as $m \to \infty$ for fixed T, and evaluate the probability $P_k(T)$ that exactly k arrivals occur in the interval of length T.

2.10. Consider a population of bacteria of size $N(t)$ at time t for which $N(0) = 1$. We consider this to be a pure birth process in which any member of the population will split into two new members in the interval $(t, t + \Delta t)$ with probability $\lambda\,\Delta t + o(\Delta t)$ or will remain unchanged in this interval with probability $1 - \lambda\,\Delta t + o(\Delta t)$ as $\Delta t \to 0$.

(a) Let $P_k(t) = P[N(t) = k]$ and write down the set of differential-difference equations that must be satisfied by these probabilities.

(b) From part (a) show that the z-transform $P(z, t)$ for $N(t)$ must satisfy

$$P(z, t) = \frac{ze^{-\lambda t}}{1 - z + ze^{-\lambda t}}$$

(c) Find $E[N(t)]$.

(d) Solve for $P_k(t)$.

(e) Solve for $P(z, t)$, $E[N(t)]$ and $P_k(t)$ that satisfy the initial condition $N(0) = n \geq 1$.

(f) Consider the corresponding deterministic problem in which each bacterium splits into two every $1/\lambda$ sec and compare with the answer in part (c).

2.11. Consider a birth–death process with coefficients

$$\lambda_k = \begin{cases} \lambda & k = 0 \\ 0 & k \neq 0 \end{cases} \qquad \mu_k = \begin{cases} \mu & k = 1 \\ 0 & k \neq 1 \end{cases}$$

which corresponds to an M/M/1 queueing system where there is no room for waiting customers.

(a) Give the differential-difference equations for $P_k(t)$ $(k = 0, 1)$.

(b) Solve these equations and express the answers in terms of $P_0(0)$ and $P_1(0)$.

2.12. Consider a birth–death queueing system in which

$$\lambda_k = \lambda \qquad k \geq 0$$
$$\mu_k = k\mu \qquad k \geq 0$$

(a) For all k, find the differential-difference equations for

$$P_k(t) = P[k \text{ in system at time } t]$$

(b) Define the z-transform

$$P(z, t) = \sum_{k=0}^{\infty} P_k(t) z^k$$

and find the *partial* differential equation that $P(z, t)$ must satisfy.

(c) Show that the solution to this equation is

$$P(z, t) = \exp\left(\frac{\lambda}{\mu}(1 - e^{-\mu t})(z - 1)\right)$$

with the initial condition $P_0(0) = 1$.

(d) Comparing the solution in part (c) with Eq. (2.134), give the expression for $P_k(t)$ by inspection.

(e) Find the limiting values for these probabilities as $t \to \infty$.

2.13. Consider a system in which the birth rate decreases and the death rate increases as the number in the system k increases, that is,

$$\lambda_k = \begin{cases} (K - k)\lambda & k \leq K \\ 0 & k \geq K \end{cases} \qquad \mu_k = \begin{cases} k\mu & k \leq K \\ 0 & k > K \end{cases}$$

Write down the differential-difference equations for

$$P_k(t) = P[k \text{ in system at time } t].$$

2.14. Consider the case of a linear birth–death process in which $\lambda_k = k\lambda$ and $\mu_k = k\mu$.

(a) Find the partial-differential equation that must be satisfied by $P(z, t)$ as defined in Eq. (2.153).

(b) Assuming that the population size is one at time zero, show that the function that satisfies the equation in part (a) is

$$P(z, t) = \frac{\mu(1 - e^{(\lambda-\mu)t}) - (\lambda - \mu e^{(\lambda-\mu)t})z}{\mu - \lambda e^{(\lambda-\mu)t} - \lambda(1 - e^{(\lambda-\mu)t})z}$$

(c) Expanding $P(z, t)$ in a power series show that

$$P_k(t) = [1 - \alpha(t)][1 - \beta(t)][\beta(t)]^{k-1} \qquad k = 1, 2, \ldots$$
$$P_0(t) = \alpha(t)$$

and find $\alpha(t)$ and $\beta(t)$.

(d) Find the mean and variance for the number in system at time t.

(e) Find the limiting probability that the population dies out by time t for $t \to \infty$.

2.15. Consider a linear birth–death process for which $\lambda_k = k\lambda + \alpha$ and $\mu_k = k\mu$.

(a) Find the differential-difference equations that must be satisfied by $P_k(t)$.

(b) From (a) find the partial-differential equation that must be satisfied by the time-dependent transform defined as

$$P(z, t) = \sum_{k=0}^{\infty} P_k(t) z^k$$

(c) What is the value of $P(1, t)$? Give a verbal interpretation for the expression

$$\bar{N}(t) = \lim_{z \to 1} \frac{\partial}{\partial z} P(z, t)$$

(d) Assuming that the population size begins with i members at time 0, find an ordinary differential equation for $\bar{N}(t)$ and then solve for $\bar{N}(t)$. Consider the case $\lambda = \mu$ as well as $\lambda \neq \mu$.

(e) Find the limiting value for $\bar{N}(t)$ in the case $\lambda < \mu$ (as $t \to \infty$).

2.16. Consider the equations of motion in Eq. (2.148) and define the Laplace transform

$$P_k^*(s) = \int_0^{\infty} P_k(t) e^{-st}\, dt$$

For our initial condition we will assume $P_0(t) = 1$ for $t = 0$. Transform Eq. (2.148) to obtain a set of linear difference equations in $\{P_k^*(s)\}$.

(a) Show that the solution to the set of equations is

$$P_k^*(s) = \frac{\prod_{i=0}^{k-1} \lambda_i}{\prod_{i=0}^{k} (s + \lambda_i)}$$

(b) From (a) find $P_k(t)$ for the case $\lambda_i = \lambda$ $(i = 0, 1, 2, \ldots)$.

2.17. Consider a time interval $(0, t)$ during which a Poisson process generates arrivals at an average rate λ. Derive Eq. (2.147) by considering the two events: exactly $k - 1$ arrivals occur in the interval $(0, t - \Delta t)$ and the event that exactly one arrival occurs in the interval $(t - \Delta t, t)$. Considering the limit as $\Delta t \to 0$ we immediately arrive at our desired result.

2.18. A barber opens up for business at $t = 0$. Customers arrive at random in a Poisson fashion; that is, the pdf of interarrival time is $u(t) - \lambda e^{-\lambda t}$. Each haircut takes X sec (where X is some random variable). Find the probability P that the second arriving customer will not have to wait and also find W, the average value of his waiting time for the two following cases:

i. $X = c =$ constant.

ii. X is exponentially distributed with pdf:

$$b(x) = \mu e^{-\mu x}$$

2.19. At $t = 0$ customer A places a request for service and finds all m servers busy and n other customers waiting for service in an M/M/m queueing system. All customers wait as long as necessary for service, waiting customers are served in order of arrival, and no new requests for service are permitted after $t = 0$. Service times are assumed to be mutually independent, identical, exponentially distributed random variables, each with mean duration $1/\mu$.

(a) Find the expected length of time customer A spends waiting for service in the queue.

(b) Find the expected length of time from the arrival of customer A at $t = 0$ until the system becomes completely empty (all customers complete service).

(c) Let X be the order of completion of service of customer A; that is, $X = k$ if A is the kth customer to complete service after $t = 0$. Find $P[X = k]$ $(k = 1, 2, \ldots, m + n + 1)$.

(d) Find the probability that customer A completes service before the customer immediately ahead of him in the queue.

(e) Let $\tilde{w}$ be the amount of time customer A waits for service. Find $P[\tilde{w} > x]$.

2.20. In this problem we wish to proceed from Eq. (2.162) to the transient solution in Eq. (2.163). Since $P^*(z, s)$ must converge in the region $|z| \leq 1$ for $\text{Re}(s) > 0$, then, in this region, the zeros of the denominator in Eq. (2.162) must also be zeros of the numerator.

(a) Find those two values of z that give the denominator zeros, and denote them by $\alpha_1(s)$, $\alpha_2(s)$ where $|\alpha_2(s)| < |\alpha_1(s)|$.

(b) Using Rouche's theorem (see Appendix I) show that the denominator of $P^*(z, s)$ has a single zero within the unit disk $|z| \leq 1$.

(c) Requiring that the numerator of $P^*(z, s)$ vanish at $z = \alpha_2(s)$ from our earlier considerations, find an explicit expression for $P_0^*(s)$.

(d) Write $P^*(z, s)$ in terms of $\alpha_1(s) = \alpha_1$ and $\alpha_2(s) = \alpha_2$. Then show that this equation may be reduced to

$$P^*(z, s) = \frac{(z^i + \alpha_2 z^{i-1} + \cdots + \alpha_2^i) + \alpha_2^{i+1}/(1 - \alpha_2)}{\lambda\alpha_1(1 - z/\alpha_1)}$$

(e) Using the fact that $|\alpha_2| < 1$ and that $\alpha_1\alpha_2 = \mu/\lambda$ show that the inversion on z yields the following expression for $P_k^*(s)$, which is the Laplace transform for our transient probabilities $P_k(t)$:

$$P_k^*(s) = \frac{1}{\lambda}\left[\alpha_1^{i-n-1} + \left(\frac{\mu}{\lambda}\right)\alpha_1^{i-n-3} + \left(\frac{\mu}{\lambda}\right)^2\alpha_1^{i-n-5} + \cdots \right.$$
$$\left. + \left(\frac{\mu}{\lambda}\right)^i\alpha_1^{-i-n-1} + \left(\frac{\lambda}{\mu}\right)^{n+1}\sum_{k=n+i+2}^{\infty}\left(\frac{\mu}{\lambda\alpha_1}\right)^k\right]$$

(f) In what follows we take advantage of property 4 in Table I.3 and also we make use of the following transform pair:

$$k\rho^{k/2}t^{-1}I_k(at) \Leftrightarrow \left[\frac{s + \sqrt{s^2 - 4\lambda\mu}}{2\lambda}\right]^{-k}$$

where ρ and a are as defined in Eqs. (2.164), (2.165) and where $I_k(x)$ is the modified Bessel function of the first kind of order k as defined in Eq. (2.166). Using these facts and the simple relations among Bessel functions, namely,

$$\frac{2k}{x}I_k(x) = I_{k-1}(x) - I_{k+1}(x) \quad \text{and} \quad I_k(x) = I_{-k}(x)$$

show that Eq. (2.163) is the inverse transform for the expression shown in part (e).

2.21. The random variables $X_1, X_2, \ldots, X_i, \ldots$ are independent, identically distributed random variables each with density $f_X(x)$ and characteristic function $\Phi_X(u) = E[e^{juX}]$. Consider a Poisson process $N(t)$ with parameter λ which is independent of the random variables X_i. Consider now a second random process of the form

$$X(t) = \sum_{i=1}^{N(t)} X_i$$

This second random process is clearly a family of staircase functions where the jumps occur at the discontinuities of the random process $N(t)$; the magnitudes of such jumps are given by the random variables X_i. Show that the characteristic function of this second random process is given by

$$\phi_{X(t)}(u) = e^{\lambda t[\phi_X(u)-1]}$$

2.22. Passengers and taxis arrive at a service point from independent Poisson processes at rates λ, μ, respectively. Let the queue size at time t be q_t, a negative value denoting a line of taxis, a positive value denoting a queue of passengers. Show that, starting with $q_0 = 0$, the distribution of q_t is given by the difference between independent Poisson variables of means λt, μt. Show by using the normal approximation that if $\lambda = \mu$, the probability that $-k \leq q_t \leq k$ is, for large t, $(2k + 1)(4\pi\lambda t)^{-1/2}$.

PART II

ELEMENTARY QUEUEING THEORY

Elementary here means that all the systems we consider are pure Markovian and, therefore, our state description is convenient and manageable. In Part I we developed the time-dependent equations for the behavior of birth–death processes; here in Chapter 3 we address the equilibrium solution for these systems. The key equation in this chapter is Eq. (3.11), and the balance of the material is the simple application of that formula. It, in fact, is no more than the solution to the equation $\boldsymbol{\pi} = \boldsymbol{\pi}\mathbf{P}$ derived in Chapter 2. The key tool used here is again that which we find throughout the text, namely, the calculation of flow rates across the boundaries of a closed system. In the case of equilibrium we merely ask that the rate of flow into be equal to the rate of flow out of a system. The application of these basic results is more than just an exercise for it is here that we first obtain some equations of use in engineering and designing queueing systems. The classical M/M/1 queue is studied and some of its important performance measures are evaluated. More complex models involving finite storage, multiple servers, finite customer population, and the like, are developed in the balance of this chapter. In Chapter 4 we leave the birth–death systems and allow more general Markovian queues, once again to be studied in equilibrium. We find that the techniques here are similar to our earlier ones, but find that no general solution such as Eq. (3.11) is available; each system is a case unto itself and so we are rapidly led into the solutions of difference equations, which force us to look carefully at the method of z-transforms for these solutions. The ingenious method of stages introduced by Erlang is considered here and its generality discussed. At the end of the chapter we introduce (for later use in Volume II) networks of Markovian queues in which we take exquisite advantage of the memoryless properties that Markovian queues provide even in a network environment. At this point, however, we have essentially exhausted the use of the memoryless distribution and we must depart from that crutch in the following parts.

3

Birth–Death Queueing Systems in Equilibrium

In the previous chapter we studied a variety of stochastic processes. We indicated that Markov processes play a fundamental role in the study of queueing systems, and after presenting the main results from that theory, we then considered a special form of Markov process known as the birth–death process. We also showed that birth–death processes enjoy a most convenient property, namely, that the time between births and the time between deaths (when the system is nonempty) are each exponentially distributed.* We then developed Eq. (2.127), which gives the basic equations of motion for the general birth–death process with stationary birth and death rates.† The solution of this set of equations gives the transient behavior of the queueing process and some important special cases were discussed earlier. In this chapter we study the limiting form of these equations to obtain the equilibrium behavior of birth–death queueing systems.

The importance of elementary queueing theory comes from its historical influence as well as its ability to describe behavior that is to be found in more complex queueing systems. The methods of analysis to be used in this chapter in large part do *not* carry over to the more involved queueing situations; nevertheless, the obtained results *do* provide insight into the basic behavior of many of these other queueing systems.

It is necessary to keep in mind how the birth–death process describes queueing systems. As an example, consider a doctor's office made up of a waiting room (in which a queue is allowed to form, unfortunately) and a service facility consisting of the doctor's examination room. Each time a patient enters the waiting room from outside the office we consider this to be an *arrival* to the queueing system; on the other hand, this arrival may well be considered to be a *birth* of a new member of a population, where the population consists of all patients present. In a similar fashion, when a patient leaves

* This comes directly from the fact that they are Markov processes.

† In addition to these equations, one requires the conservation relation given in Eq. (2.122) and a set of initial conditions $\{P_k(0)\}$.

the office after being treated, he is considered to be a *departure* from the queueing system; in terms of a birth–death process this is considered to be a *death* of a member of the population.

We have considerable freedom in constructing a large number of queueing systems through the choice of the birth coefficients λ_k and death coefficients μ_k, as we shall see shortly. First, let us establish the general solution for the equilibrium behavior.

3.1. GENERAL EQUILIBRIUM SOLUTION

As we saw in Chapter 2 the time-dependent solution of the birth–death system quickly becomes unmanageable when we consider any sophisticated set of birth–death coefficients. Furthermore, were we always capable of solving for $P_k(t)$ it is not clear how useful that set of functions would be in aiding our understanding of the behavior of these queueing systems (too much information is sometimes a curse!). Consequently, it is natural for us to ask whether the probabilities $P_k(t)$ eventually settle down as t gets large and display no more "transient" behavior. This inquiry on our part is analogous to the questions we asked regarding the existence of π_k in the limit of $\pi_k(t)$ as $t \to \infty$. For our queueing studies here we choose to denote the limiting probability as p_k rather than π_k, purely for convenience. Accordingly, let

$$p_k \triangleq \lim_{t \to \infty} P_k(t) \tag{3.1}$$

where p_k is interpreted as the limiting probability that the system contains k members (or equivalently is in state E_k) at some arbitrary time in the distant future. The question regarding the existence of these limiting probabilities is of concern to us, but will be deferred at this point until we obtain the general steady-state or limiting solution. It is important to understand that whereas p_k (assuming it exists) is no longer a function of t, we are not claiming that the process does not move from state to state in this limiting case; certainly, the number of members in the population will change with time, but the *long-run* probability of finding the system with k members will be properly described by p_k.

Accepting the existence of the limit in Eq. (3.1), we may then set lim $dP_k(t)/dt$ as $t \to \infty$ equal to zero in the Kolmogorov forward equations (of motion) for the birth–death system [given in Eqs. (2.127)] and immediately obtain the result

$$0 = -(\lambda_k + \mu_k)p_k + \lambda_{k-1}p_{k-1} + \mu_{k+1}p_{k+1} \qquad k \geq 1 \tag{3.2}$$

$$0 = -\lambda_0 p_0 + \mu_1 p_1 \qquad k = 0 \tag{3.3}$$

The annoying task of providing a separate equation for $k = 0$ may be overcome by agreeing once and for all that the following birth and death coefficients are identically equal to 0:

$$\lambda_{-1} = \lambda_{-2} = \lambda_{-3} = \cdots = 0$$
$$\mu_0 = \mu_{-1} = \mu_{-2} = \cdots = 0$$

Furthermore, since it is perfectly clear that we cannot have a negative number of members in our population, we will, in most cases, adopt the convention that

$$p_{-1} = p_{-2} = p_{-3} = \cdots = 0$$

Thus, for all values of k, we may reformulate Eqs. (3.2) and (3.3) into the following set of difference equations for $k = \ldots, -2, -1, 0, 1, 2, \ldots$

$$0 = -(\lambda_k + \mu_k)p_k + \lambda_{k-1}p_{k-1} + \mu_{k+1}p_{k+1} \tag{3.4}$$

We also require the conservation relation

$$\sum_{k=0}^{\infty} p_k = 1 \tag{3.5}$$

Recall from the previous chapter that the limit given in the Eq. (3.1) is independent of the initial conditions.

Just as we used the state-transition-rate diagram as an inspection technique for writing down the equations of motion in Chapter 2, so may we use the same concept in writing down the *equilibrium* equations [Eqs. (3.2) and (3.3)] directly from that diagram. In this equilibrium case it is clear that flow must be *conserved* in the sense that the input flow must equal the output flow from a given state. For example, if we look at Figure 2.9 once again and concentrate on state E_k in equilibrium, we observe that

$$\text{Flow rate into } E_k = \lambda_{k-1}p_{k-1} + \mu_{k+1}p_{k+1}$$

and

$$\text{Flow rate out of } E_k = (\lambda_k + \mu_k)p_k$$

In equilibrium these two must be the same and so we have immediately

$$\lambda_{k-1}p_{k-1} + \mu_{k+1}p_{k+1} = (\lambda_k + \mu_k)p_k \tag{3.6}$$

But this last is just Eq. (3.4) again! *By inspection we have established the equilibrium difference equations for our system.* The same comments apply here as applied earlier regarding the conservation of flow across *any* closed boundary; for example, rather than surrounding each state and writing down its equation we could choose a sequence of boundaries the first of which surrounds E_0, the second of which surrounds E_0 and E_1, and so on, each time adding the next higher-numbered state to get a new boundary. In such an example the kth boundary (which surrounds states $E_0, E_1, \ldots, E_{k-1}$) would lead to the following simple conservation of flow relationship:

$$\lambda_{k-1}p_{k-1} = \mu_k p_k \tag{3.7}$$

This last set of equations is equivalent to drawing a vertical line separating adjacent states and equating flows across this boundary; this set of difference equations is equivalent to our earlier set.

The solution for p_k in Eq. (3.4) may be obtained by at least two methods. One way is first to solve for p_1 in terms of p_0 by considering the case $k = 0$, that is,

$$p_1 = \frac{\lambda_0}{\mu_1} p_0 \tag{3.8}$$

We may then consider Eq. (3.4) for the case $k = 1$ and using Eq. (3.8) obtain

$$0 = -(\lambda_1 + \mu_1)p_1 + \lambda_0 p_0 + \mu_2 p_2$$

$$0 = -(\lambda_1 + \mu_1)\frac{\lambda_0}{\mu_1} p_0 + \lambda_0 p_0 + \mu_2 p_2$$

$$0 = -\frac{\lambda_1\lambda_0}{\mu_1} p_0 - \lambda_0 p_0 + \lambda_0 p_0 + \mu_2 p_2$$

and so

$$p_2 = \frac{\lambda_0\lambda_1}{\mu_1\mu_2} p_0 \tag{3.9}$$

If we examine Eqs. (3.8) and (3.9) we may justifiably guess that the general solution to Eq. (3.4) must be

$$p_k = \frac{\lambda_0\lambda_1 \cdots \lambda_{k-1}}{\mu_1\mu_2 \cdots \mu_k} p_0 \tag{3.10}$$

To validate this assertion we need merely use the inductive argument and apply Eq. (3.10) to Eq. (3.4) solving for p_{k+1}. Carrying out this operation we do, in fact, find that (3.10) is the solution to the general birth–death process in this steady-state or limiting case. We have thus expressed all equilibrium probabilities p_k in terms of a single unknown constant p_0:

$$p_k = p_0 \prod_{i=0}^{k-1} \frac{\lambda_i}{\mu_{i+1}} \qquad k = 0, 1, 2, \ldots \qquad \blacksquare \tag{3.11}$$

(Recall the usual convention that an empty product is unity by definition.) Equation (3.5) provides the additional condition that allows us to determine p_0; thus, summing over all k, we obtain

$$p_0 = \frac{1}{1 + \sum_{k=1}^{\infty} \prod_{i=0}^{k-1} \frac{\lambda_i}{\mu_{i+1}}} \qquad \blacksquare \tag{3.12}$$

This "product" solution for p_k $(k = 0, 1, 2, \ldots)$ simply obtained, is a *principal* equation in elementary queueing theory and, in fact, is the point of departure for all of our further solutions in this chapter.

A second easy way to obtain the solution to Eq. (3.4) is to rewrite that equation as follows:

$$\lambda_{k-1}p_{k-1} - \mu_k p_k = \lambda_k p_k - \mu_{k+1}p_{k+1} \tag{3.13}$$

Defining

$$g_k = \lambda_k p_k - \mu_{k+1}p_{k+1} \tag{3.14}$$

we have from Eq. (3.13) that

$$g_{k-1} = g_k \tag{3.15}$$

Clearly Eq. (3.15) implies that

$$g_k = \text{constant with respect to } k \tag{3.16}$$

However, since $\lambda_{-1} = \mu_0 = 0$, Eq. (3.14) gives

$$g_{-1} = 0$$

and so the constant in Eq. (3.16) must be 0. Setting g_k equal to 0, we immediately obtain from Eq. (3.14)

$$p_{k+1} = \frac{\lambda_k}{\mu_{k+1}} p_k \tag{3.17}$$

Solving Eq. (3.17) successively beginning with $k = 0$ we obtain the earlier solution, namely, Eqs. (3.11) and (3.12).

We now address ourselves to the *existence* of the steady-state probabilities p_k given by Eqs. (3.11) and (3.12). Simply stated, in order for those expressions to represent a probability distribution, we usually require that $p_0 > 0$. This clearly places a condition upon the birth and death coefficients in those equations. Essentially, what we are requiring is that the system occasionally empties; that this is a condition for stability seems quite reasonable when one interprets it in terms of real life situations.* More precisely, we may classify the possibilities by first defining the two sums

$$S_1 \triangleq \sum_{k=0}^{\infty} \prod_{i=0}^{k-1} \frac{\lambda_i}{\mu_{i+1}} \tag{3.18}$$

$$S_2 \triangleq \sum_{k=0}^{\infty} \left(1 \bigg/ \left(\lambda_k \prod_{i=0}^{k-1} \frac{\lambda_i}{\mu_{i+1}} \right) \right) \tag{3.19}$$

* It is easy to construct counterexamples to this case, and so we require the precise arguments which follow.

All states E_k of our birth–death process will be *ergodic* if and only if

$$\text{Ergodic:} \qquad S_1 < \infty \qquad S_2 = \infty$$

On the other hand, all states will be *recurrent null* if and only if

$$\text{Recurrent null:} \qquad S_1 = \infty \qquad S_2 = \infty$$

Also, all states will be *transient* if and only if

$$\text{Transient:} \qquad S_1 = \infty \qquad S_2 < \infty$$

It is the ergodic case that gives rise to the equilibrium probabilities $\{p_k\}$ and that is of most interest to our studies. We note that the condition for ergodicity is met whenever the sequence $\{\lambda_k/\mu_k\}$ remains below unity from some k onwards, that is, if there exists some k_0 such that for all $k \geq k_0$ we have

$$\frac{\lambda_k}{\mu_k} < 1 \tag{3.20}$$

We will find this to be true in most of the queueing systems we study.

We are now ready to apply our general solution as given in Eqs. (3.11) and (3.12) to some very important special cases. Before we launch headlong into that discussion, let us put at ease those readers who feel that the birth–death constraints of permitting only nearest-neighbor transitions are too confining. It is true that the solution given in Eqs. (3.11) and (3.12) applies only to nearest-neighbor birth–death processes. However, rest assured that the equilibrium methods we have described can be extended to more general than nearest-neighbor systems; these generalizations are considered in Chapter 4.

3.2. M/M/1: THE CLASSICAL QUEUEING SYSTEM

As mentioned in Chapter 2, the celebrated M/M/1 queue is the simplest nontrivial interesting system and may be described by selecting the birth–death coefficients as follows:

$$\lambda_k = \lambda \qquad k = 0, 1, 2, \ldots$$

$$\mu_k = \mu \qquad k = 1, 2, 3, \ldots$$

That is, we set all birth* coefficients equal to a constant λ and all death*

* In this case, the average interarrival time is $\bar{t} = 1/\lambda$ and the average service time is $\bar{x} = 1/\mu$; this follows since $\tilde{t}$ and $\tilde{x}$ are both exponentially distributed.

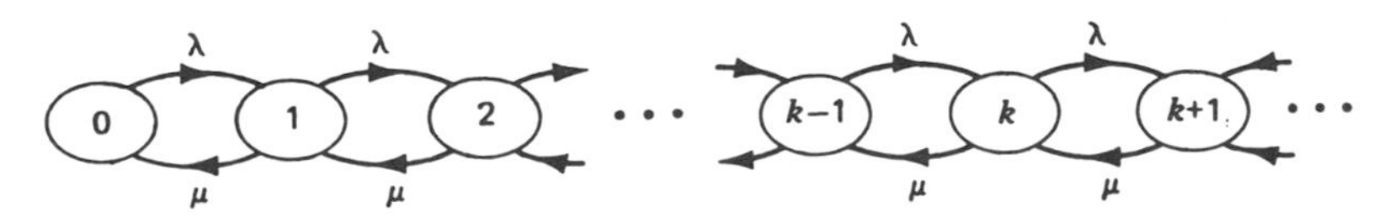

Figure 3.1 State-transition-rate diagram for M/M/1.

coefficients equal to a constant μ. We further assume that infinite queueing space is provided and that customers are served in a first-come-first-served fashion (although this last is not necessary for many of our results). For this important example the state-transition-rate diagram is as given in Figure 3.1.

Applying these coefficients to Eq. (3.11) we have

$$p_k = p_0 \prod_{i=0}^{k-1} \frac{\lambda}{\mu}$$

or

$$p_k = p_0 \left(\frac{\lambda}{\mu}\right)^k \qquad k \geq 0 \tag{3.21}$$

The result is immediate. The conditions for our system to be ergodic (and, therefore, to have an equilibrium solution $p_k > 0$) are that $S_1 < \infty$ and $S_2 = \infty$; in this case the first condition becomes

$$S_1 = \sum_{k=0}^{\infty} \frac{p_k}{p_0} = \sum_{k=0}^{\infty} \left(\frac{\lambda}{\mu}\right)^k < \infty$$

The series on the left-hand side of the inequality will converge if and only if $\lambda/\mu < 1$. The second condition for ergodicity becomes

$$S_2 = \sum_{k=0}^{\infty} \frac{1}{\lambda(p_k/p_0)} = \sum_{k=0}^{\infty} \frac{1}{\lambda}\left(\frac{\mu}{\lambda}\right)^k = \infty$$

This last condition will be satisfied if $\lambda/\mu \leq 1$; thus the necessary and sufficient condition for ergodicity in the M/M/1 queue is simply $\lambda < \mu$. In order to solve for p_0 we use Eq. (3.12) [or Eq. (3.5) as suits the reader] and obtain

$$p_0 = 1 \Bigg/ \left[1 + \sum_{k=1}^{\infty} \left(\frac{\lambda}{\mu}\right)^k\right]$$

The sum converges since $\lambda < \mu$ and so

$$p_0 = \frac{1}{1 + \dfrac{\lambda/\mu}{1 - \lambda/\mu}}$$

Thus

$$p_0 = 1 - \frac{\lambda}{\mu} \tag{3.22}$$

From Eq. (2.29) we have $\rho = \lambda/\mu$. From our stability conditions, we therefore require that $0 \le \rho < 1$; note that this insures that $p_0 > 0$. From Eq. (3.21) we have, finally,

$$p_k = (1 - \rho)\rho^k \qquad k = 0, 1, 2, \ldots \tag{3.23}$$

Equation (3.23) is indeed the solution for the steady-state probability of finding k customers in the system.* We make the important observation that p_k depends upon λ and μ only through their ratio ρ.

The solution given by Eq. (3.23) for this fundamental system is graphed in Figure 3.2 for the case of $\rho = 1/2$. Clearly, this is the geometric distribution (which shares the fundamental memoryless property with the exponential distribution). As we develop the behavior of the M/M/1 queue, we shall continue to see that almost all of its important probability distributions are of the memoryless type.

An important measure of a queueing system is the average number of customers in the system $\bar{N}$. This is clearly given by

$$\begin{aligned} \bar{N} &= \sum_{k=0}^{\infty} k p_k \\ &= (1 - \rho) \sum_{k=0}^{\infty} k \rho^k \end{aligned}$$

Using the trick similar to the one used in deriving Eq. (2.142) we have

$$\begin{aligned} \bar{N} &= (1 - \rho)\rho \frac{\partial}{\partial \rho} \sum_{k=0}^{\infty} \rho^k \\ &= (1 - \rho)\rho \frac{\partial}{\partial \rho} \frac{1}{1 - \rho} \end{aligned}$$

$$\bar{N} = \frac{\rho}{1 - \rho} \tag{3.24}$$

* If we inspect the transient solution for M/M/1 given in Eq. (2.163), we see the term $(1 - \rho)\rho^k$; the reader may verify that, for $\rho < 1$, the limit of the transient solution agrees with our solution here.

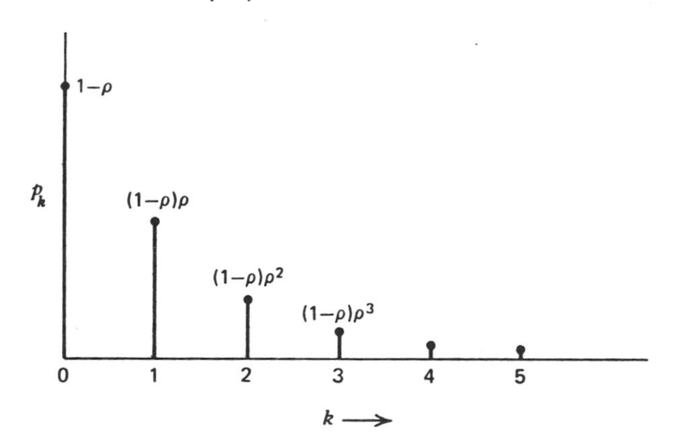

Figure 3.2 The solution for p_k in the system M/M/1.

The behavior of the expected number in the system is plotted in Figure 3.3. By similar methods we find that the variance of the number in the system is given by

$$\sigma_N^2 = \sum_{k=0}^{\infty}(k - \bar{N})^2 p_k$$

$$\sigma_N^2 = \frac{\rho}{(1 - \rho)^2} \qquad \blacksquare \; (3.25)$$

We may now apply Little's result directly from Eq. (2.25) in order to obtain

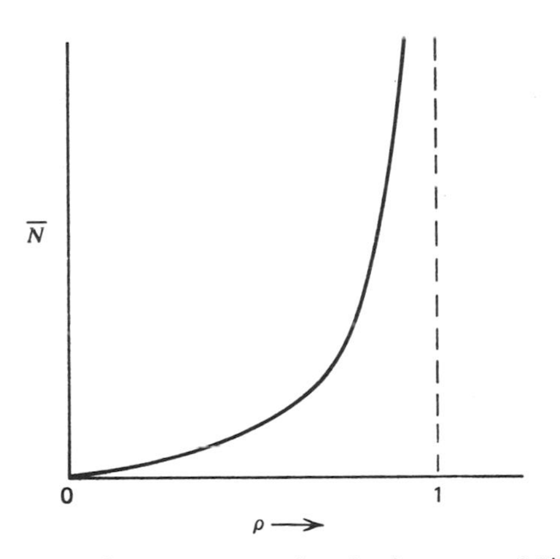

Figure 3.3 The average number in the system M/M/1.

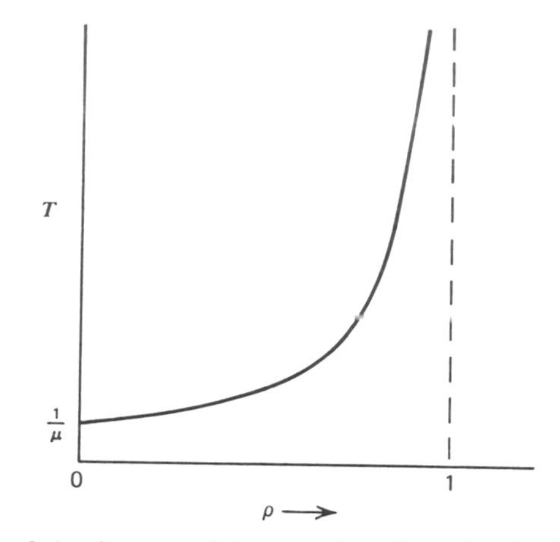

Figure 3.4 Average delay as a function of ρ for M/M/1.

T, the average *time* spent in the system as follows:

$$T = \frac{\bar{N}}{\lambda}$$

$$T = \left(\frac{\rho}{1-\rho}\right)\left(\frac{1}{\lambda}\right)$$

$$T = \frac{1/\mu}{1-\rho} \qquad \blacksquare \; (3.26)$$

This dependence of average time on the utilization factor ρ is shown in Figure 3.4. The value obtained by T when $\rho = 0$ is exactly the average service time expected by a customer; that is, he spends no time in queue and $1/\mu$ sec in service on the average.

The behavior given by Eqs. (3.24) and (3.26) is rather dramatic. As ρ approaches unity, both the average number in the system and the average time in the system grow in an unbounded fashion.* Both these quantities have a

* We observe at $\rho = 1$ that the system behavior is unstable; this is not surprising if one recalls that $\rho < 1$ was our condition for ergodicity. What is perhaps surprising is that the behavior of the average number $\bar{N}$ and of the average system time T deteriorates so badly as $\rho \to 1$ from below; we had seen for steady flow systems in Chapter 1 that so long as $R < C$ (which corresponds to the case $\rho < 1$) no queue formed and smooth, rapid flow proceeded through the system. Here in the M/M/1 queue we find this is no longer true and that we pay an extreme penalty when we attempt to run the system near (but below) its capacity. The

simple pole at $\rho = 1$. *This type of behavior with respect to ρ as ρ approaches 1 is characteristic of almost every queueing system one can encounter.* We will see it again in M/G/1 in Chapter 5 as well as in the heavy traffic behavior of G/G/1 (and also in the tight bounds on G/G/1 behavior) in Volume II, Chapter 2.

Another interesting quantity to calculate is the probability of finding at least k customers in the system:

$$P[\geq k \text{ in system}] = \sum_{i=k}^{\infty} p_i$$

$$= \sum_{i=k}^{\infty} (1 - \rho)\rho^i$$

$$P[\geq k \text{ in system}] = \rho^k \qquad (3.27)$$

Thus we see that the probability of exceeding some limit on the number of customers in the system is a geometrically decreasing function of that number and decays very rapidly.

With the tools at hand we are now in a position to develop the probability density function for the time spent in the system. However, we defer that development until we treat the more general case of M/G/1 in Chapter 5 [see Eq. (5.118)]. Meanwhile, we proceed to discuss numerous other birth–death queues in equilibrium.

3.3. DISCOURAGED ARRIVALS

This next example considers a case where arrivals tend to get discouraged when more and more people are present in the system. One possible way to model this effect is to choose the birth and death coefficients as follows:

$$\lambda_k = \frac{\alpha}{k+1} \qquad k = 0, 1, 2, \ldots$$

$$\mu_k = \mu \qquad k = 1, 2, 3, \ldots$$

We are here assuming an harmonic discouragement of arrivals with respect to the number present in the system. The state-transition-rate diagram in this

intuitive explanation here is that with random flow (e.g., M/M/1) we get occasional bursts of traffic which temporarily overwhelm the server; while it is still true that the server will be idle on the average $1 - \rho = p_0$ of the time this average idle time will not be distributed uniformly within small time intervals but will only be true in the long run. On the other hand, in the steady flow case (which corresponds to our system D/D/1) the system idle time will be distributed quite uniformly in the sense that after every service time (of exactly $1/\mu$ secs) there will be an idle time of exactly $(1/\lambda) - (1/\mu)$ sec. Thus it is the *variability* in both the interarrival time and in the service time which gives rise to the disastrous behavior near $\rho = 1$; any reduction in the variation of either random variable will lead to a reduction in the average waiting time, as we shall see again and again.

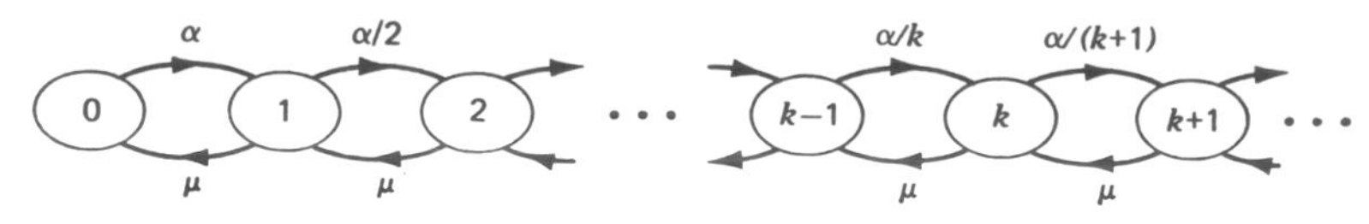

Figure 3.5 State-transition-rate diagram for discouraged arrivals.

case is as shown in Figure 3.5. We apply Eq. (3.11) immediately to obtain for p_k

$$p_k = p_0 \prod_{i=0}^{k-1} \frac{\alpha/(i+1)}{\mu} \tag{3.28}$$

$$p_k = p_0 \left(\frac{\alpha}{\mu}\right)^k \frac{1}{k!} \tag{3.29}$$

Solving for p_0 from Eq. (3.12) we have

$$p_0 = 1 \Big/ \left[1 + \sum_{k=1}^{\infty} \left(\frac{\alpha}{\mu}\right)^k \frac{1}{k!}\right]$$

$$p_0 = e^{-\alpha/\mu}$$

From Eq. (2.32) we have therefore,

$$\rho = 1 - e^{-\alpha/\mu} \tag{3.30}$$

Note that the ergodic condition here is merely $\alpha/\mu < \infty$. Going back to Eq. (3.29) we have the final solution

$$p_k = \frac{(\alpha/\mu)^k}{k!} e^{-\alpha/\mu} \qquad k = 0, 1, 2, \ldots \tag{3.31}$$

We thus have a Poisson distribution for the number of customers in the system of discouraged arrivals! From Eqs. (2.131) and (2.132) we have that the expected number in the system is

$$\bar{N} = \frac{\alpha}{\mu}$$

In order to calculate T, the average time spent in the system, we may use Little's result again. For this we require λ, which is directly calculated from $\rho = \lambda\bar{x} = \lambda/\mu$; thus from Eq. (3.30)

$$\lambda = \mu\rho = \mu(1 - e^{-\alpha/\mu})$$

Using this* and Little's result we then obtain

$$T = \frac{\alpha}{\mu^2(1 - e^{-\alpha/\mu})} \tag{3.32}$$

* Note that this result could have been obtained from $\lambda = \sum_k \lambda_k p_k$. The reader should verify this last calculation.

3.4. M/M/∞: RESPONSIVE SERVERS (INFINITE NUMBER OF SERVERS)

Here we consider the case that may be interpreted either as that of a responsive server who accelerates her service rate linearly when more customers are waiting or may be interpreted as the case where there is always a new clerk or server available for each arriving customer. In particular, we set

$$\lambda_k = \lambda \qquad k = 0, 1, 2, \ldots$$
$$\mu_k = k\mu \qquad k = 1, 2, 3, \ldots$$

Here the state-transition-rate diagram is that shown in Figure 3.6. Going directly to Eq. (3.11) for the solution we obtain

$$p_k = p_0 \prod_{i=0}^{k-1} \frac{\lambda}{(i+1)\mu} \tag{3.33}$$

Need we go any further? The reader should compare Eq. (3.33) with Eq. (3.28). These two are in fact equivalent for $\alpha = \lambda$, and so we immediately have the solutions for p_k and $\bar{N}$,

$$p_k = \frac{(\lambda/\mu)^k}{k!} e^{-\lambda/\mu} \qquad k = 0, 1, 2, \ldots \tag{3.34}$$

$$\bar{N} = \frac{\lambda}{\mu}$$

Here, too, the ergodic condition is simply $\lambda/\mu < \infty$. It appears then that a system of discouraged arrivals behaves exactly the same as a system that includes a responsive server. However, Little's result provides a different (and simpler) form for T here than that given in Eq. (3.32); thus

$$T = \frac{1}{\mu}$$

This answer is, of course, obvious since if we use the interpretation where each arriving customer is granted his own server, then his time in system will be merely his service time which clearly equals $1/\mu$ on the average.

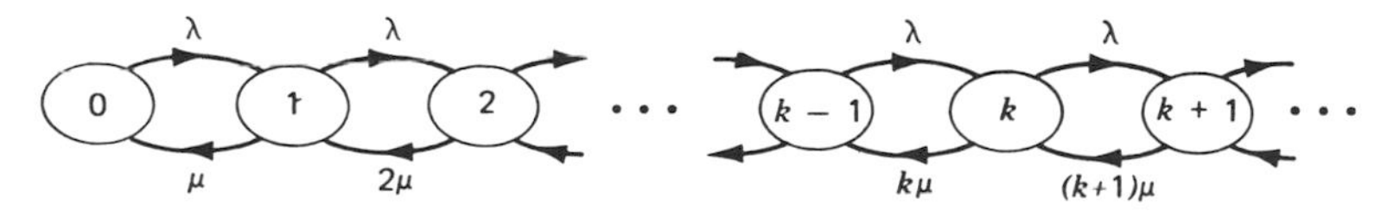

Figure 3.6 State-transition-rate diagram for the infinite-server case M/M/∞.

3.5. M/M/m: THE *m*-SERVER CASE

Here again we consider a system with an unlimited waiting room and with a constant arrival rate λ. The system provides for a *maximum* of m servers. This is within the reach of our birth–death formulation and leads to

$$\lambda_k = \lambda \qquad k = 0, 1, 2, \ldots$$
$$\mu_k = \min [k\mu, m\mu]$$
$$= \begin{cases} k\mu & 0 \le k \le m \\ m\mu & m \le k \end{cases}$$

From Eq. (3.20) it is easily seen that the condition for ergodicity is $\lambda/m\mu < 1$. The state-transition-rate diagram is shown in Figure 3.7. When we go to solve for p_k from Eq. (3.11) we find that we must separate the solution into two parts, since the dependence of μ_k upon k is also in two parts. Accordingly, for $k \le m$,

$$p_k = p_0 \prod_{i=0}^{k-1} \frac{\lambda}{(i+1)\mu}$$
$$= p_0 \left(\frac{\lambda}{\mu}\right)^k \frac{1}{k!} \tag{3.35}$$

Similarly, for $k \ge m$,

$$p_k = p_0 \prod_{i=0}^{m-1} \frac{\lambda}{(i+1)\mu} \prod_{j=m}^{k-1} \frac{\lambda}{m\mu}$$
$$= p_0 \left(\frac{\lambda}{\mu}\right)^k \frac{1}{m!\, m^{k-m}} \tag{3.36}$$

Collecting together the results from Eqs. (3.35) and (3.36) we have

$$p_k = \begin{cases} p_0 \dfrac{(m\rho)^k}{k!} & k \le m \\[2ex] p_0 \dfrac{(\rho)^k m^m}{m!} & k \ge m \end{cases} \tag{3.37}$$

where

$$\rho = \frac{\lambda}{m\mu} < 1 \tag{3.38}$$

This expression for ρ follows that in Eq. (2.30) and is consistent with our

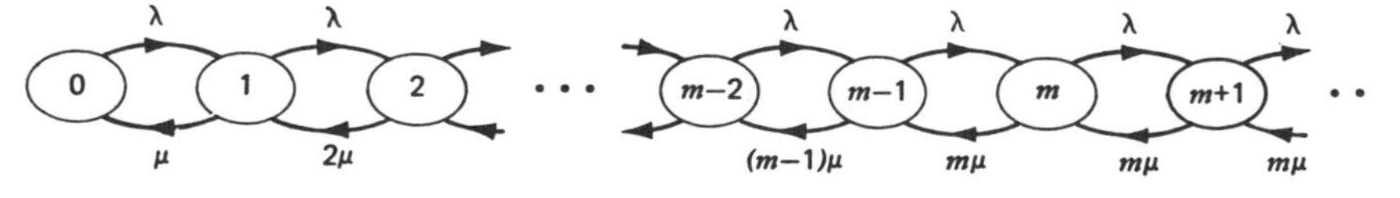

Figure 3.7 State-transition-rate diagram for M/M/m.

definition in terms of the expected fraction of busy servers. We may now solve for p_0 from Eq. (3.12), which gives us

$$p_0 = \left[1 + \sum_{k=1}^{m-1} \frac{(m\rho)^k}{k!} + \sum_{k=m}^{\infty} \frac{(m\rho)^k}{m!} \frac{1}{m^{k-m}}\right]^{-1}$$

and so

$$p_0 = \left[\sum_{k=0}^{m-1} \frac{(m\rho)^k}{k!} + \left(\frac{(m\rho)^m}{m!}\right)\left(\frac{1}{1-\rho}\right)\right]^{-1} \qquad \blacksquare (3.39)$$

The probability that an arriving customer is forced to join the queue is given by

$$\begin{aligned} P[\text{queueing}] &= \sum_{k=m}^{\infty} p_k \\ &= \sum_{k=m}^{\infty} p_0 \frac{(m\rho)^k}{m!} \frac{1}{m^{k-m}} \end{aligned}$$

Thus

$$P[\text{queueing}] = \frac{\left(\frac{(m\rho)^m}{m!}\right)\left(\frac{1}{1-\rho}\right)}{\left[\sum_{k=0}^{m-1} \frac{(m\rho)^k}{k!} + \left(\frac{(m\rho)^m}{m!}\right)\left(\frac{1}{1-\rho}\right)\right]} \qquad \blacksquare (3.40)$$

This probability is of wide use in telephony and gives the probability that no trunk (i.e., server) is available for an arriving call (customer) in a system of m trunks; it is referred to as *Erlang's C formula* and is often denoted* by $C(m, \lambda/\mu)$.

3.6. M/M/1/K: FINITE STORAGE

We now consider for the first time the case of a queueing system in which there is a maximum number of customers that may be stored; in particular, we assume the system can hold at most a total of K customers (including the customer in service) and that any further arriving customers will in fact be refused entry to the system and will depart immediately without service. Newly arriving customers will continue to be generated according to a Poisson process but only those who find the system with strictly less than K customers will be allowed entry. In telephony the refused customers are considered to be "lost"; for the system in which $K = 1$ (i.e., no waiting room at all) this is referred to as a "blocked calls cleared" system with a single server.

* Europeans use the symbol $E_{2,m}(\lambda/\mu)$.

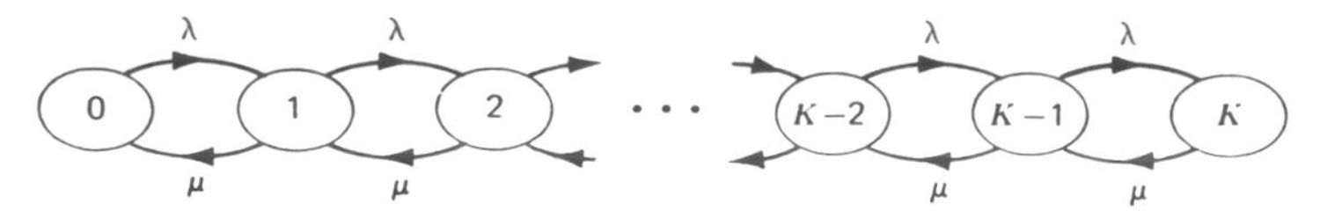

Figure 3.8 State-transition-rate diagram for the case of finite storage room M/M/1/K.

It is interesting that we are capable of accommodating this seemingly complex system description with our birth–death model. In particular, we accomplish this by effectively "turning off" the Poisson input as soon as the systems fills up, as follows:

$$\lambda_k = \begin{cases} \lambda & k < K \\ 0 & k \geq K \end{cases}$$

$$\mu_k = \mu \qquad k = 1, 2, \ldots, K$$

From Eq. (3.20), we see that this system is always ergodic. The state-transition-rate diagram for this finite Markov chain is shown in Figure 3.8. Proceeding directly with Eq. (3.11) we obtain

$$p_k = p_0 \prod_{i=0}^{k-1} \frac{\lambda}{\mu} \qquad k \leq K$$

or

$$p_k = p_0\left(\frac{\lambda}{\mu}\right)^k \qquad k \leq K \tag{3.41}$$

Of course, we also have

$$p_k = 0 \qquad k > K \tag{3.42}$$

In order to solve for p_0 we use Eqs. (3.41) and (3.42) in Eq. (3.12) to obtain

$$p_0 = \left[1 + \sum_{k=1}^{K}\left(\frac{\lambda}{\mu}\right)^k\right]^{-1}$$

$$= \left[1 + \frac{(\lambda/\mu)(1 - (\lambda/\mu)^K)}{1 - \lambda/\mu}\right]^{-1}$$

and so

$$p_0 = \frac{1 - \lambda/\mu}{1 - (\lambda/\mu)^{K+1}}$$

Thus, finally,

$$p_k = \begin{cases} \dfrac{1 - \lambda/\mu}{1 - (\lambda/\mu)^{K+1}}\left(\dfrac{\lambda}{\mu}\right)^k & 0 \leq k \leq K \\ 0 & \text{otherwise} \end{cases} \tag{3.43}$$

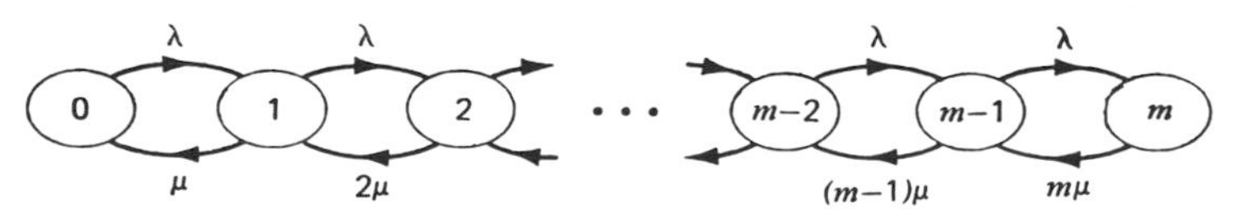

Figure 3.9 State-transition-rate diagram for *m*-server loss system M/M/m/m.

For the case of blocked calls cleared ($K = 1$) we have

$$p_k = \begin{cases} \dfrac{1}{1 + \lambda/\mu} & k = 0 \\ \dfrac{\lambda/\mu}{1 + \lambda/\mu} & k = 1 = K \\ 0 & \text{otherwise} \end{cases} \tag{3.44}$$

3.7. M/M/m/m: *m*-SERVER LOSS SYSTEMS

Here we have again a blocked calls cleared situation in which there are available m servers. Each newly arriving customer is given his private server; however, if a customer arrives when all servers are occupied, that customer is lost. We create this artifact as above by choosing the following birth and death coefficients:

$$\lambda_k = \begin{cases} \lambda & k < m \\ 0 & k \geq m \end{cases}$$

$$\mu_k = k\mu \qquad k = 1, 2, \ldots, m$$

Here again, ergodicity is always assured. This finite state-transition-rate diagram is shown in Figure 3.9.

Applying Eq. (3.11) we obtain

$$p_k = p_0 \prod_{i=0}^{k-1} \frac{\lambda}{(i + 1)\mu} \qquad k \leq m$$

or

$$p_k = \begin{cases} p_0 \left(\dfrac{\lambda}{\mu}\right)^k \dfrac{1}{k!} & k \leq m \\ 0 & k > m \end{cases} \tag{3.45}$$

Solving for p_0 we have

$$p_0 = \left[\sum_{k=0}^{m} \left(\frac{\lambda}{\mu}\right)^k \frac{1}{k!}\right]^{-1}$$

This particular system is of great interest to those in telephony [so much so that a special case of Eq. (3.45) has been tabulated and graphed in many books on telephony]. Specifically, p_m describes the fraction of time that all m servers are busy. The name given to this probability expression is *Erlang's loss formula* and it is given by

$$p_m = \frac{(\lambda/\mu)^m/m!}{\sum_{k=0}^{m}(\lambda/\mu)^k/k!} \qquad (3.46)$$

This equation is also referred to as *Erlang's B formula* and is commonly denoted* by $B(m, \lambda/\mu)$. Formula (3.46) was first derived by Erlang in 1917!

3.8. M/M/1//M†: FINITE CUSTOMER POPULATION—SINGLE SERVER

Here we consider the case where we no longer have a Poisson input process with an infinite user population, but rather have a *finite* population of possible users. The system structure is such that we have a total of M users; a customer is either in the system (consisting of a queue and a single server) or outside the system and in some sense "arriving." In particular, when a customer is in the "arriving" condition then the time it takes him to arrive is a random variable with an exponential distribution whose mean is $1/\lambda$ sec. All customers act independently of each other. As a result, when there are k customers in the system (queue plus service) then there are $M - k$ customers in the arriving state and, therefore, the total average arrival rate in this state is $\lambda(M - k)$. We see that this system is in a strong sense self-regulating. By this we mean that when the system gets busy, with many of these customers in the queue, then the rate at which additional customers arrive is in fact reduced, thus lowering the further congestion of the system. We model this quite appropriately with our birth–death process choosing for parameters

$$\lambda_k = \begin{cases} \lambda(M - k) & 0 \le k \le M \\ 0 & \text{otherwise} \end{cases}$$

$$\mu_k = \mu \qquad k = 1, 2, \ldots$$

The system is ergodic. We assume that we have sufficient room to contain M customers in the system. The finite state-transition-rate diagram is shown in Figure 3.10. Using Eq. (3.11) we solve for p_k as follows:

$$p_k = p_0 \prod_{i=0}^{k-1} \frac{\lambda(M - i)}{\mu} \qquad 0 \le k \le M$$

* Europeans use the notation $E_{1,m}(\lambda/\mu)$.

† Recall that a blank entry in either of the last two optional positions in this notation means an entry of ∞; thus here we have the system M/M/1/∞/M.

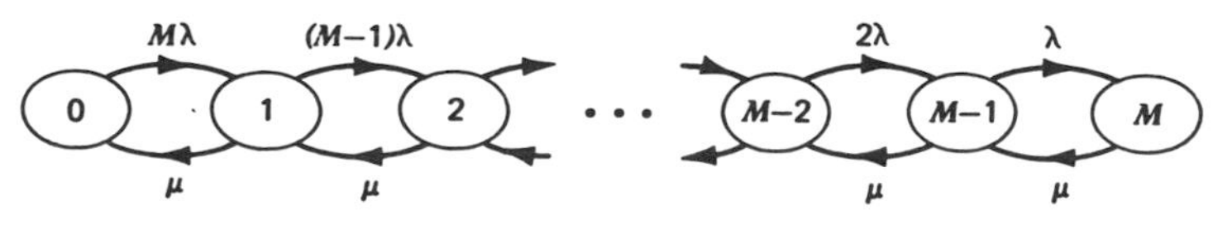

Figure 3.10 State-transition-rate diagram for single-server finite population system M/M/1//M.

Thus

$$p_k = \begin{cases} p_0 \left(\dfrac{\lambda}{\mu}\right)^k \dfrac{M!}{(M-k)!} & 0 \le k \le M \\ 0 & k > M \end{cases} \qquad (3.47)$$

In addition, we obtain for p_0

$$p_0 = \left[\sum_{k=0}^{M} \left(\frac{\lambda}{\mu}\right)^k \frac{M!}{(M-k)!}\right]^{-1} \qquad (3.48)$$

3.9. M/M/∞//M: FINITE CUSTOMER POPULATION—"INFINITE" NUMBER OF SERVERS

We again consider the finite population case, but now provide a separate server for each customer in the system. We model this as follows:

$$\lambda_k = \begin{cases} \lambda(M-k) & 0 \le k \le M \\ 0 & \text{otherwise} \end{cases}$$

$$\mu_k = k\mu \qquad k = 1, 2, \ldots$$

Clearly, this too is an ergodic system. The finite state-transition-rate diagram is shown in Figure 3.11. Solving this system, we have from Eq. (3.11)

$$p_k = p_0 \prod_{i=0}^{k-1} \frac{\lambda(M-i)}{(i+1)\mu}$$

$$= p_0 \left(\frac{\lambda}{\mu}\right)^k \binom{M}{k} \qquad 0 \le k \le M \qquad (3.49)$$

where the binomial coefficient is defined in the usual way,

$$\binom{M}{k} \triangleq \frac{M!}{k!\,(M-k)!}$$

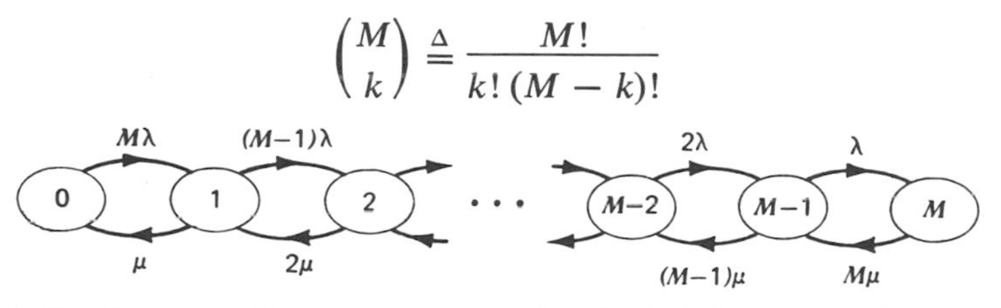

Figure 3.11 State-transition-rate diagram for "infinite"-server finite population system M/M/∞//M.

Solving for p_0 we have

$$p_0 = \left[\sum_{k=0}^{M} \left(\frac{\lambda}{\mu}\right)^k \binom{M}{k}\right]^{-1}$$

and so

$$p_0 = \frac{1}{(1 + \lambda/\mu)^M}$$

Thus

$$p_k = \begin{cases} \dfrac{\left(\dfrac{\lambda}{\mu}\right)^k \dbinom{M}{k}}{(1 + \lambda/\mu)^M} & 0 \le k \le M \\ 0 & \text{otherwise} \end{cases} \tag{3.50}$$

We may easily calculate the expected number of people in the system from

$$\bar{N} = \sum_{k=0}^{M} k p_k = \frac{\sum_{k=0}^{M} k \left(\dfrac{\lambda}{\mu}\right)^k \dbinom{M}{k}}{(1 + \lambda/\mu)^M}$$

Using the partial-differentiation trick such as for obtaining Eq. (3.24) we then have

$$\bar{N} = \frac{M\lambda/\mu}{1 + \lambda/\mu}$$

3.10. M/M/m/K/M: FINITE POPULATION, *m*-SERVER CASE, FINITE STORAGE

This rather general system is the most complicated we have so far considered and will reduce to all of the previous cases (except the example of discouraged arrivals) as we permit the parameters of this system to vary. We assume we have a finite population of M customers, each with an "arriving" parameter λ. In addition, the system has m servers, each with parameter μ. The system also has finite storage room such that the total number of customers in the system (queueing plus those in service) is no more than K. We assume $M \ge K \ge m$; customers arriving to find K already in the system are "lost" and return immediately to the arriving state as if they had just completed service. This leads to the following set of birth–death coefficients:

$$\lambda_k = \begin{cases} \lambda(M - k) & 0 \le k \le K - 1 \\ 0 & \text{otherwise} \end{cases}$$

$$\mu_k = \begin{cases} k\mu & 0 \le k \le m \\ m\mu & k \ge m \end{cases}$$

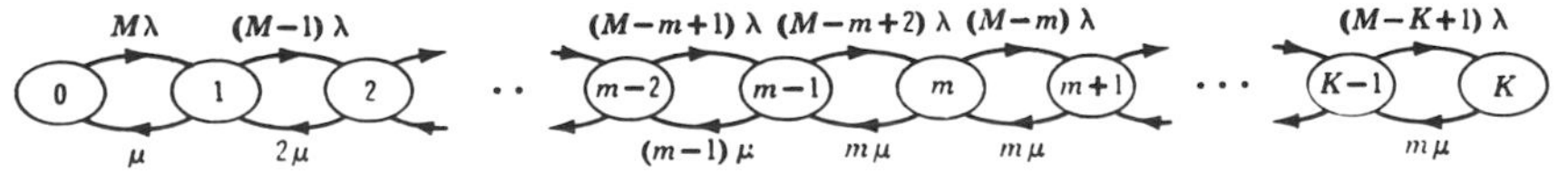

Figure 3.12 State-transition-rate diagram for m-server, finite storage, finite population system M/M/m/K/M.

In Figure 3.12 we see the most complicated of our finite state-transition-rate diagrams. In order to apply Eq. (3.11) we must consider two regions. First, for the range $0 \le k \le m - 1$ we have

$$\begin{aligned} p_k &= p_0 \prod_{i=0}^{k-1} \frac{\lambda(M-i)}{(i+1)\mu} \\ &= p_0 \left(\frac{\lambda}{\mu}\right)^k \binom{M}{k} \qquad 0 \le k \le m-1 \end{aligned} \tag{3.51}$$

For the region $m \le k \le K$ we have

$$\begin{aligned} p_k &= p_0 \prod_{i=0}^{m-1} \frac{\lambda(M-i)}{(i+1)\mu} \prod_{i=m}^{k-1} \frac{\lambda(M-i)}{m\mu} \\ &= p_0 \left(\frac{\lambda}{\mu}\right)^k \binom{M}{k} \frac{k!}{m!} m^{m-k} \qquad m \le k \le K \end{aligned} \tag{3.52}$$

The expression for p_0 is rather complex and will not be given here, although it may be computed in a straightforward manner. In the case of a pure loss system (i.e., $M \ge K = m$), the stationary state probabilities are given by

$$p_k = \frac{\binom{M}{k}\left(\frac{\lambda}{\mu}\right)^k}{\sum_{i=0}^{m} \binom{M}{i}\left(\frac{\lambda}{\mu}\right)^i} \qquad k = 0, 1, \ldots, m \tag{3.53}$$

This is known as the *Engset* distribution.

We could continue these examples ad nauseam but we will instead take a benevolent approach and terminate the set of examples here. Additional examples are given in the exercises. It should be clear to the reader by now that a large number of interesting queueing structures can be modeled with the birth–death process. In particular, we have demonstrated the ability to model the multiple-server case, the finite-population case, the finite-storage case and combinations thereof. The common element in all of these is that the solution for the equilibrium probabilities $\{p_k\}$ is given in Eqs. (3.11) and (3.12). Only systems whose solutions are given by these equations have been considered in this chapter. However, there are many other Markovian systems that lend themselves to simple solution and which are important in queueing theory. In the next chapter (4) we consider the equilibrium solution for Markovian queues; in Chapter 5 we will generalize to semi-Markov processes in which the service time distribution $B(x)$ is permitted to be general, and in Chapter 6 we revert back to the exponential service time case, but permit the interarrival time distribution $A(t)$ to be general; in both of these cases an imbedded Markov chain will be identified and solved. Only when both $A(t)$ and $B(x)$ are nonexponential do we require the methods of advanced queueing theory discussed in Chapter 8. (There are some special nonexponential distributions that may be described with the theory of Markov processes and these too are discussed in Chapter 4.)

EXERCISES

3.1. Consider a pure Markovian queueing system in which

$$\lambda_k = \begin{cases} \lambda & 0 \le k \le K \\ 2\lambda & K < k \end{cases}$$

$$\mu_k = \mu \qquad k = 1, 2, \ldots$$

(a) Find the equilibrium probabilities p_k for the number in the system.

(b) What relationship must exist among the parameters of the problem in order that the system be stable and, therefore, that this equilibrium solution in fact be reached? Interpret this answer in terms of the possible dynamics of the system.

3.2. Consider a Markovian queueing system in which

$$\lambda_k = \alpha^k \lambda \qquad k \ge 0, 0 \le \alpha < 1$$

$$\mu_k = \mu \qquad k \ge 1$$

(a) Find the equilibrium probability p_k of having k customers in the system. Express your answer in terms of p_0.

(b) Give an expression for p_0.

3.3. Consider an M/M/2 queueing system where the average arrival rate is λ customers per second and the average service time is $1/\mu$ sec, where $\lambda < 2\mu$.

(a) Find the differential equations that govern the time-dependent probabilities $P_k(t)$.

(b) Find the equilibrium probabilities

$$p_k = \lim_{t \to \infty} P_k(t)$$

3.4. Consider an M/M/1 system with parameters λ, μ in which customers are impatient. Specifically, upon arrival, customers estimate their queueing time w and then join the queue with probability $e^{-\alpha w}$ (or leave with probability $1 - e^{-\alpha w}$). The estimate is $w = k/\mu$ when the new arrival finds k in the system. Assume $0 \leq \alpha$.

(a) In terms of p_0, find the equilibrium probabilities p_k of finding k in the system. Give an expression for p_0 in terms of the system parameters.

(b) For $0 < \alpha$, $0 < \mu$ under what conditions will the equilibrium solution hold?

(c) For $\alpha \to \infty$, find p_k explicitly and find the average number in the system.

3.5. Consider a birth–death system with the following birth and death coefficients:

$$\lambda_k = (k + 2)\lambda \qquad k = 0, 1, 2, \ldots$$
$$\mu_k = k\mu \qquad k = 1, 2, 3, \ldots$$

All other coefficients are zero.

(a) Solve for p_k. Be sure to express your answer explicitly in terms of λ, k, and μ only.

(b) Find the average number of customers in the system.

3.6. Consider a birth–death process with the following coefficients:

$$\lambda_k = \alpha k(K_2 - k) \qquad k = K_1, K_1 + 1, \ldots, K_2$$
$$\mu_k = \beta k(k - K_1) \qquad k = K_1, K_1 + 1, \ldots K_2$$

where $K_1 \leq K_2$ and where these coefficients are zero outside the range $K_1 \leq k \leq K_2$. Solve for p_k (assuming that the system initially contains $K_1 \leq k \leq K_2$ customers).

3.7. Consider an M/M/m system that is to serve the pooled sum of two Poisson arrival streams; the ith stream has an average arrival rate given by λ_i and exponentially distributed service times with mean $1/\mu_i$ $(i = 1, 2)$. The first stream is an ordinary stream whereby each arrival requires exactly one of the m servers; if all m servers are busy then any newly arriving customer of type 1 is lost. Customers from the second class each require the simultaneous use of m_0 servers (and will occupy them all simultaneously for the same exponentially distributed amount of time whose mean is $1/\mu_2$ sec); if a customer from this class finds less than m_0 idle servers then he too is lost to the system. Find the fraction of type 1 customers and the fraction of type 2 customers that are lost.

3.8. Consider a finite customer population system with a single server such as that considered in Section 3.8; let the parameters M, λ be replaced by M, λ'. It can be shown that if $M \to \infty$ and $\lambda' \to 0$ such that lim $M\lambda' = \lambda$ then the finite population system becomes an infinite population system with exponential interarrival times (at a mean rate of λ customers per second). Now consider the case of Section 3.10; the parameters of that case are now to be denoted M, λ', m, μ, K in the obvious way. Show what value these parameters must take on if they are to represent the earlier cases described in Sections 3.2, 3.4, 3.5, 3.6, 3.7, 3.8, or 3.9.

3.9. Using the definition for $B(m, \lambda/\mu)$ in Section 3.7 and the definition of $C(m, \lambda/\mu)$ given in Section 3.5 establish the following for $\lambda/\mu > 0$, $m = 1, 2, \ldots$

(a) $$B\left(m, \frac{\lambda}{\mu}\right) < \sum_{k=m}^{\infty} \frac{(\lambda/\mu)^k}{k!} e^{-\lambda/\mu} < C\left(m, \frac{\lambda}{\mu}\right)$$

(b) $$C\left(m, \frac{\lambda}{\mu}\right) = \frac{B\left(m, \frac{\lambda}{\mu}\right)}{1 - \frac{\lambda}{\mu}\left[1 - B\left(m, \frac{\lambda}{\mu}\right)\right]}$$

(c) $$B\left(m+1, \frac{\lambda}{\mu}\right) = \frac{\frac{\mu}{\lambda} B\left(m, \frac{\lambda}{\mu}\right)}{m + 1 + \frac{\lambda}{\mu} B\left(m, \frac{\lambda}{\mu}\right)}$$

3.10. Here we consider an M/M/1 queue in discrete time where time is segmented into intervals of length q sec each. We assume that events can only occur at the ends of these discrete time intervals. In particular the probability of a single arrival at the end of such an interval is given by λq and the probability of no arrival at that point is $1 - \lambda q$ (thus at most one arrival may occur). Similarly the departure process is such that if a customer is in service during an interval he will complete service at the end of that interval with probability $1 - \sigma$ or will require at least one more interval with probability σ.

(a) Derive the form for $a(t)$ and $b(x)$, the interarrival time and service time pdf's, respectively.

(b) Assuming FCFS, write down the equilibrium equations that govern the behavior of $p_k = P[k$ customers in system at the end of a discrete time interval] where k includes any arrivals who have occurred at the end of this interval as well as any customers who are about to leave at this point.

(c) Solve for the expected value of the number of customers at these points.

3.11. Consider an M/M/1 system with "feedback"; by this we mean that when a customer departs from service he has probability σ of rejoining the tail of the queue after a random feedback time, which is exponentially distributed (with mean $1/\gamma$ sec); on the other hand, with probability $1 - \sigma$ he will depart forever after completing service. It is clear that a customer may return many times to the tail of the queue before making his eventual final departure. Let p_{kj} be the equilibrium probability that there are k customers in the "system" (that is, in the queue and the service facility) and that there are j customers in the process of returning to the system.

(a) Write down the set of difference equations for the equilibrium probabilities p_{kj}.

(b) Defining the double z-transform

$$P(z_1, z_2) = \sum_{k=0}^{\infty} \sum_{j=0}^{\infty} p_{kj} z_1^k z_2^j$$

show that

$$\gamma(z_2 - z_1) \frac{\partial P(z_1, z_2)}{\partial z_2} + \left\{\lambda(1 - z_1) + \mu\left[1 - \frac{1-\sigma}{z_1} - \sigma \frac{z_2}{z_1}\right]\right\} P(z_1, z_2) = \mu\left[1 - \frac{1-\sigma}{z_1} - \sigma\frac{z_2}{z_1}\right] P(0, z_2)$$

(c) By taking advantage of the moment-generating properties of our z-transforms, show that the mean number in the "system" (queue plus server) is given by $\rho/(1 - \rho)$ and that the mean number returning to the tail of the queue is given by $\mu\sigma\rho/\gamma$, where $\rho = \lambda/(1 - \sigma)\mu$.

3.12. Consider a "cyclic queue" in which M customers circulate around through two queueing facilities as shown below.

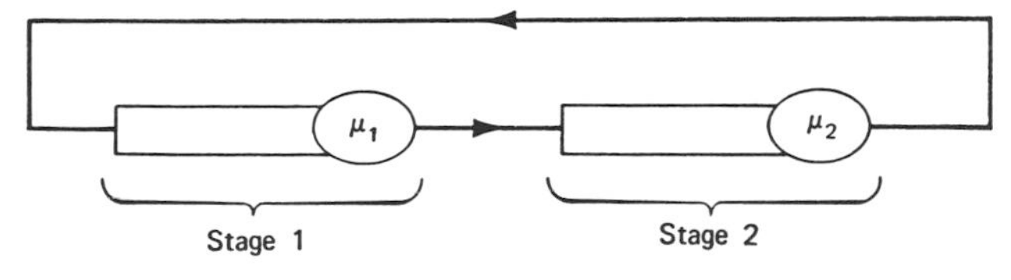

Both servers are of the exponential type with rates μ_1 and μ_2, respectively. Let

$$p_k = P[k \text{ customers in stage 1 and } M\text{-}k \text{ in stage 2}]$$

(a) Draw the state-transition-rate diagram.
(b) Write down the relationship among $\{p_k\}$.
(c) Find

$$P(z) = \sum_{k=0}^{M} p_k z^k$$

(d) Find p_k.

3.13. Consider an M/M/1 queue with parameters λ and μ. A customer in the queue will defect (depart without service) with probability $\alpha \, \Delta t + o(\Delta t)$ in any interval of duration Δt.
(a) Draw the state-transition-rate diagram.
(b) Express p_{k+1} in terms of p_k.
(c) For $\alpha = \mu$, solve for p_k $(k = 0, 1, 2, \ldots)$.

3.14. Let us elaborate on the $M/M/1/K$ system of Section 3.6.
(a) Evaluate p_k when $\lambda = \mu$.
(b) Find $\bar{N}$ for $\lambda \neq \mu$ and for $\lambda = \mu$.
(c) Find T by carefully solving for the average arrival rate to the system.

4

Markovian Queues in Equilibrium

The previous chapter was devoted to the study of the birth–death product solution given in Eq. (3.11). The beauty of that solution lies not only in its simplicity but also in its broad range of application to queueing systems, as we have discussed. When we venture beyond the birth–death process into the more general Markov process, then the product solution mentioned above no longer applies; however, one seeks and often finds some other form of product solution for the pure Markovian systems. In this chapter we intend to investigate some of these Markov processes that are of direct interest to queueing systems. Most of what we say will apply to random walks of the Markovian type; we may think of these as somewhat more general birth–death processes where steps beyond nearest neighbors are permitted, but which nevertheless contain sufficient structure so as to permit explicit solutions. All of the underlying distributions are, of course, exponential.

Our concern here again is with equilibrium results. We begin by outlining a general method for finding the equilibrium equations by inspection. Then we consider the special Erlangian distribution E_r, which is applied to the queueing systems $M/E_r/1$ and $E_r/M/1$. We find that the system $M/E_r/1$ has an interpretation as a bulk arrival process whose general form we study further; similarly the system $E_r/M/1$ may be interpreted as a bulk service system, which we also investigate separately. We then consider the more general systems $E_{r_a}/E_{r_b}/1$ and step beyond that to mention a broad class of G/G/1 systems that are derivable from the Erlangian by "series-parallel" combinations. Finally, we consider the case of queueing networks in which all the underlying distributions once again are of the memoryless type. As we shall see in most of these cases we obtain a product form of solution.

4.1. THE EQUILIBRIUM EQUATIONS

Our point of departure is Eq. (2.116), namely, $\boldsymbol{\pi}\mathbf{Q} = 0$, which expresses the equilibrium conditions for a general ergodic discrete-state continuous-time Markov process; recall that $\boldsymbol{\pi}$ is the row vector of equilibrium state probabilities and that $\mathbf{Q}$ is the infinitesimal generator whose elements are the infinitesimal transition rates of our Markov process. As discussed in the previous chapter, we adopt the more standard queueing-theory notation and replace the vector $\boldsymbol{\pi}$ with the row vector $\mathbf{p}$ whose kth element is the equilibrium probability p_k of finding the system in state E_k. Our task then is to solve

$$\mathbf{pQ} = 0$$

with the additional conservation relation given in Eq. (2.117), namely,

$$\sum_k p_k = 1$$

This vector equation describes the "equations of motion" in equilibrium.

In Chapter 3 we presented a graphical inspection method for writing down equations of motion making use of the state-transition-rate diagram. For the equilibrium case that method was based on the observation that the probabilistic flow rate into a state must equal the probabilistic flow rate out of that state. It is clear that this notion of flow conservation applies more generally than only to the birth–death process, but in fact to any Markov chain. Thus we may construct "non-nearest-neighbor" systems and still expect that our flow conservation technique should work; this in fact is the case. Our approach then is to describe our Markov chain in terms of a state diagram and then apply conservation of flow to each state in turn. This graphical representation is often easier for this purpose than, in fact, is the verbal, mathematical, or matrix description of the system. Once we have this graphical representation we can, by inspection, write down the equations that govern the system dynamics. As an example, let us consider the very simple three-state Markov chain (which clearly is not a birth–death process since the transition $E_0 \to E_2$ is permitted), as shown in Figure 4.1. Writing down the flow conservation law for each state yields

$$\tfrac{3}{2}\lambda p_0 = \mu p_1 \tag{4.1}$$

$$(\lambda + \mu)p_1 = \lambda p_0 + \mu p_2 \tag{4.2}$$

$$\mu p_2 = \frac{\lambda}{2} p_0 + \lambda p_1 \tag{4.3}$$

where Eqs. (4.1), (4.2), and (4.3) correspond to the flow conservation for states E_0, E_1, and E_2, respectively. Observe also that the last equation is exactly the sum of the first two; we always have exactly one redundant equation in these finite Markov chains. We know that the additional equation required is

$$p_0 + p_1 + p_2 = 1$$

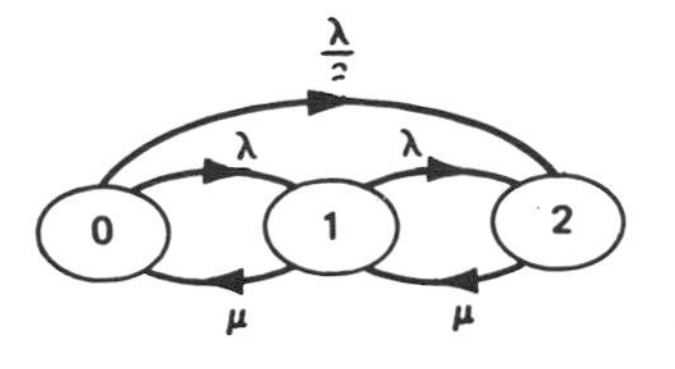

Figure 4.1 Example of non-nearest-neighbor system.

The solution to this system of equations gives

$$p_0 = \left[1 + 2\frac{\lambda}{\mu} + \frac{3}{2}\left(\frac{\lambda}{\mu}\right)^2\right]^{-1}$$

$$p_1 = \frac{3}{2}\frac{\lambda}{\mu} p_0 \tag{4.4}$$

$$p_2 = \left[\frac{1}{2}\frac{\lambda}{\mu} + \frac{3}{2}\left(\frac{\lambda}{\mu}\right)^2\right] p_0$$

Voilà! Simple as pie. In fact, it is as "simple" as inverting a set of simultaneous linear equations.

We take advantage of this inspection technique in solving a number of Markov chains in equilibrium in the balance of this chapter.*

As in the previous chapter we are here concerned with the limiting probability defined as $p_k = \lim P[N(t) = k]$ as $t \to \infty$, assuming it exists. This probability may be interpreted as giving the proportion of *time* that the system spends in state E_k. One could, in fact, estimate this probability by measuring how often the system contained k customers as compared to the total measurement time. Another quantity of interest (perhaps of greater interest) in queueing systems is the probability that an arriving customer finds the system in state E_k; that is, we consider the equilibrium probability

$$r_k = P[\text{arriving customer finds the system in state } E_k]$$

in the case of an ergodic system. One might intuitively feel that in all cases $p_k = r_k$, but it is easy to show that this is not generally true. For example, let us consider the (non-Markovian) system D/D/1 in which arrivals are uniformly spaced in time such that we get one arrival every $\bar{t}$ sec exactly; the service-time requirements are identical for all customers and equal, say

* It should also be clear that this inspection technique permits us to write down the time-dependent state probabilities $P_k(t)$ directly as we have already seen for the case of birth–death processes; these time-dependent equations will in fact be exactly Eq. (2.114).

to $\bar{x}$ sec. We recognize this single-server system as an instance of steady flow through a single channel (remember the pineapple factory). For stability we require that $\bar{x} < \bar{t}$. Now it is clear that no arrival will ever have to wait once equilibrium is reached and, therefore, $r_0 = 1$ and $r_k = 0$ for $k = 1, 2, \ldots$. Moreover, it is clear that the fraction of time that the system contains one customer (in service) is exactly equal to $\rho = \bar{x}/\bar{t}$, and the remainder of the time the system will be empty; therefore, we have $p_0 = 1 - \rho$, $p_1 = \rho$, $p_k = 0$ for $k = 2, 3, 4, \ldots$. So we have a trivial example in which $p_k \neq r_k$. However, as is often the case, one's intuition has a basis in fact, and we find that there is a large class of queueing systems for which $p_k = r_k$ for all k. This, in fact, is the class of stable queueing systems with Poisson arrivals!

Actually, we can prove more, and as we show below for any queueing system with Poisson arrivals we must have

$$P_k(t) = R_k(t)$$

where $P_k(t)$ is, as before, the probability that the system is in state E_k at time t and where $R_k(t)$ is the probability that a customer arriving at time t finds the system in state E_k. Specifically, for our system with Poisson arrivals we define $A(t, t + \Delta t)$ to be the *event* that an arrival occurs in the interval $(t, t + \Delta t)$; then we have

$$R_k(t) \triangleq \lim_{\Delta t \to 0} P[N(t) = k \mid A(t, t + \Delta t)] \tag{4.5}$$

[where $N(t)$ gives the number in system at time t]. Using our definition of conditional probability we may rewrite $R_k(t)$ as

$$R_k(t) = \lim_{\Delta t \to 0} \frac{P[N(t) = k, A(t, t + \Delta t)]}{P[A(t, t + \Delta t)]}$$

$$= \lim_{\Delta t \to 0} \frac{P[A(t, t + \Delta t) \mid N(t) = k]P[N(t) = k]}{P[A(t, t + \Delta t)]}$$

Now for the case of Poisson arrivals we know (due to the memoryless property) that the event $A(t, t + \Delta t)$ must be independent of the number in the system at time t (and also of the time t itself); consequently $P[A(t, t + \Delta t) \mid N(t) = k] = P[A(t, t + \Delta t)]$, and so we have

$$R_k(t) = \lim_{\Delta t \to 0} P[N(t) = k]$$

or

$$R_k(t) = P_k(t) \tag{4.6}$$

This is what we set out to prove, namely, that the time-dependent probability of an arrival finding the system in state E_k is exactly equal to the time-dependent probability of the system being in state E_k. Clearly this also applies to the equilibrium probability r_k that an arrival finds k customers in the system and the proportion of time p_k that the system finds itself with k customers. This equivalence does not surprise us in view of the memoryless property of the Poisson process, which as we have just shown generates a sequence of arrivals that take a really "random look" at the system.

4.2. THE METHOD OF STAGES—ERLANGIAN DISTRIBUTION E_r

The "method of stages" permits one to study queueing systems that are more general than the birth–death systems. This ingenious method is a further testimonial to the brilliance of A. K. Erlang, who developed it early in this century long before our tools of modern probability theory were available. Erlang recognized the extreme simplicity of the exponential distribution and its great power in solving Markovian queueing systems. However, he also recognized that the exponential distribution was not always an appropriate candidate for representing the true situation with regard to service times (and interarrival times). He must also have observed that to allow a more general service distribution would have destroyed the Markovian property and then would have required some more complicated solution method.* The inherent beauty of the Markov chain was not to be given up so easily. What Erlang conceived was the notion of *decomposing* the service† time distribution into a collection of structured exponential distributions.

The principle on which the method of stages is based is the memoryless property of the exponential distribution; again we repeat that this lack of memory is reflected by the fact that the distribution of time remaining for an exponentially distributed random variable is independent of the acquired "age" of that random variable.

Consider the diagram of Figure 4.2. In this figure we are defining a service facility with an exponentially distributed service time pdf given by

$$b(x) \triangleq \frac{dB(x)}{dx} = \mu e^{-\mu x} \qquad x \geq 0 \tag{4.7}$$

The notation of the figure shows an oval which represents the service facility and is labeled with the symbol μ, which represents the service-rate parameter

* As we shall see in Chapter 5, a newer approach to this problem, the "method of imbedded Markov chains," was not available at the time of Erlang.

† Identical observations apply also to the interarrival time distribution.

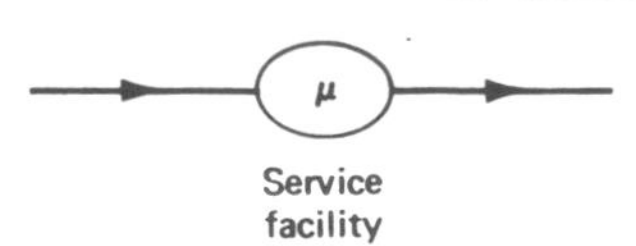

Figure 4.2 The single-stage exponential server.

as in Eq. (4.7). The reader will recall from Chapter 2 that the exponential distribution has a mean and variance given by

$$E[\tilde{x}] = \frac{1}{\mu}$$

$$\sigma_b^{\,2} = \frac{1}{\mu^2}$$

where the subscript b on $\sigma_b^{\,2}$ identifies this as the service time variance.

Now consider the system shown in Figure 4.3. In this figure the large oval represents the service facility. The internal structure of this service facility is revealed as a series or tandem connection of two smaller ovals. Each of these small ovals represents a single exponential server such as that depicted in Figure 4.2; in Figure 4.3, however, the small ovals are labeled internally with the parameter 2μ indicating that they each have a pdf given by

$$h(y) = 2\mu e^{-2\mu y} \qquad y \geq 0 \tag{4.8}$$

Thus the mean and variance for $h(y)$ are $E(\tilde{y}) = 1/2\mu$ and $\sigma_h^{\,2} = (1/2\mu)^2$. The fashion in which this two-stage service facility functions is that upon departure of a customer from this facility a new customer is allowed to enter from the left. This new customer enters stage 1 and remains there for an amount of time randomly chosen from $h(y)$. Upon his departure from this first stage he then proceeds immediately into the second stage and spends an amount of time there equal to a random variable drawn independently once again from $h(y)$. After this second random interval expires he then departs from the service facility and at this point only may a new customer enter the facility from the left. We see then, that one, and only one, customer is

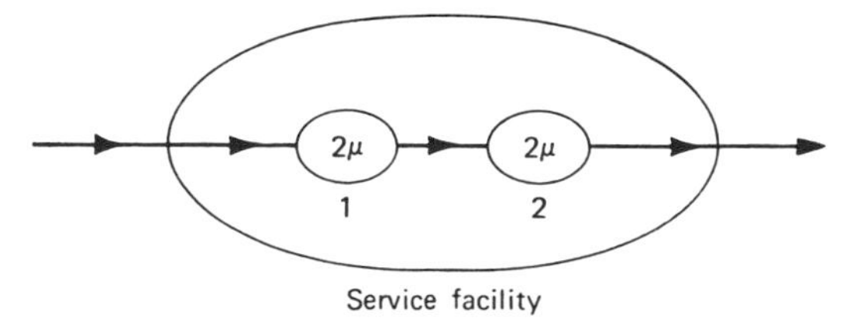

Figure 4.3 The two-stage Erlangian server E_2.

allowed into the box entitled "service facility" at any time.* This implies that at least one of the two service stages must always be empty. We now inquire as to the specific distribution of total time spent in the service facility. Clearly this is a random variable, which is the sum of two independent and identically distributed random variables. Thus, as shown in Appendix II, we must form the convolution of the density function associated with each of the two summands. Alternatively, we may calculate the Laplace transform of the service time pdf as being equal to the product of the Laplace transform of the pdf's associated with each of the summands. Since both random variables are (independent and) identically distributed we must form the product of a function with itself. First, as always, we define the appropriate transforms as

$$B^*(s) \triangleq \int_0^\infty e^{-sx} b(x)\, dx \tag{4.9}$$

$$H^*(s) \triangleq \int_0^\infty e^{-sy} h(y)\, dy \tag{4.10}$$

From our earlier statements we have

$$B^*(s) = [H^*(s)]^2$$

But, we already know the transform of the exponential from Eq. (2.144) and so

$$H^*(s) = \frac{2\mu}{s + 2\mu}$$

Thus

$$B^*(s) = \left(\frac{2\mu}{s + 2\mu}\right)^2 \tag{4.11}$$

We must now invert Eq. (4.11). However, the reader may recall that we already have seen this form in Eq. (2.146) with its inverse in Eq. (2.147). Applying that result we have

$$b(x) = 2\mu(2\mu x)e^{-2\mu x} \qquad x \geq 0 \tag{4.12}$$

We may now calculate the mean and variance of this two-stage system in one of three possible ways: by arguing on the basis of the structure in Figure 4.3; by using the moment generating properties of $B^*(s)$; or by direct calculation

* As an example of a two-stage service facility in which only one stage may be active at a time, consider a courtroom in a small town. A queue of defendants forms, waiting for trial. The judge tries a case (the first service stage) and then fines the defendant. The second stage consists of paying the fine to the court clerk. However, in this small town, the judge is also the clerk and so he moves over to the clerk's desk, collects the fine, releases the defendant, goes back to his bench, and then accepts the next defendant into "service."

from the density function given in Eq. (4.12). We choose the first of these three methods since it is most straightforward (the reader may verify the other two for his own satisfaction). Since the time spent in service is the sum of two random variables, then it is clear that the expected time in service is the sum of the expectations of each. Thus we have

$$E[\tilde{x}] = 2E[\tilde{y}] = \frac{1}{\mu}$$

Similarly, since the two random variables being summed are independent, we may, therefore, sum their variances to find the variance of the sum:

$$\sigma_b^{\,2} = \sigma_h^{\,2} + \sigma_h^{\,2} = \frac{1}{2\mu^2}$$

Note that we have arranged matters such that the mean time in service in the single-stage system of Figure 4.2 and the two-stage system of Figure 4.3 is the same. We accomplished this by speeding up each of the two-stage service stations by a factor of 2. Note further that the variance of the two-stage system is one-half the variance of the one-stage system.

The previous paragraph introduced the notion of a two-stage service facility but we have yet to discuss the crucial point. Let us consider the state variable for a queueing system with Poisson arrivals and a two-stage exponential server as given in Figure 4.3. As always, as part of our state description, we must record the number of customers waiting in the queue. In addition we must supply sufficient information about the service facility so as to summarize the relevant past history. Owing to the memoryless property of the exponential distribution it is enough to indicate which of the following three possible situations may be found within the service facility: either both stages are idle (indicating an empty service facility); or the first stage is busy and the second stage is idle; or the first stage is idle and the second stage is busy. This service-facility state information may be supplied by identifying the stage of service in which the customer may be found. Our state description then becomes a two-dimensional vector that specifies the number of customers in queue and the number of stages yet to be completed by our customer in service. The time this customer has already spent in his current stage of service is irrelevant in calculating the future behavior of the system. Once again we have a Markov process with a discrete (two-dimensional) state space!

The method generalizes and so now we consider the case in which we provide an r-stage service facility, as shown in Figure 4.4. In this system, of course, when a customer departs by exiting from the right side of the oval service facility a new customer may then enter from the left side and proceed one stage at a time through the sequence of r stages. Upon his departure from

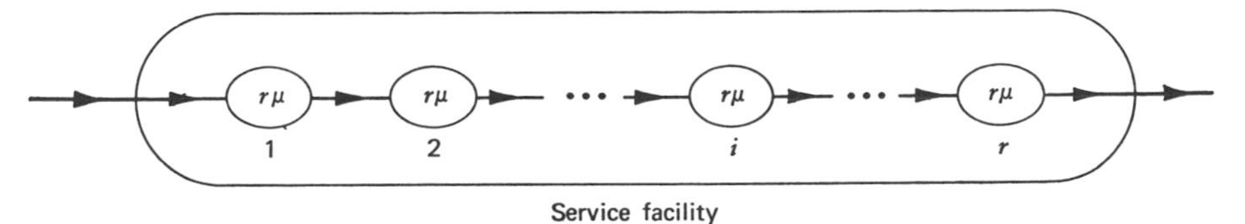

Figure 4.4 The r-stage Erlangian server E_r.

the rth stage a new customer again may then enter, and so on. The time that he spends in the ith stage is drawn from the density function

$$h(y) = r\mu e^{-r\mu y} \qquad y \geq 0 \tag{4.13}$$

The total time that a customer spends in this service facility is the sum of r independent identically distributed random variables, each chosen from the distribution given in Eq. (4.13). We have the following expectation and variance associated with each stage:

$$E[\tilde{y}] = \frac{1}{r\mu}$$

$$\sigma_h^{\,2} = \left(\frac{1}{r\mu}\right)^2$$

It should be clear to the reader that we have chosen each stage in this system to have a service rate equal to $r\mu$ in order that the mean service time remain constant:

$$E[\tilde{x}] = r\left(\frac{1}{r\mu}\right) = \frac{1}{\mu}$$

Similarly, since the stage times are independent we may add the variances to obtain

$$\sigma_b^{\,2} = r\left(\frac{1}{r\mu}\right)^2 = \frac{1}{r\mu^2}$$

Also, we observe that the coefficient of variation [see Eq. (II.23)] is

$$C_b = \frac{1}{\sqrt{r}} \tag{4.14}$$

Once again we wish to solve for the pdf of the service time. This we do by generalizing the notions leading up to Eq. (4.11) to obtain

$$B^*(s) = \left(\frac{r\mu}{s + r\mu}\right)^r \tag{4.15}$$

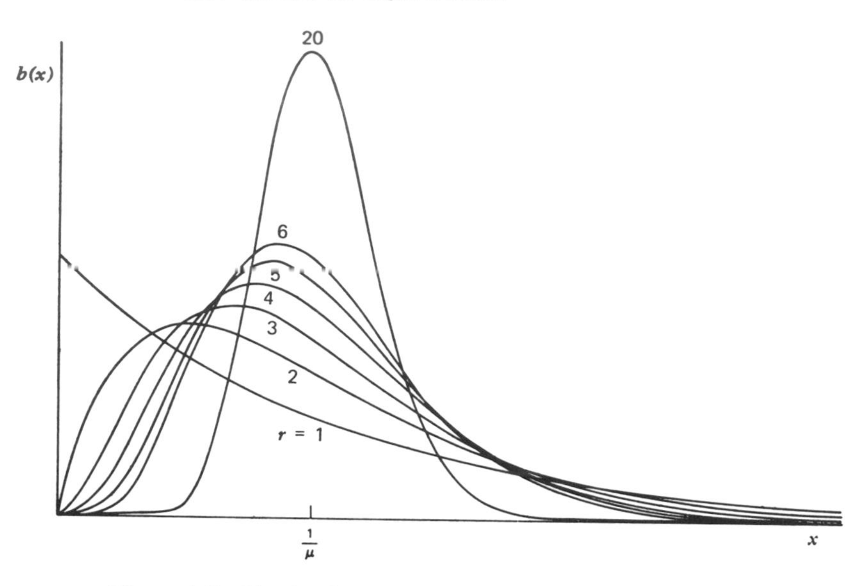

Figure 4.5 The family of r-stage Erlangian distributions E_r.

Equation (4.15) is easily inverted as earlier to give

$$b(x) = \frac{r\mu(r\mu x)^{r-1}e^{-r\mu x}}{(r-1)!} \qquad x \geq 0 \qquad \blacksquare \; (4.16)$$

This we recognize as the Erlangian distribution given in Eq. (2.147). We have carefully adjusted the mean of this density function to be independent of r. In order to obtain an indication of its width we must examine the standard deviation as given by

$$\sigma_b = \frac{1}{\sqrt{r}}\left(\frac{1}{\mu}\right) \qquad \blacksquare$$

Thus we see that the standard deviation for the r-stage Erlangian distribution is $1/\sqrt{r}$ times the standard deviation for the single stage. It should be clear to the sophisticated reader that as r increases, the density function given by Eq. (4.16) must approach that of the normal or Gaussian distribution due to the central limit theorem. This is indeed true but we give more in Eq. (4.16) by specifying the actual sequence of distributions as r increases to show the fashion in which the limit is approached. In Figure 4.5 we show the family of r-stage Erlangian distributions (compare with Figure 2.10). From this figure we observe that the mean holds constant as the width or standard deviation of the density shrinks by $1/\sqrt{r}$. Below, we show that the limit (as r goes to infinity) for this density function must, in fact, be a unit impulse function (see Appendix I) at the point $x = 1/\mu$; this implies that the time spent in an infinite-stage Erlangian service facility approaches a constant with probability 1 (this constant, of course, equals the mean $1/\mu$). We see further that the peak of the family shown moves to the right in a regular fashion. To calculate the location of the peak, we differentiate the density function as given in Eq. (4.16) and set this derivative equal to zero to obtain

$$\frac{db(x)}{dx} = \frac{(r\mu)^2(r-1)(r\mu x)^{r-2}e^{-r\mu x}}{(r-1)!} - \frac{(r\mu x)^{r-1}e^{-r\mu x}(r\mu)^2}{(r-1)!} = 0$$

or

$$(r-1) = r\mu x$$

and so we have

$$x_{\text{peak}} = \left(\frac{r-1}{r}\right)\frac{1}{\mu} \tag{4.17}$$

Thus we see that the location of the peak moves rather quickly toward its final location at $1/\mu$.

We now show that the limiting distribution is, in fact, a unit impulse by considering the limit of the Laplace transform given in Eq. (4.15):

$$\lim_{r\to\infty} B^*(s) = \lim_{r\to\infty}\left(\frac{r\mu}{s+r\mu}\right)^r$$

$$= \lim_{r\to\infty}\left(\frac{1}{1+s/r\mu}\right)^r$$

$$\lim_{r\to\infty} B^*(s) = e^{-s/\mu} \tag{4.18}$$

We recognize the inverse transform of this limiting distribution from entry 3 in Table I.4 of Appendix I; it is merely a unit impulse located at $x = 1/\mu$.

Thus the family of Erlangian distributions varies over a fairly broad range; as such, it is extremely useful for approximating empirical (and even theoretical) distributions. For example, if one had measured a service-time operation and had sufficient data to give acceptable estimates of its mean and variance only, then one could select one member of this two-parameter family such that $1/\mu$ matched the mean and $1/r\mu^2$ matched the variance; this would then be a method for approximating $B(x)$ in a way that permits solution of the queueing system (as we shall see below). If the measured coefficient of variation exceeds unity, we see from Eq. (4.14) that this procedure fails, and we must use the hyperexponential distribution described later or some other distribution.

It is clear for each member of this family of density functions that we may describe the state of the service facility by merely giving the number of stages yet to be completed by a customer in service. We denote the r-stage Erlangian distribution by the symbol E_r (not to be confused with the notation for the state of a random process). Since our state variable is discrete, we are in a position to analyze the queueing system* $\text{M/E}_r/1$. This we do in the following section. Moreover, we will use the same technique in Section 4.4 to decompose the interarrival time distribution $A(t)$ into an r-stage Erlangian distribution. Note in these next two sections that we neurotically require at least one of our distributions to be a pure exponential (this is also true for Chapters 5 and 6).

4.3. THE QUEUE $\text{M/E}_r/1$

Here we consider the system for which

$$a(t) = \lambda e^{-\lambda t} \qquad t \geq 0$$

$$b(x) = \frac{r\mu(r\mu x)^{r-1}e^{-r\mu x}}{(r-1)!} \qquad x \geq 0$$

Since in addition to specifying the number of customers in the system (as in Chapter 3), we must also specify the number of stages remaining in the service facility for the man in service, it behooves us to represent each customer in the queue as possessing r stages of service yet to be completed for him. Thus we agree to take the state variable as the total number of service stages yet to be completed by all customers in the system at the time the state is described.† In particular, if we consider the state at a time when the system contains k customers and when the ith stage of service contains the customer in service we then have that the number of stages contained in the total system is

$$\begin{aligned} j &\triangleq \text{number of stages left in total system} \\ &= (k-1)r + (r-i+1) \end{aligned}$$

Thus

$$j = rk - i + 1 \tag{4.19}$$

As usual, p_k is defined as the equilibrium probability for the number of customers in the system; we further define

$$P_j \triangleq P[j \text{ stages in system}] \tag{4.20}$$

The relationship between customers and stages allows us to write

$$p_k = \sum_{j=(k-1)r+1}^{kr} P_j \qquad k = 1, 2, 3, \ldots$$

* Clearly this is a special case of the system M/G/1 which we will analyze in Chapter 5 using the imbedded Markov chain approach.

† Note that this converts our proposed two-dimensional state vector into a one-dimensional description.

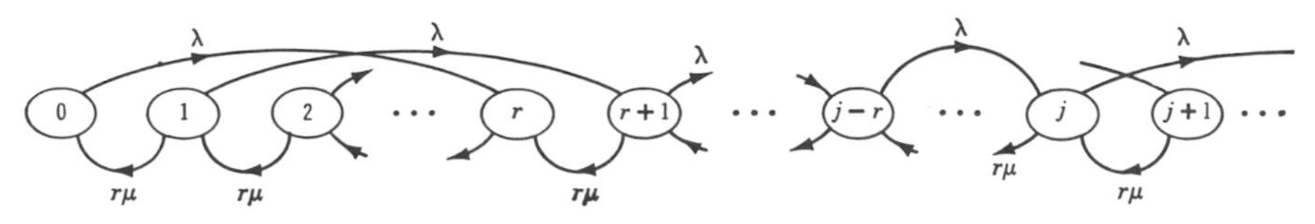

Figure 4.6 State-transition-rate diagram for number of stages: $M/E_r/1$.

And now for the beauty of Erlang's approach: We may represent the state-transition-rate diagram for stages in our system as shown in Figure 4.6. Focusing on state E_j we see that it is entered from below by a state which is r positions to its left and also entered from above by state E_{j+1}; the former transition is due to the arrival of r new stages when a new customer enters, and the latter is due to the completion of one stage within the r-stage service facility. Furthermore we may leave state E_j at a rate λ due to an arrival and at a rate $r\mu$ due to a service completion. Of course, we have special boundary conditions for states $E_0, E_1, \ldots, E_{r-1}$. In order to handle the boundary situation simply let us agree, as in Chapter 3, that state probabilities with negative subscripts are in fact zero. We thus define

$$P_j = 0 \qquad j < 0 \tag{4.21}$$

We may now write down the system state equations immediately by using our flow conservation inspection method. (Note that we are writing the forward equations in equilibrium.) Thus we have

$$\lambda P_0 = r\mu P_1 \tag{4.22}$$

$$(\lambda + r\mu)P_j = \lambda P_{j-r} + r\mu P_{j+1} \qquad j = 1, 2, \ldots \tag{4.23}$$

Let us now use our "familiar" method of solving difference equations, namely the z-transform. Thus we define

$$P(z) = \sum_{j=0}^{\infty} P_j z^j$$

As usual, we multiply the jth equation given in Eq. (4.23) by z^j and then sum over all applicable j. This yields

$$\sum_{j=1}^{\infty}(\lambda + r\mu)P_j z^j = \sum_{j=1}^{\infty}\lambda P_{j-r} z^j + \sum_{j=1}^{\infty} r\mu P_{j+1} z^j$$

Rewriting we have

$$(\lambda + r\mu)\left[\sum_{j=0}^{\infty} P_j z^j - P_0\right] = \lambda z^r \sum_{j=1}^{\infty} P_{j-r} z^{j-r} + \frac{r\mu}{z}\sum_{j=1}^{\infty} P_{j+1} z^{j+1}$$

Recognizing $P(z)$, we then have

$$(\lambda + r\mu)[P(z) - P_0] = \lambda z^r P(z) + \frac{r\mu}{z}[P(z) - P_0 - P_1 z]$$

The first term on the right-hand side of this last equation is obtained by taking special note of Eq. (4.21). Simplifying we have

$$P(z) = \frac{P_0[\lambda + r\mu - (r\mu/z)] - r\mu P_1}{\lambda + r\mu - \lambda z^r - (r\mu/z)}$$

We may now use Eq. (4.22) to simplify this last further:

$$P(z) = \frac{r\mu P_0[1 - (1/z)]}{\lambda + r\mu - \lambda z^r - (r\mu/z)}$$

yielding finally

$$P(z) = \frac{r\mu P_0(1 - z)}{r\mu + \lambda z^{r+1} - (\lambda + r\mu)z} \tag{4.24}$$

We may evaluate the constant P_0 by recognizing that $P(1) = 1$ and using L'Hospital's rule, thus

$$P(1) = 1 = \frac{r\mu P_0}{r\mu - \lambda r}$$

giving (observe that $p_0 = P_0$)

$$P_0 = 1 - \frac{\lambda}{\mu}$$

In this system the arrival rate is λ and the average service time is held fixed at $1/\mu$ independent of r. Thus we recognize that our utilization factor is

$$\rho \triangleq \lambda\bar{x} = \frac{\lambda}{\mu} \tag{4.25}$$

Substituting back into Eq. (4.24) we find

$$P(z) = \frac{r\mu(1 - \rho)(1 - z)}{r\mu + \lambda z^{r+1} - (\lambda + r\mu)z} \tag{4.26}$$

We must now invert this z-transform to find the distribution of the number of stages in the system.

The case $r = 1$, which is clearly the system M/M/1, presents no difficulties; this case yields

$$P(z) = \frac{\mu(1 - \rho)(1 - z)}{\mu + \lambda z^2 - (\lambda + \mu)z}$$
$$= \frac{(1 - \rho)(1 - z)}{1 + \rho z^2 - (1 + \rho)z}$$

The denominator factors into $(1 - z)(1 - \rho z)$ and so canceling the common term $(1 - z)$ we obtain

$$P(z) = \frac{1 - \rho}{1 - \rho z}$$

We recognize this function as entry 6 in Table I.2 of Appendix I, and so we have immediately

$$P_k = (1 - \rho)\rho^k \qquad k = 0, 1, 2, \ldots \tag{4.27}$$

Now in the case $r = 1$ it is clear that $p_k = P_k$ and so Eq. (4.27) gives us the distribution of the number of customers in the system M/M/1, as we had seen previously in Eq. (3.23).

For arbitrary values of r, things are a bit more complex. The usual approach to inverting a z-transform such as that given in Eq. (4.26) is to make a partial fraction expansion and then to invert each term by inspection; let us follow this approach. Before we can carry out this expansion we must identify the $r + 1$ zeroes of the denominator polynomial. Unity is easily seen to be one such. The denominator may therefore be written as $(1 - z)[r\mu - \lambda(z + z^2 + \cdots + z^r)]$, where the remaining r zeroes (which we choose to denote by $z_1, z_2, \ldots, z_r$) are the roots of the bracketed expression. Once we have found these roots* (which are unique) we may then write the denominator as $r\mu(1 - z)(1 - z/z_1)\cdots(1 - z/z_r)$. Substituting this back into Eq. (4.26) we find

$$P(z) = \frac{1 - \rho}{(1 - z/z_1)(1 - z/z_2)\cdots(1 - z/z_r)}$$

Our partial fraction expansion now yields

$$P(z) = (1 - \rho)\sum_{i=1}^{r}\frac{A_i}{(1 - z/z_i)} \tag{4.28}$$

where

$$A_i = \prod_{\substack{n=1 \\ n \neq i}}^{r}\frac{1}{(1 - z_i/z_n)}$$

We may now invert Eq. (4.28) by inspection (from entry 6 in Table I.2) to obtain the final solution for the distribution of the number of stages in the system, namely,

$$P_j = (1 - \rho)\sum_{i=1}^{r} A_i(z_i)^{-j} \qquad j = 1, 2, \ldots, r \qquad \blacksquare \tag{4.29}$$

* Many of the analytic problems in queueing theory reduce to the (difficult) task of locating the roots of a function.

and where as before $P_0 = 1 - \rho$. Thus we see for the system M/E$_r$/1 that the distribution of the number of stages in the system is a weighted sum of *geometric* distributions. The waiting-time distribution may be calculated using the methods developed later in Chapter 5.

4.4. THE QUEUE E_r/M/1

Let us now consider the queueing system E$_r$/M/1 for which

$$a(t) = \frac{r\lambda(r\lambda t)^{r-1}e^{-r\lambda t}}{(r-1)!} \qquad t \geq 0 \tag{4.30}$$

$$b(x) = \mu e^{-\mu x} \qquad x \geq 0 \tag{4.31}$$

Here the roles of interarrival time and service time are interchanged from those of the previous section; in many ways these two systems are duals of each other. The system operates as follows: Given that an arrival has just occurred, then one immediately introduces a new "arriving" customer into an r-stage Erlangian facility much like that in Figure 4.4; however, rather than consider this to be a service facility we consider it to be an "arriving" facility. When this arriving customer is inserted from the left side he must then pass through r exponential stages each with parameter $r\lambda$. It is clear that the pdf of the time spent in the arriving facility will be given by Eq. (4.30). When he exits from the right side of the arriving facility he is then said to "arrive" to the queueing system E$_r$/M/1. Immediately upon his arrival, a new customer (taken from an infinite pool of available customers) is inserted into the left side of the arriving box and the process is repeated. Once having arrived, the customer joins the queue, waits for service, and is then served according to the distribution given in Eq. (4.31). It is clear that an appropriate state description for this system is to specify not only the number of customers in the system, but also to identify which stage in the arriving facility the arriving customer now occupies. We will consider that each customer who has already arrived (but not yet departed) is contributing r stages of "arrival"; in addition we will count the number of stages so far completed by the arriving customer as a further contribution to the number of arrival stages in the system. Thus our state description will consist of the total number of stages of arrival currently in the system; when we find k customers in the system and when our arriving customer is in the ith stage of arrival ($1 \leq i \leq r$) then the total number of stages of arrival in the system is given by

$$j = rk + i - 1$$

Once again let us use the definition given in Eq. (4.20) so that P_j is defined to be the number of *arrival* stages in the system; as always p_k will be the

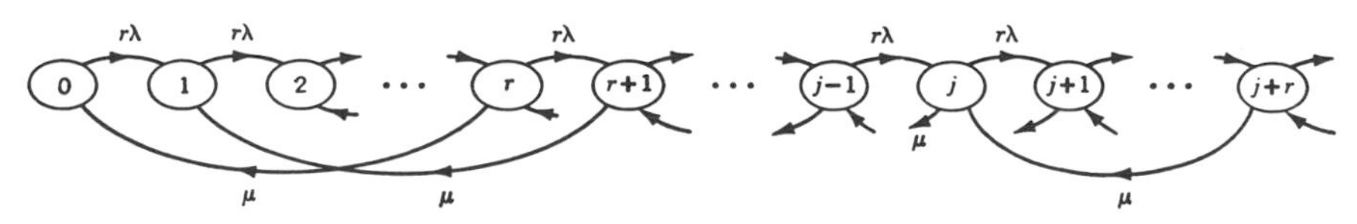

Figure 4.7 State-transition-rate diagram for number of stages: $E_r/M/1$.

equilibrium probability for number of *customers* in the system, and clearly they are related through

$$p_k = \sum_{j=rk}^{r(k+1)-1} P_j$$

The system we have defined is an irreducible ergodic Markov chain with its state-transition-rate diagram for stages given in Figure 4.7. Note that when a customer departs from service, he "removes" r stages of "arrival" from the system. Using our inspection method, we may write down the equilibrium equations as

$$r\lambda P_0 = \mu P_r \tag{4.32}$$

$$r\lambda P_j = r\lambda P_{j-1} + \mu P_{j+r} \qquad 1 \le j \le r-1 \tag{4.33}$$

$$(r\lambda + \mu)P_j = r\lambda P_{j-1} + \mu P_{j+r} \qquad r \le j \tag{4.34}$$

Again we define the z-transform for these probabilities as

$$P(z) = \sum_{j=0}^{\infty} P_j z^j$$

Let us now apply our transform method to the equilibrium equations. Equations (4.33) and (4.34) are almost identical except that the former is missing the term μP_j; consequently let us operate upon the equations in the range $j \ge 1$, adding and subtracting the missing terms as appropriate. Thus we obtain

$$\sum_{j=1}^{\infty}(\mu + r\lambda)P_j z^j - \sum_{j=1}^{r-1} \mu P_j z^j = \sum_{j=1}^{\infty} r\lambda P_{j-1} z^j + \sum_{j=1}^{\infty} \mu P_{j+r} z^j$$

Identifying the transform in this last equation we have

$$(\mu + r\lambda)[P(z) - P_0] - \sum_{j=1}^{r-1} \mu P_j z^j = r\lambda z P(z) + \frac{\mu}{z^r}\left[P(z) - \sum_{j=0}^{r} P_j z^j\right]$$

We may now use Eq. (4.32) to eliminate the term P_r and then finally solve for our transform to obtain

$$P(z) = \frac{(1 - z^r)\sum_{j=0}^{r-1} P_j z^j}{r\rho z^{r+1} - (1 + r\rho)z^r + 1} \tag{4.35}$$

where as always we have defined $\rho = \lambda\bar{x} = \lambda/\mu$. We must now study the poles (zeroes of the denominator) for this function. The denominator polynomial has $r + 1$ zeroes of which unity is one such [the factor $(1 - z)$ is almost always present in the denominator]. Of the remaining r zeroes it can be shown (see Exercise 4.10) that exactly $r - 1$ of them lie in the range $|z| < 1$ and the last, which we shall denote by z_0, is such that $|z_0| > 1$. We are still faced with the numerator summation that contains the unknown probabilities P_j; we must now appeal to the second footnote in step 5 of our z-transform procedure (see Chapter 2, pp. 74–75), which takes advantage of the observation that the z-transform of a probability distribution must be analytic in the range $|z| < 1$ in the following way. Since $P(z)$ must be bounded in the range $|z| < 1$ [see Eq. (II.28)] and since the denominator has $r - 1$ zeroes in this range, then certainly the numerator must also have zeroes at the same $r - 1$ points. The numerator consists of two factors; the first of the form $(1 - z^r)$ all of whose zeroes have absolute value equal to unity; and the second in the form of a summation. Consequently, the "compensating" zeroes in the numerator must come from the summation itself (the summation is a polynomial of degree $r - 1$ and therefore has exactly $r - 1$ zeroes). These observations, therefore, permit us to equate the numerator sum to the denominator (after its two roots at $z = 1$ and $z = z_0$ are factored out) as follows:

$$\frac{r\rho z^{r+1} - (1 + r\rho)z^r + 1}{(1 - z)(1 - z/z_0)} = K \sum_{j=0}^{r-1} P_j z^j$$

where K is a constant to be evaluated below. This computation permits us to rewrite Eq. (4.35) as

$$P(z) = \frac{(1 - z^r)}{K(1 - z)(1 - z/z_0)}$$

But since $P(1) = 1$ we find that

$$K = r/(1 - 1/z_0)$$

and so we have

$$P(z) = \frac{(1 - z^r)(1 - 1/z_0)}{r(1 - z)(1 - z/z_0)} \tag{4.36}$$

We now know all there is to know about the poles and zeroes of $P(z)$; we are, therefore, in a position to make a partial fraction expansion so that we may invert on z. Unfortunately $P(z)$ as expressed in Eq. (4.36) is not in the proper form for the partial fraction expansion, since the numerator degree is not less than the denominator degree. However, we will take advantage of property 8 in Table I.1 of Appendix I, which states that if $F(z) \Leftrightarrow f_n$ then

$z^r F(z) \Leftrightarrow f_{n-r}$, where we recall that the notation $\Leftrightarrow$ indicates a transform pair. With this observation then, we carry out the following partial fraction expansion

$$P(z) = (1 - z^r)\left[\frac{1/r}{1 - z} + \frac{-1/rz_0}{1 - z/z_0}\right]$$

If we denote the inverse transform of the quantity in square brackets by f_j then it is clear that the inverse transform for $P(z)$ must be

$$P_j = f_j - f_{j-r} \tag{4.37}$$

By inspection we see that

$$f_j = \begin{cases} \dfrac{1}{r}(1 - z_0^{-j-1}) & j \geq 0 \\ 0 & j < 0 \end{cases} \tag{4.38}$$

First we solve for P_j in the range $j \geq r$; from Eqs. (4.37) and (4.38) we, therefore, have

$$P_j = \frac{1}{r} z_0^{r-j-1}(1 - z_0^{-r}) \qquad j \geq r \tag{4.39}$$

We may simplify this last expression by recognizing that the denominator of Eq. (4.35) must equal zero for $z = z_0$; this observation leads to the equality $r\rho(z_0 - 1) = 1 - z_0^{-r}$, and so Eq. (4.39) becomes

$$P_j = \rho(z_0 - 1)z_0^{r-j-1} \qquad j \geq r \tag{4.40}$$

On the other hand, in the range $0 \leq j < r$ we have that $f_{j-r} = 0$, and so P_j is easily found for the rest of our range. Combining this and Eq. (4.40) we finally obtain the distribution for the number of arrival stages in our system:

$$P_j = \begin{cases} \dfrac{1}{r}(1 - z_0^{-j-1}) & 0 \leq j < r \\ \rho(z_0 - 1)z_0^{r-j-1} & j \geq r \end{cases} \tag{4.41}$$

Using our earlier relationship between p_k and P_j we find (the reader should check this algebra for himself) that the distribution of the number of customers in the system is given by

$$p_k = \begin{cases} 1 - \rho & k = 0 \\ \rho(z_0^r - 1)z_0^{-rk} & k > 0 \end{cases} \tag{4.42}$$

We note that this distribution for number of customers is *geometric* with a slightly modified first term. We could at this point calculate the waiting time distribution, but we will postpone that until we study the system G/M/1 in Chapter 6.

4.5. BULK ARRIVAL SYSTEMS

In Section 4.3 we studied the system $M/E_r/1$ in which each customer had to pass through r stages of service to complete his total service. The key to the solution of that system was to count the number of service stages remaining in the system, each customer contributing r stages to that number upon his arrival into the system. We may look at the system from another point of view in which we consider each "customer" arrival to be in reality the arrival of r customers. Each of these r customers will require only a single stage of service (that is, the service time distribution is an exponential*). Clearly, these two points of view define identical systems: The former is the system $M/E_r/1$ and the latter is an M/M/1 system with "bulk" arrivals of size r. In fact, if we were to draw the state-transition-rate diagram for the number of *customers* in the system, then the bulk arrival system would lead to the diagram given in Figure 4.6; of course, that diagram was for the number of stages in the system $M/E_r/1$. As a consequence, we see that the generating function for the number of customers in the bulk arrival system must be given by Eq. (4.26) and that the distribution of number of customers in the system is given by Eq. (4.29) since we are equating stages in the original system to customers in the current system.

Since we are considering bulk arrival systems, we may as well be more generous and permit other than a fixed-size bulk to arrive at each (Poisson) arrival instant. What we have in mind is to permit a bulk (or group) at each arrival instant to be of random size where

$$g_i \triangleq P[\text{bulk size is } i] \tag{4.43}$$

(As an example, one may think of random-size families arriving at the doctor's office for individual vaccinations.) As usual, we will assume that the arrival rate (of bulks) is λ. Taking the number of customers in the system as our state variable, we have the state-transition-rate diagram of Figure 4.8. In this figure we have shown details only for state E_k for clarity. Thus we find that we can enter E_k from any state below it (since we permit bulks of any size to arrive); similarly, we can move from state E_k to any state above it, the net rate at which we leave E_k being $\lambda g_1 + \lambda g_2 + \cdots = \lambda \sum_{i=1}^{\infty} g_i = \lambda$. If, as usual we define p_k to be the equilibrium probability for the number of customers in the system, then we may write down the following equilibrium

* To make the correspondence complete, the parameter for this exponential distribution should indeed be $r\mu$. However, in the following development, we will choose the parameter merely to be μ and recall this fact whenever we compare the bulk arrival system to the system $M/E_r/1$.

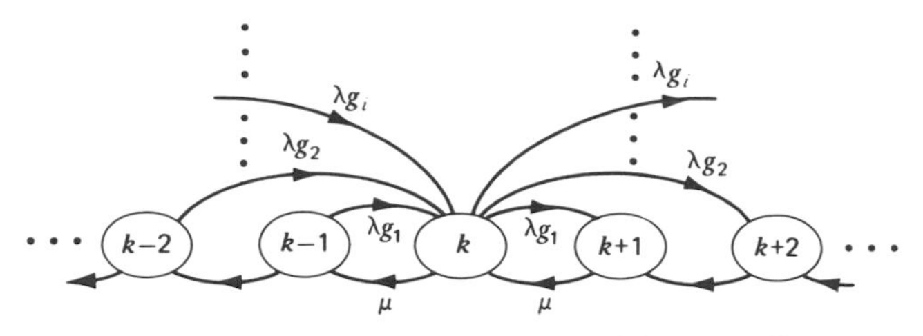

Figure 4.8 The bulk arrival state-transition-rate diagram.

equations using our inspection method:

$$(\lambda + \mu)p_k = \mu p_{k+1} + \sum_{i=0}^{k-1} p_i \lambda g_{k-i} \qquad k \geq 1 \tag{4.44}$$

$$\lambda p_0 = \mu p_1 \tag{4.45}$$

Equation (4.44) has equated the rate out of state E_k (the left-hand side) to the rate into that state, where the first term refers to a service completion and the second term (the sum) refers to all possible ways that arrivals may occur and drive us into state E_k from below. Equation (4.45) is the single boundary equation for the state E_0. As usual, we shall solve these equations using the method of z-transforms; thus we have

$$(\lambda + \mu)\sum_{k=1}^{\infty} p_k z^k = \frac{\mu}{z}\sum_{k=1}^{\infty} p_{k+1} z^{k+1} + \sum_{k=1}^{\infty}\sum_{i=0}^{k-1} p_i \lambda g_{k-i} z^k \tag{4.46}$$

We may interchange the order of summation for the double sum such that

$$\sum_{k=1}^{\infty}\sum_{i=0}^{k-1} = \sum_{i=0}^{\infty}\sum_{k=i+1}^{\infty}$$

and regrouping the terms, we have

$$\begin{aligned}\sum_{k=1}^{\infty}\sum_{i=0}^{k-1} p_i \lambda g_{k-i} z^k &= \lambda \sum_{i=0}^{\infty} p_i z^i \sum_{k=i+1}^{\infty} g_{k-i} z^{k-i} \\ &= \lambda \sum_{i=0}^{\infty} p_i z^i \sum_{j=1}^{\infty} g_j z^j \end{aligned} \tag{4.47}$$

The z-transform we are seeking is

$$P(z) \triangleq \sum_{k=0}^{\infty} p_k z^k$$

and we see from Eq. (4.47) that we should define the z-transform for the distribution of bulk size as*

$$G(z) \triangleq \sum_{k=1}^{\infty} g_k z^k \tag{4.48}$$

* We could just as well have permitted $g_0 > 0$, which would then have allowed zero-size bulks to arrive, and this would have put self-loops in our state-transition diagram corresponding to null arrivals. Had we done so, then the definition for $G(z)$ would have ranged from zero to infinity, and everything we say below applies for this case as well.

Extracting these transforms from Eq. (4.46) we have

$$(\lambda + \mu)[P(z) - p_0] = \frac{\mu}{z}[P(z) - p_0 - p_1 z] + \lambda P(z)G(z)$$

Note that the product $P(z)G(z)$ is a manifestation of property 11 in Table I.1 of Appendix I, since we have in effect formed the transform of the convolution of the sequence $\{p_k\}$ with that of $\{g_k\}$ in Eq. (4.44). Applying the boundary equation (4.45) and simplifying, we have

$$P(z) = \frac{\mu p_0(1 - z)}{\mu(1 - z) - \lambda z[1 - G(z)]}$$

To eliminate p_0 we use $P(1) = 1$; direct application yields the indeterminate form 0/0 and so we must use L'Hospital's rule, which gives $p_0 = 1 - \rho$. We obtain

$$P(z) = \frac{\mu(1 - \rho)(1 - z)}{\mu(1 - z) - \lambda z[1 - G(z)]} \qquad (4.49)$$

This is the final solution for the transform of number of customers in the bulk arrival M/M/1 system. Once the sequence $\{g_k\}$ is given, we may then face the problem of inverting this transform. One may calculate the mean and variance of the number of customers in the system in terms of the system parameters directly from $P(z)$ (see Exercise 4.8). Let us note that the appropriate definition for the utilization factor ρ must be carefully defined here. Recall that ρ is the average arrival rate of customers times the average service time. In our case, the average arrival rate of customers is the product of the average arrival rate of bulks and the average bulk size. From Eq. (II.29) we have immediately that the average bulk size must be $G'(1)$. Thus we naturally conclude that the appropriate definition for ρ in this system is

$$\rho = \frac{\lambda G'(1)}{\mu} \qquad (4.50)$$

It is instructive to consider the special case where all bulk sizes are the same, namely,

$$g_k = \begin{cases} 1 & k = r \\ 0 & k \neq r \end{cases}$$

Clearly, this is the simplified bulk system discussed in the beginning of this section; it corresponds exactly to the system M/E_r/1 (where we must make the minor modification as indicated in our earlier footnote that μ must now be replaced by $r\mu$). We find immediately that $G(z) = z^r$ and after substituting this into our solution Eq. (4.49) we find that it corresponds exactly to our earlier solution Eq. (4.26) as, of course, it must.

4.6. BULK SERVICE SYSTEMS

In Section 4.4 we studied the system $E_r/M/1$ in which arrivals were considered to have passed through r stages of "arrival." We found it expedient in that case to take as our state variable the number of "arrival stages" that were in the system (where each fully arrived customer still in the system contributed r stages to that count). As we found an analogy between bulk *arrival* systems and the Erlangian service systems of Section 4.3, here also we find an analogy between bulk *service* systems and the Erlangian arrival systems studied in Section 4.4. Thus let us consider an M/M/1 system which provides service to groups of size r. That is, when the server becomes free he will accept a "bulk" of exactly r customers from the queue and administer service to them collectively; the service time for this group is drawn from an exponential distribution with parameter μ. If, upon becoming free, the server finds less than r customers in the queue, he then waits until a total of r accumulate and then accepts them for bulk service, and so on.* Customers arrive from a simple Poisson process, at a rate λ, one at a time. It should be clear to the reader that this bulk service system and the $E_r/M/1$ are identical. Were we to draw the state-transition-rate diagram for the number of customers in the bulk service system, then we would find exactly the diagram of Figure 4.7 (with the parameter $r\lambda$ replaced by λ; we must account for this parameter change, however, whenever we compare our bulk service system with the system $E_r/M/1$). Since the two systems are equivalent, then the solution for the distribution of number of customers in the bulk service system must be given by Eq. (4.41) (since stages in the original system correspond to customers in the current system).

It certainly seems a waste for our server to remain idle when less than r customers are available for bulk service. Therefore let us now consider a system in which the server will, upon becoming free, accept r customers for bulk service if they are available, or if not will accept less than r if any are available. We take the number of customers in the system as our state variable and find Figure 4.9 to be the state-transition-rate diagram. In this figure we see that all states (except for state E_0) behave in the same way in that they are entered from their left-hand neighbor by an arrival, and from their neighbor r units to the right by a group departure, and they are exited by either an arrival or a group departure; on the other hand, state E_0 can be entered from any one of the r states immediately to its right and can be exited only by an arrival. These considerations lead directly to the following set of equations for the equilibrium probability p_k of finding k customers in

* For example, the shared taxis in Israel do not (usually) depart until they have collected a full load of customers, all of whom receive service simultaneously.

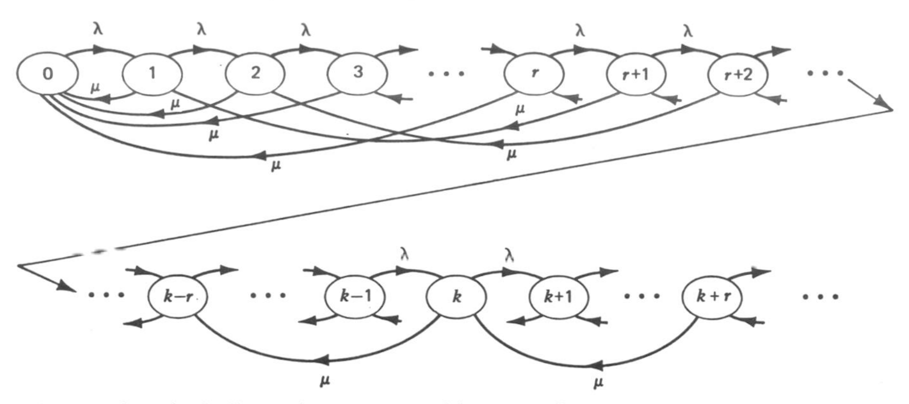

Figure 4.9 The bulk service state-transition-rate diagram.

the system:

$$(\lambda + \mu)p_k = \mu p_{k+r} + \lambda p_{k-1} \qquad k \geq 1$$

$$\lambda p_0 = \mu(p_1 + p_2 + \cdots + p_r) \tag{4.51}$$

Let us now apply our z-transform method; as usual we define

$$P(z) = \sum_{k=0}^{\infty} p_k z^k$$

We then multiply by z^k, sum, and then identify $P(z)$ to obtain in the usual way

$$(\lambda + \mu)[P(z) - p_0] = \frac{\mu}{z^r}\left[P(z) - \sum_{k=0}^{r} p_k z^k\right] + \lambda z P(z)$$

Solving for $P(z)$ we have

$$P(z) = \frac{\mu \sum_{k=0}^{r} p_k z^k - (\lambda + \mu)p_0 z^r}{\lambda z^{r+1} - (\lambda + \mu)z^r + \mu}$$

From our boundary Eq. (4.51) we see that the negative term in the numerator of this last equation may be written as

$$-z^r(\lambda p_0 + \mu p_0) = -\mu z^r \sum_{k=0}^{r} p_k$$

and so we have

$$P(z) = \frac{\sum_{k=0}^{r-1} p_k(z^k - z^r)}{r\rho z^{r+1} - (1 + r\rho)z^r + 1} \tag{4.52}$$

where we have defined $\rho = \lambda/\mu r$ since, for this system, up to r customers may be served simultaneously in an interval whose average length is $1/\mu$ sec. We immediately observe that the denominator of this last equation is precisely the same as in Eq. (4.35) from our study of the system $E_r/M/1$. Thus we may give the same arguments regarding the location of the denominator roots; in particular, of the $r + 1$ denominator zeroes, exactly one will occur at the point $z = 1$, exactly $r - 1$ will be such that $|z| < 1$, and only one will be found, which we will denote by z_0, such that $|z_0| > 1$. Now let us study the numerator of Eq. (4.52). We note that this is a polynomial in z of degree r. Clearly one root occurs at $z = 1$. By arguments now familiar to us, $P(z)$ must remain bounded in the region $|z| < 1$, and so the $r - 1$ remaining zeroes of the numerator must exactly match the $r - 1$ zeroes of the denominator for which $|z| < 1$; as a consequence of this the two polynomials of degree $r - 1$ must be proportional, that is,

$$\frac{K \sum_{k=0}^{r-1} p_k(z^k - z^r)}{1 - z} = \frac{r\rho z^{r+1} - (1 + r\rho)z^r + 1}{(1 - z)(1 - z/z_0)}$$

Taking advantage of this last equation we may then cancel common factors in the numerator and denominator of Eq. (4.52) to obtain

$$P(z) = \frac{1}{K(1 - z/z_0)}$$

The constant K may be evaluated in the usual way by requiring that $P(1) = 1$, which provides the following simple form for our generating function:

$$P(z) = \frac{1 - 1/z_0}{1 - z/z_0} \tag{4.53}$$

This last we may invert by inspection to obtain finally the distribution for the number of customers in our bulk service system

$$p_k = \left(1 - \frac{1}{z_0}\right)\left(\frac{1}{z_0}\right)^k \qquad k = 0, 1, 2, \ldots \qquad \blacksquare \tag{4.54}$$

Once again we see the familiar geometric distribution appear in the solution of our Markovian queueing systems!

4.7. SERIES-PARALLEL STAGES: GENERALIZATIONS

How general is the method of stages studied in Section 4.3 for the system $M/E_r/1$ and studied in Section 4.4 for the system $E_r/M/1$? The Erlangian distribution is shown in Figure 4.5; recall that we may select its mean by appropriate choice of μ and may select a range of standard deviations by adjusting r. Note, however, that we are restricted to accept a coefficient of variation that is less than that of the exponential distribution [from Eq. (4.14) we see that $C_b = 1/\sqrt{r}$ whereas for $r = 1$, the exponential gives $C_b = 1$] and so in some sense Erlangian random variables are "more regular" than exponential variables. This situation is certainly less than completely general.

One direction for generalization would be to remove the restriction that one of our two basic queueing distributions must be exponential; that is, we certainly could consider the system $E_{r_a}/E_{r_b}/1$ in which we have an r_a-stage Erlangian distribution for the interarrival times and an r_b-stage Erlangian distribution for the service times.* On the other hand, we could attempt to generalize by broadening the class of distributions we consider beyond that of the Erlangian. This we do next.

We wish to find a stage-type arrangement that gives larger coefficients of variation than the exponential. One might consider a generalization of the r-stage Erlangian in which we permit each stage to have a *different* service rate (say, the ith stage has rate μ_i). Perhaps this will extend the range of C_b above unity. In this case we will have instead of Eq. (4.15) a Laplace transform for the service-time pdf given by

$$B^*(s) = \left(\frac{\mu_1}{s + \mu_1}\right)\left(\frac{\mu_2}{s + \mu_2}\right) \cdots \left(\frac{\mu_r}{s + \mu_r}\right) \tag{4.55}$$

The service time density $b(x)$ will merely be the convolution of r exponential densities each with its own parameter μ_i. The squared coefficient of variation in this case is easily shown [see Eq. (II.26), Appendix II] to be

$$C_b^2 = \left(\sum_i \frac{1}{\mu_i^2}\right) \bigg/ \left(\sum_i \frac{1}{\mu_i}\right)^2$$

But for real $a_i \geq 0$, it is always true that $\sum_i a_i^2 \leq (\sum_i a_i)^2$ since the right-hand side contains the left-hand side plus the sum of all the nonnegative cross terms. Choosing $a_i = 1/\mu_i$, we find that $C_b^2 \leq 1$. Thus, unfortunately, no generalization to larger coefficients of variation is obtained this way.

We previously found that sending a customer through an increasing sequence of faster exponential stages in *series* tended to reduce the variability of the service time, and so one might expect that sending him through a *parallel* arrangement would increase the variability. This in fact is true. Let us therefore consider the two-stage parallel service system shown in Figure 4.10. The situation may be contrasted to the service structure shown in Figure 4.3. In Figure 4.10 an entering customer approaches the large oval (which represents the service facility) from the left. Upon entry into the

* We consider this shortly.

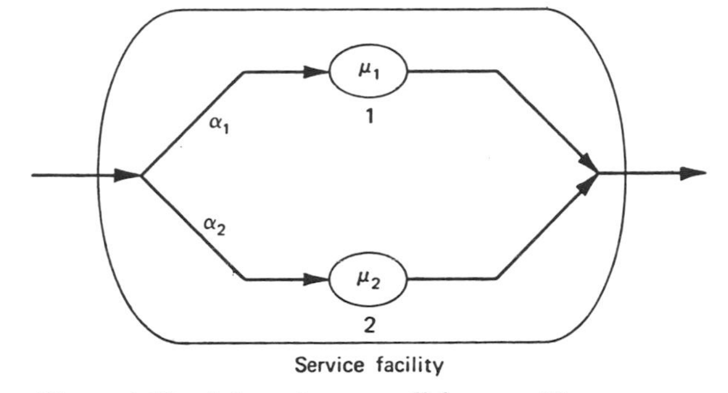

Figure 4.10 A two-stage parallel server H_2.

facility he will proceed to service stage 1 with probability α_1 or will proceed to service stage 2 with probability α_2, where $\alpha_1 + \alpha_2 = 1$. He will then spend an exponentially distributed interval of time in the ith such stage whose mean is $1/\mu_i$ sec. After that interval the customer departs and only then is a new customer allowed into the service facility. It is clear from this description that the service time pdf will be given by

$$b(x) = \alpha_1\mu_1 e^{-\mu_1 x} + \alpha_2\mu_2 e^{-\mu_2 x} \qquad x \geq 0$$

and also we have

$$B^*(s) = \alpha_1 \frac{\mu_1}{s + \mu_1} + \alpha_2 \frac{\mu_2}{s + \mu_2}$$

Of course the more general case with R parallel stages is shown in Figure 4.11. (Contrast this with Figure 4.4.) In this case, as always, at most one customer at any one time is permitted within the large oval representing the service facility. Here we assume that

$$\sum_{i=1}^{R} \alpha_i = 1 \tag{4.56}$$

Clearly,

$$b(x) = \sum_{i=1}^{R} \alpha_i \mu_i e^{-\mu_i x} \qquad x \geq 0 \tag{4.57}$$

and

$$B^*(s) = \sum_{i=1}^{R} \alpha_i \frac{\mu_i}{s + \mu_i}$$

The pdf given in Eq. (4.57) is referred to as the *hyperexponential* distribution and is denoted by H_R. Hopefully, the coefficient of variation ($C_b \triangleq \sigma_b/\bar{x}$) is now greater than unity and therefore represents a wider variation than

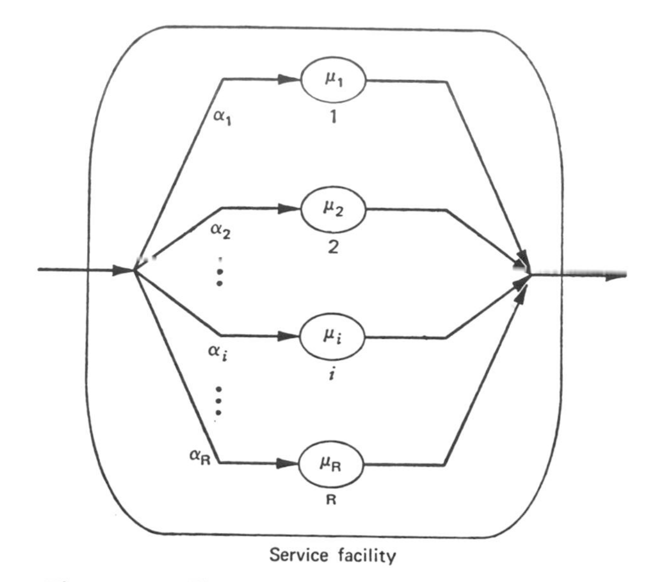

Figure 4.11 The R-stage parallel server H_R.

that of the exponential. Let us prove this. From Eq. (II.26) we find immediately that

$$\bar{x} = \sum_{i=1}^{R} \frac{\alpha_i}{\mu_i}$$

$$\overline{x^2} = 2\sum_{i=1}^{R} \frac{\alpha_i}{\mu_i^2}$$

Forming the square of the coefficient of variation we then have

$$C_b^2 \triangleq \frac{\sigma_b^2}{\bar{x}^2} = \frac{\overline{x^2} - \bar{x}^2}{\bar{x}^2}$$

$$= \frac{2\sum_{i=1}^{R} \frac{\alpha_i}{\mu_i^2}}{\left(\sum_{i=1}^{R} \frac{\alpha_i}{\mu_i}\right)^2} - 1 \tag{4.58}$$

Now, Eq. (II.35), the Cauchy–Schwarz inequality, may also be expressed as follows (for a_i, b_i real):

$$\left(\sum_i a_i b_i\right)^2 \le \left(\sum_i a_i^2\right)\left(\sum_i b_i^2\right) \tag{4.59}$$

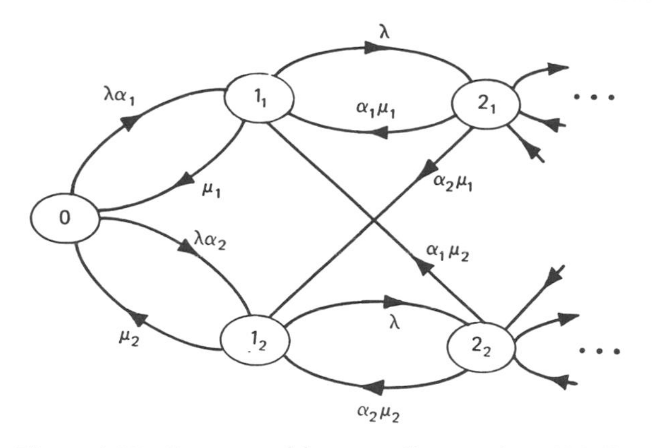

Figure 4.12 State-transition-rate diagram for $M/H_2/1$.

(This is often referred to as the *Cauchy* inequality.) If we make the association $a_i = \sqrt{\alpha_i}$, $b_i = \sqrt{\alpha_i}/\mu_i$, then Eq. (4.59) shows

$$\left(\sum_i \frac{\alpha_i}{\mu_i}\right)^2 \leq \left(\sum_i \alpha_i\right)\left(\sum_i \frac{\alpha_i}{\mu_i^2}\right)$$

But from Eq. (4.56) the first factor on the right-hand side of this inequality is just unity; this result along with Eq. (4.58) permits us to write

$$C_b^2 \geq 1 \qquad \blacksquare(4.60)$$

which proves the desired result.

One might expect that an analysis by the method of stages exists for the systems $M/H_R/1$, $H_R/M/1$, $H_{R_a}/H_{R_b}/1$, and this is indeed true. The reason that the analysis can proceed is that we may take account of the nonexponential character of the service (or arrival) facility merely by specifying which stage within the service (or arrival) facility the customer currently occupies. This information along with a statement regarding the number of customers in the system creates a Markov chain, which may then be studied much as was done earlier in this chapter.

For example, the system $M/H_2/1$ would have the state-transition-rate diagram shown in Figure 4.12. In this figure the designation k_i implies that the system contains k customers and that the customer in service is located in stage i ($i = 1, 2$). The transitions for higher numbered states are identical to the transitions between states 1_i and 2_i.

We are now led directly into the following generalization of series stages and parallel stages; specifically we are free to combine series and parallel

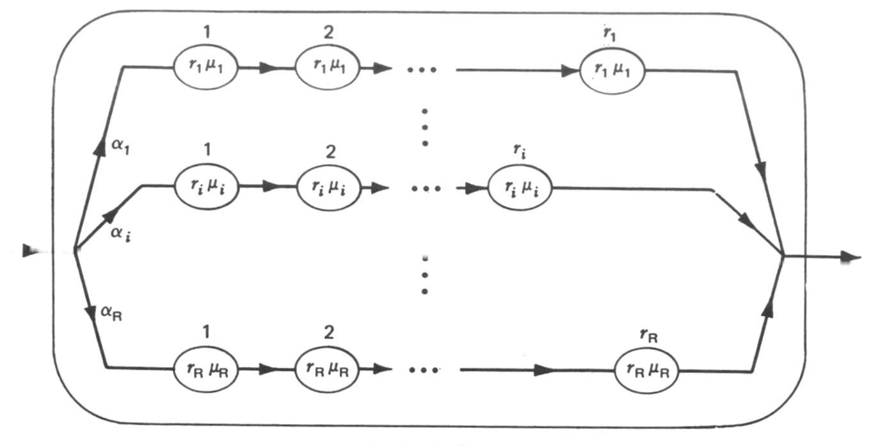

Figure 4.13 Series-parallel server.

stages into arbitrarily complex structures such as shown in Figure 4.13. This diagram shows R parallel "stages," the ith "stage" consisting of an r_i-stage series system ($i = 1, 2, \ldots, R$); each stage in the ith series branch is an exponential service facility with parameter $r_i\mu_i$. It is clear that great generality can be built into such series–parallel systems. Within the service facility one and only one of the multitude of stages may be occupied by a customer and no new customer may enter the large oval (representing the service facility) until the previous customer departs. In all cases, however, we note that the state of the service facility is completely contained in the specification of the particular single stage of service in which the customer may currently be found. Clearly the pdf for the service time is calculable directly as above to give

$$b(x) = \sum_{i=1}^{R} \alpha_i \frac{r_i\mu_i(r_i\mu_i x)^{r_i-1}}{(r_i - 1)!} e^{-r_i\mu_i x} \qquad x \geq 0 \tag{4.61}$$

and has a transform given by

$$B^*(s) = \sum_{i=1}^{R} \alpha_i \left(\frac{r_i\mu_i}{s + r_i\mu_i} \right)^{r_i} \tag{4.62}$$

One further way in which we may generalize our series–parallel server is to remove the restriction that each stage within the same series branch has the same service rate ($r_i\mu_i$); if indeed we permit the jth series stage in the ith

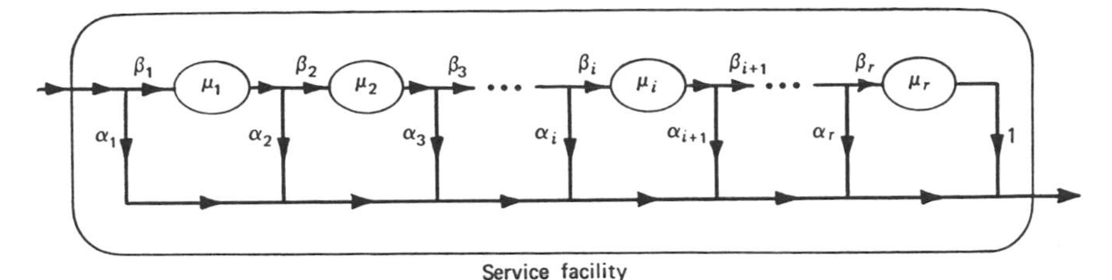

Figure 4.14 Another stage-type server.

parallel branch to have a service rate given by μ_{ij}, then we find that the Laplace transform of the service time density will be generalized to

$$B^*(s) = \sum_{j=1}^{R} \alpha_i \prod_{i=1}^{r_i} \left(\frac{\mu_{ij}}{s + \mu_{ij}} \right) \tag{4.63}$$

These generalities lead to rather complex system equations.

Another way to create the series–parallel effect is as follows. Consider the service facility shown in Figure 4.14. In this system there are r service stages only one of which may be occupied at a given time. Customers enter from the left and depart to the right. Before entering the ith stage an independent choice is made such that with probability β_i the customer will proceed into the ith exponential service stage and with probability α_i he will depart from the system immediately; clearly we require $\beta_i + \alpha_i = 1$ for $i = 1, 2, \ldots, r$. After completing the rth stage he will depart from the system with probability 1. One may immediately write down the Laplace transform of the pdf for the system as follows:

$$B^*(s) = \alpha_1 + \sum_{i=1}^{r} \beta_1\beta_2 \cdots \beta_i\alpha_{i+1} \prod_{j=1}^{i} \left(\frac{\mu_j}{s + \mu_j} \right) \tag{4.64}$$

where $\alpha_{r+1} = 1$. One is tempted to consider more general transitions among stages than that shown in this last figure; for example, rather than choosing only between immediate departure and entry into the next stage one might consider feedback or feedforward to other stages. Cox [COX 55] has shown that no further generality is introduced with this feedback and feedforward concept over that of the system shown in Figure 4.14.

It is clear that each of these last three expressions for $B^*(s)$ may be rewritten as a rational function of s, that is, as a ratio of polynomials in s. The positions of the poles (zeroes of the denominator polynomial) for $B^*(s)$ will of necessity be located on the negative real axis of the complex s-plane. This is not quite as general as we would like, since an arbitrary pdf for service time may have poles located anywhere in the negative half s-plane [that is, for $R_e(s) < 0$]. Cox [COX 55] has studied this problem and suggests that complex values for the exponential parameters $r_i\mu_i$ be permitted; the argument is that whereas this corresponds to no physically realizable exponential stage, so long as we provide poles in complex conjugate pairs then the entire service facility will have a real pdf, which corresponds to the feasible cases. If we permit complex-conjugate pairs of poles then we have *complete generality* in synthesizing any rational function of s for our service-time transform $B^*(s)$. In addition, we have in effect outlined a method of solving these systems by keeping track of the state of the service facility. Moreover, we can similarly construct an interarrival time distribution from series–parallel stages, and thereby we are capable of considering any G/G/1 system where the distributions have transforms that are rational functions of s. It is further true that any nonrational function of s may be approximated arbitrarily closely with rational functions.* Thus in principle we have solved a very general problem. Let us discuss this method of solution. The state description clearly will be the number of customers in the system, the stage in which the arriving customer finds himself within the (stage-type) arriving box and the stage in which the customer finds himself in service. From this we may draw a (horribly complicated) state-transition diagram. Once we have this diagram we may (by inspection) write down the equilibrium equations in a rather straightforward manner; this large set of equations will typically have many boundary conditions. However, these equations will all be linear in the unknowns and so the solution method is straightforward (albeit extremely tedious). What more natural setup for a computer solution could one ask for? Indeed, a digital computer is extremely adept at solving large sets of linear equations (such a task is much easier for a digital computer to handle than is a small set of nonlinear equations). In carrying out the digital solution of this (typically infinite) set of linear equations, we must reduce it to a finite set; this can only be done in an approximate way by first deciding at what point we are satisfied in truncating the sequence $p_0, p_1, p_2, \ldots$. Then we may solve the finite set and perhaps extrapolate the

* In a real sense, then, we are faced with an approximation problem: how may we "best" approximate a given distribution by one that has a rational transform. If we are given a pdf in numerical form then Prony's method [WHIT 44] is one acceptable procedure. On the other hand, if the pdf is given analytically it is difficult to describe a general procedure for suitable approximation. Of course one would like to make these approximations with the fewest number of stages possible. We comment that if one wishes to fit the first and second moments of a given distribution by the method of stages then the number of stages cannot be significantly less than $1/C_b^2$; unfortunately, this implies that when the distribution tends to concentrate around a fixed value, then the number of stages required grows rather quickly.

solution to the infinite set; all this is in way of approximation and hopefully we are able to carry out the computation far enough so that the neglected terms are indeed negligible.

One must not overemphasize the usefulness of this procedure; this solution method is not as yet automated but does at least in principle provide a method of approach. Other analytic methods for handling the more complex queueing situations are discussed in the balance of this book.

4.8. NETWORKS OF MARKOVIAN QUEUES

We have so far considered Markovian systems in which each customer was demanding a single service operation from the system. We may refer to this as a "single-node" system. In this section we are concerned with multiple-node systems in which a customer requires service at more than one station (node). Thus we may think of a *network of nodes*, each of which is a service center (perhaps with multiple servers at some of the nodes) and each with storage room for queues to form. Customers enter the system at various points, queue for service, and upon departure from a given node then proceed to some other node, there to receive additional service. We are now describing the last category of flow system discussed in Chapter 1, namely, stochastic flow in a network.

A number of new considerations emerge when one considers networks. For example, the topological structure of the network is important since it describes the permissible transitions between nodes. Also the paths taken by individual customers must somehow be described. Of great significance is the nature of the stochastic flow in terms of the basic stochastic processes describing that flow; for example, in the case of a tandem queue where customers departing from node i immediately enter node $i + 1$, we see that the interdeparture times from the former generate the interarrival times to the latter. Let us for the moment consider the simple two-node tandem network shown in Figure 4.15. Each oval in that figure describes a queueing *system* consisting of a queue and server(s); within each oval is given the node number. (It is important not to confuse these physical *network* diagrams with the abstract *state-transition-rate* diagrams we have seen earlier.) For the moment let us assume that a Poisson process generates the arrivals to the system at a rate λ, all of which enter node one; further assume that node one consists of a single exponential server at rate μ. Thus node one is exactly an M/M/1 queueing system. Also we will assume that node two has a single

Figure 4.15 A two-node tandem network.

exponential server also of rate μ. The basic question is to solve for the interarrival time distribution feeding node two; this certainly will be equivalent to the interdeparture time distribution from node one. Let $d(t)$ be the pdf describing the interdeparture process from node one and as usual let its Laplace transform be denoted by $D^*(s)$. Let us now calculate $D^*(s)$. When a customer departs from node one either a second customer is available in the queue and ready to be taken into service immediately or the queue is empty. In the first case, the time until this next customer departs from node one will be distributed exactly as a service time and in that case we will have

$$D^*(s)\Big|_{\text{node one nonempty}} = B^*(s)$$

On the other hand, if the node is empty upon this first customer's departure then we must wait for the sum of two intervals, the first being the time until the second customer arrives and the next being his service time; since these two intervals are independently distributed then the pdf of the sum must be the convolution of the pdf's for each. Certainly then the transform of the sum pdf will be the product of the transforms of the individual pdfs and so we have

$$D^*(s)\Big|_{\text{node one empty}} = \frac{\lambda}{s+\lambda} B^*(s)$$

where we have given the explicit expression for the transform of the interarrival time density. Since we have an exponential server we may also write $B^*(s) = \mu/(s + \mu)$; furthermore, as we shall discuss in Chapter 5 the probability of a departure leaving behind an empty system is the same as the probability of an arrival finding an empty system, namely, $1 - \rho$. This permits us to write down the unconditional transform for the interdeparture time density as

$$D^*(s) = (1-\rho)D^*(s)\Big|_{\text{node one empty}} + \rho D^*(s)\Big|_{\text{node one nonempty}}$$

Using our above calculations we then have

$$D^*(s) = (1-\rho)\left(\frac{\lambda}{s+\lambda}\right)\left(\frac{\mu}{s+\mu}\right) + \rho\left(\frac{\mu}{s+\mu}\right)$$

A little algebra gives

$$D^*(s) = \frac{\lambda}{s+\lambda} \tag{4.65}$$

and so the interdeparture time distribution is given by

$$D(t) = 1 - e^{-\lambda t} \qquad t \geq 0 \qquad ■$$

Thus we find the remarkable conclusion that the interdeparture times are exponentially distributed with the same parameter as the interarrival times! In other words (in the case of a stable stationary queueing system), a Poisson process driving an exponential server generates a Poisson process for departures. This startling result is usually referred to as *Burke's* theorem [BURK 56]; a number of others also studied the problem (see, for example, the discussion in [SAAT 65]). In fact, Burke's theorem says more, namely, that the steady-state output of a stable M/M/m queue with input parameter λ and service-time parameter μ for each of the m channels is in fact a Poisson process at the same rate λ. Burke also established that the output process was independent of the other processes in the system. It has also been shown that the M/M/m system is the only such FCFS system with this property. Returning now to Figure 4.15 we see therefore that node two is driven by an independent Poisson arrival process and therefore it too behaves like an M/M/1 system and so may be analyzed independently of node one. In fact Burke's theorem tells us that we may connect many multiple-server nodes (each server with exponential pdf) together in a feedforward* network fashion and still preserve this node-by-node decomposition.

Jackson [JACK 57] addressed himself to this question by considering an arbitrary network of queues. The system he studied consists of N nodes where the ith node consists of m_i exponential servers each with parameter μ_i; further the ith node receives arrivals from outside the system in the form of a Poisson process at rate γ_i. Thus if $N = 1$ then we have an M/M/m system. Upon leaving the ith node a customer then proceeds to the jth node with probability r_{ij}; this formulation permits the case where $r_{ii} \geq 0$. On the other hand, after completing service in the ith node the probability that the customer departs from the network (never to return again) is given by $1 - \sum_{j=1}^{N} r_{ij}$. We must calculate the total average arrival rate of customers to a given node. To do so, we must sum the (Poisson) arrivals from outside the system plus arrivals (not necessarily Poisson) from all internal nodes; that is, denoting the total average arrival rate to node i by λ_i we easily find that this set of parameters must satisfy the following equations:

$$\lambda_i = \gamma_i + \sum_{j=1}^{N} \lambda_j r_{ji} \qquad i = 1, 2, \ldots, N \tag{4.66}$$

In order for all nodes in this system to represent ergodic Markov chains we require that $\lambda_i < m_i\mu_i$ for all i; again we caution the reader not to confuse the nodes in this discussion with the system states of each node from our

* Specifically we do not permit feedback paths since this may destroy the Poisson nature of the feedback departure stream. In spite of this, the following discussion of Jackson's work points out that even networks with feedback are such that the individual nodes behave *as if* they were fed totally by Poisson arrivals, when in fact they are not.

previous discussions. What is amazing is that Jackson was able to show that each node (say the ith) in the network behaves as if it were an independent M/M/m system with a Poisson input rate λ_i. In general, the total input will *not* be a Poisson process. The state variable for this N-node system consists of the vector $(k_1, k_2, \ldots, k_N)$, where k_i is the number of customers in the ith node [including the customer(s) in service]. Let the equilibrium probability associated with this state be denoted by $p(k_1, k_2, \ldots, k_N)$. Similarly we denote the marginal distribution of finding k_i customers in the ith node by $p_i(k_i)$. Jackson was able to show that the joint distribution for all nodes factored into the product of each of the marginal distributions, that is,

$$p(k_1, k_2, \ldots, k_N) = p_1(k_1)p_2(k_2)\cdots p_N(k_N) \qquad \blacksquare(4.67)$$

and $p_i(k_i)$ is given as the solution to the classical M/M/m system [see, for example, Eqs. (3.37)–(3.39) with the obvious change in notation]! This last result is commonly referred to as *Jackson's theorem*. Once again we see the "product" form of solution for Markovian queues in equilibrium.

A modification of Jackson's network of queues was considered by Gordon and Newell [GORD 67]. The modification they investigated was that of a *closed* Markovian network in the sense that a fixed and finite number of customers, say K, are considered to be in the system and are trapped in that system in the sense that no others may enter and none of these may leave; this corresponds to Jackson's case in which $\sum_{j=1}^{N} r_{ij} = 1$ and $\gamma_i = 0$ for all i. (An interesting example of this class of systems known as cyclic queues had been considered earlier by Koenigsberg [KOEN 58]; a cyclic queue is a tandem queue in which the last stage is connected back to the first.) In the general case considered by Gordon and Newell we do not quite expect a product solution since there is a dependency among the elements of the state vector $(k_1, k_2, \ldots, k_N)$ as follows:

$$\sum_{i=1}^{N} k_i = K \qquad (4.68)$$

As is the case for Jackson's model we assume that this discrete-state Markov process is irreducible and therefore a unique equilibrium probability distribution exists for $p(k_1, k_2, \ldots, k_N)$. In this model, however, there is a finite number of states; in particular it is easy to see that the number of distinguishable states of the system is equal to the number of ways in which one can place K customers among the N nodes, and is equal to the binomial coefficient

$$\binom{N+K-1}{N-1}$$

The following equations describe the behavior of the equilibrium distribution of customers in this closed system and may be written by inspection as

$$p(k_1, k_2, \ldots, k_N) \sum_{i=1}^{N} \delta_{k_i-1}\alpha_i(k_i)\mu_i$$
$$= \sum_{i=1}^{N} \sum_{j=1}^{N} \delta_{k_j-1}\alpha_i(k_i + 1)\mu_i r_{ij} p(k_1, k_2, \ldots, k_j - 1, \ldots, k_i + 1, \ldots, k_N) \tag{4.69}$$

where the discrete unit step-function defined in Appendix I takes the form

$$\delta_k \triangleq \begin{cases} 1 & k = 0, 1, 2, \ldots \\ 0 & k < 0 \end{cases} \tag{4.70}$$

and is included in the equilibrium equations to indicate the fact that the service rate must be zero when a given node is empty; furthermore we define

$$\alpha_i(k_i) = \begin{cases} k_i & k_i \le m_i \\ m_i & k_i \ge m_i \end{cases}$$

which merely gives the number of customers in service in the ith node when there are k_i customers at that node. As usual the left-hand side of Eq. (4.69) describes the flow of probability out of state $(k_1, k_2, \ldots, k_N)$ whereas the right-hand side accounts for the flow of probability into that state from neighboring states. Let us proceed to write down the solution to these equations. We define the function $\beta_i(k_i)$ as follows:

$$\beta_i(k_i) = \begin{cases} k_i! & k_i \le m_i \\ m_i!\, m_i^{k_i - m_i} & k_i \ge m_i \end{cases}$$

Consider a set of numbers $\{x_i\}$, which are solutions to the following set of linear equations:

$$\mu_i x_i = \sum_{j=1}^{N} \mu_j x_j r_{ji} \qquad i = 1, 2, \ldots, N \tag{4.71}$$

Note that this set of equations is in the same form as $\boldsymbol{\pi} = \boldsymbol{\pi}\mathbf{P}$ where now the vector $\boldsymbol{\pi}$ may be considered to be $(\mu_1 x_1, \ldots, \mu_N x_N)$ and the elements of the matrix $\mathbf{P}$ are considered to be the elements r_{ij}.* Since we assume that the

* Again the reader is cautioned that, on the one hand, we have been considering Markov chains in which the quantities p_{ij} refer to the transition probabilities among the possible states that the system may take on, whereas, on the other hand, we have in this section in addition been considering a network of queueing systems in which the probabilities r_{ij} refer to transitions that customers make between nodes in that network.

matrix of transition probabilities (whose elements are r_{ij}) is irreducible, then by our previous studies we know that there must be a solution to Eqs. (4.71), all of whose components are positive; of course, they will only be determined to within a multiplicative constant since there are only $N - 1$ independent equations there. With these definitions the solution to Eq. (4.69) can be shown to equal

$$p(k_1,k_2,\ldots,k_N) = \frac{1}{G(K)}\prod_{i=1}^{N}\frac{x_i^{k_i}}{\beta_i(k_i)} \qquad \blacksquare(4.72)$$

where the normalization constant is given by

$$G(K) = \sum_{\mathbf{k}\in A}\prod_{i=1}^{N}\frac{x_i^{k_i}}{\beta_i(k_i)} \qquad (4.73)$$

Here we imply that the summation is taken over all state vectors $\mathbf{k} \triangleq (k_1, \ldots, k_N)$ that lie in the set A, and this is the set of all state vectors for which Eq. (4.68) holds. This then is the solution to the closed finite queueing network problem, and we observe once again that it has the product form.

We may expose the product formulation somewhat further by considering the case where $K \to \infty$. As it turns out, the quantities x_i/m_i are critical in this calculation; we will assume that there exists a unique such ratio that is largest and we will renumber the nodes such that $x_1/m_1 > x_i/m_i$ $(i \neq 1)$. It can then be shown that $p(k_1, k_2, \ldots, k_N) \to 0$ for any state in which $k_1 < \infty$. This implies that an infinite number of customers will form in node one, and this node is often referred to as the "bottleneck" for the given network. On the other hand, however, the marginal distribution $p(k_2, \ldots, k_N)$ is well-defined in the limit and takes the form

$$p(k_2, k_3, \ldots, k_N) = p_2(k_2)p_3(k_3)\cdots p_N(k_N) \qquad (4.74)$$

Thus we see the product solution directly for this marginal distribution and, of course, it is similar to Jackson's theorem in Eq. (4.67); note that in one case we have an open system (one that permits external arrivals) and in the other case we have a closed system. As we shall see in Chapter 4, Volume II, this model has significant applications in time-shared and multi-access computer systems.

Jackson [JACK 63] earlier considered an even more general open queueing system, which includes the closed system just considered as a special case. The new wrinkles introduced by Jackson are, first, that the customer arrival process is permitted to depend upon the total number of customers in the system (using this, he easily creates closed networks) and, second, that the service rate at any node may be a function of the number of customers in that node. Thus defining

$$S(\mathbf{k}) \triangleq k_1 + k_2 + \cdots + k_N$$

we then permit the total arrival rate to be a function of $S(\mathbf{k})$ when the system state is given by the vector $\mathbf{k}$. Similarly we define the exponential service rate at node i to be μ_{k_i} when there are k_i customers at that node (including those in service). As earlier, we have the node transition probabilities r_{ij} $(i, j = 1, 2, \ldots, N)$ with the following additional definitions: r_{0i} is the probability that the next externally generated arrival will enter the network at node i; $r_{i,N+1}$ is the probability that a customer leaving node i departs from the system; and $r_{0,N+1}$ is the probability that the next arrival will require no service from the system and leave immediately upon arrival. Thus we see that in this case $\gamma_i = r_{0i}\gamma(S(\mathbf{k}))$, where $\gamma(S(\mathbf{k}))$ is the total external arrival rate to the system [conditioned on the number of customers $S(\mathbf{k})$ at the moment] from our external Poisson process. It can be seen that the probability of a customer arriving at node i_1 and then passing through the node sequence $i_2, i_3, \ldots, i_n$ and then departing is given by $r_{0i_1}r_{i_1i_2}r_{i_2i_3}\cdots r_{i_{n-1}i_n}r_{i_nN+1}$. Rather than seek the solution of Eq. (4.66) for the traffic rates, since they are functions of the total number of customers in the system we rather seek the solution for the following equivalent set:

$$e_i = r_{0i} + \sum_{j=1}^{N} e_j r_{ji} \tag{4.75}$$

[In the case where the arrival rates are independent of the number in the system then Eqs. (4.66) and (4.75) differ by a multiplicative factor equal to the total arrival rate of customers to the system.] We assume that the solution to Eq. (4.75) exists, is unique, and is such that $e_i \geq 0$ for all i; this is equivalent to assuming that with probability 1 a customer's journey through the network is of finite length. e_i is, in fact, the expected number of times a customer will visit node i in passing through the network.

Let us define the time-dependent state probabilities as

$$P_{\mathbf{k}}(t) = P[\text{system (vector) state at time } t \text{ is } \mathbf{k}] \tag{4.76}$$

By our usual methods we may write down the differential-difference equations governing these probabilities as follows:

$$\frac{d}{dt}P_{\mathbf{k}}(t) = -\left[\gamma(S(\mathbf{k})) + \sum_{i=1}^{N}\mu_{k_i}(1 - r_{ii})\right]P_{\mathbf{k}}(t) + \sum_{i=1}^{N}\gamma(S(\mathbf{k}) - 1)r_{0i}P_{\mathbf{k}(i^-)}(t)$$
$$+ \sum_{i=1}^{N}\mu_{k_i+1}r_{i,N+1}P_{\mathbf{k}(i^+)}(t) + \sum_{\substack{i=1\\ i\neq j}}^{N}\sum_{j=1}^{N}\mu_{k_j+1}r_{ji}P_{\mathbf{k}(i,j)}(t) \tag{4.77}$$

where terms are omitted when any component of the vector argument goes negative; $\mathbf{k}(i^-) = \mathbf{k}$ except for its ith component, which takes on the value

$k_i - 1$; $\mathbf{k}(i^+) = \mathbf{k}$ except for its ith component, which takes on the value $k_i + 1$; and $\mathbf{k}(i, j) = \mathbf{k}$ except that its ith component is $k_i - 1$ and its jth component is $k_j + 1$ where $i \neq j$. Complex as this notation appears its interpretation should be rather straightforward for the reader. Jackson shows that the equilibrium distribution is unique (if it exists) and defines it in our earlier notation to be $\lim P_{\mathbf{k}}(t) \triangleq p_{\mathbf{k}} \triangleq p(k_1, k_2, \ldots, k_N)$ as $t \to \infty$. In order to give the equilibrium solution for $p_{\mathbf{k}}$ we must unfortunately define the following further notation:

$$F(K) \triangleq \prod_{S(\mathbf{k})=0}^{K-1} \gamma(S(\mathbf{k})) \qquad K = 0, 1, 2, \ldots \tag{4.78}$$

$$f(\mathbf{k}) \triangleq \prod_{i=1}^{N} \prod_{j_i=1}^{k_i} \frac{e_i}{\mu_{j_i}} \tag{4.79}$$

$$H(K) \triangleq \sum_{\mathbf{k} \in A} f(\mathbf{k}) \tag{4.80}$$

$$G \triangleq \begin{cases} \sum_{K=0}^{\infty} F(K)H(K) & \text{if the sum converges} \\ \infty & \text{otherwise} \end{cases} \tag{4.81}$$

where the set A shown in Eq. (4.80) is the same as that defined for Eq. (4.73). In terms of these definitions then Jackson's more general theorem states that if $G < \infty$ then a unique equilibrium-state probability distribution exists for the general state-dependent networks and is given by

$$p_{\mathbf{k}} = \frac{1}{G} f(\mathbf{k}) F(S(\mathbf{k})) \tag{4.82}$$

Again we detect the product form of solution. It is also possible to show that in the case when arrivals are independent of the total number in the system [that is, $\gamma \triangleq \gamma(S(\mathbf{k}))$] then even in the case of state-dependent service rates Jackson's first theorem applies, namely, that the joint pdf factors into the product of the individual pdf's given in Eq. (4.67). In fact $p_i(k_i)$ turns out to be the same as the probability distribution for the number of customers in a single-node system where arrivals come from a Poisson process at rate γe_i and with the state-dependent service rates μ_{k_i} such as we have derived for our general birth–death process in Chapter 3. Thus one impact of Jackson's second theorem is that for the constant-arrival-rate case, the equilibrium probability distributions of number of customers in the system at individual centers are independent of other centers; in addition, each of these distributions is identical to the well-known single-node service center with the same parameters.* A remarkable result!

This last theorem is perhaps as far as one can go† with simple Markovian networks, since it seems to extend Burke's theorem in its most general sense. When one relaxes the Markovian assumption on arrivals and/or service times, then extreme complexity in the interdeparture process arises not only from its marginal distribution, but also from its lack of independence on other state variables.

These Markovian queueing networks lead to rather depressing sets of (linear) system equations; this is due to the enormous (yet finite) state description. It is indeed remarkable that such systems do possess reasonably straightforward solutions. The key to solution lies in the observation that these systems may be represented as Markovian population processes, as neatly described by Kingman [KING 69] and as recently pursued by Chandy [CHAN 72]. In particular, a Markov population process is a continuous-time Markov chain over the set of finite-dimensional state vectors $\mathbf{k} = (k_1, k_2, \ldots, k_N)$ for which transitions are permitted only between states‡: $\mathbf{k}$ and $\mathbf{k}(i^+)$ (an external arrival at node i); $\mathbf{k}$ and $\mathbf{k}(i^-)$ (an external departure from node i); and $\mathbf{k}$ and $\mathbf{k}(i, j)$ (an internal transfer from node i to node j). Kingman gives an elegant discussion of the interesting classes and properties of these processes (using the notion and properties of reversible Markov chains). Chandy discusses some of these issues by observing that the equilibrium probabilities for the system states obey not only the global-balance equations that we have so far seen (and typically which lead to product form solutions) but also that this system of equations may be decomposed into many sets of smaller systems of equations, each of which is simpler to solve. This transformed set is referred to as the set of "local"-balance equations, which we now proceed to discuss.

The concept of local balance is most valuable when one deals with a network of queues. However, the concept does apply to single-node Markovian queues, and in fact we have already seen an example of local balance at play.

* This model also permits one to handle the closed queueing systems studied by Gordon and Newell. In order to create the constant total number of customers one need merely set $\gamma(k) = 0$ for $k \geq K$ and $\gamma(K-1) = \infty$, where K is the fixed number one wishes to contain within the system. In order to keep the node transition probabilities identical in the open and closed systems, let us denote the former as earlier by r_{ij} and the latter now by r_{ij}'; to make the limit of Jackson's general system equivalent to the closed system of Gordon and Newell we then require $r_{ij}' = r_{ij} + (r_{i,N+1})(r_{0i})$.

† In Chapter 4, Volume II, we describe some recent results that do in fact extend the model to handle different customer classes and different service disciplines at each node (permitting, in some cases, more general service-time distributions).

‡ See the definitions following Eq. (4.77).

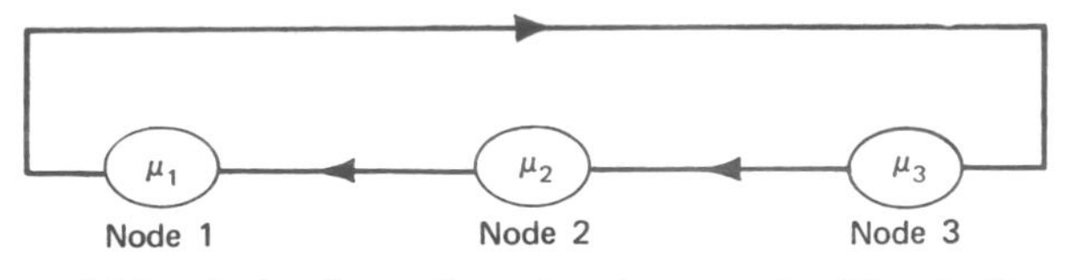

Figure 4.16 A simple cyclic network example: $N = 3$, $K = 2$.

Let us recall the global-balance equations (the flow-conservation equations) for the general birth–death process as exemplified in Eq. (3.6). This equation was obtained by balancing flow into and out of state E_k in Figure 2.9. We also commented at that time that a different boundary could be considered across which flow must be conserved, and this led to the set of equations (3.7). These latter equations are in fact local-balance equations and have the extremely interesting property that they *match* terms from the left-hand side of Eq. (3.6) with corresponding terms on the right-hand side; for example, the term $\lambda_{k-1}p_{k-1}$ on the left-hand side of Eq. (3.6) is seen to be equal to $\mu_k p_k$ on the right-hand side of that equation directly from Eq. (3.7), and by a second application of Eq. (3.7) we see that the two remaining terms in Eq. (3.6) must be equal. This is precisely the way in which local balance operates, namely, to observe that certain sets of terms in the global-balance equation must balance by themselves giving rise to a number of "local"-balance equations.

The significant observation is that, if we are dealing with an ergodic Markov process, then we know for sure that there is a unique solution for the equilibrium probabilities as defined by the generic equation $\boldsymbol{\pi} = \boldsymbol{\pi}\mathbf{P}$. Second, if we decompose the global-balance equations for such a process by matching terms of the large global-balance equations into sets of smaller local-balance equations (and of course account for all the terms in the global balance), then any solution satisfied by this large set of local-balance equations must also satisfy the global-balance equations; the converse is not generally true. Thus any solution for the local-balance equations will yield the unique solution for our Markov process.

In the interesting case of a network of queues we define a *local-balance equation* (with respect to a given network state and a network node i) as one that equates the rate of flow out of that network state due to the departure of a customer from node i to the rate of flow into that network state due to the arrival of a customer to node i.* This notion in the case of networks is best illustrated by the simple example shown in Figure 4.16. Here we show the case of a three-node network where the service rate in the ith node is given as

* When service is nonexponential but rather given in terms of a stage-type service distribution, then one equates arrivals to and departures from a given stage of service (rather than to and from the node itself).

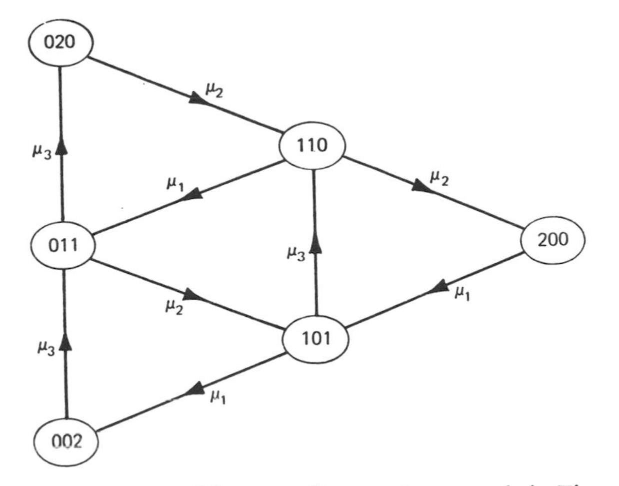

Figure 4.17 State-transition-rate diagram for example in Figure 4.16.

μ_i and is independent of the number of customers at that node; we assume there are exactly $K = 2$ customers circulating in this closed cyclic network. Clearly we have $r_{13} = r_{32} = r_{21} = 1$ and $r_{ij} = 0$ otherwise. Our state description is merely the triplet (k_1, k_2, k_3), where as usual k_i gives the number of customers in node i and where we require, of course, that $k_1 + k_2 + k_3 = 2$. For this network we will therefore have exactly

$$\binom{N + K - 1}{N - 1} = 6$$

states with state-transition rates as shown in Figure 4.17.

For this system we have six global-balance equations (one of which will be redundant as usual; the extra condition comes from the conservation of probability); these are

$$\mu_1 p(2, 0, 0) = \mu_2 p(1, 1, 0) \tag{4.83}$$

$$\mu_2 p(0, 2, 0) = \mu_3 p(0, 1, 1) \tag{4.84}$$

$$\mu_3 p(0, 0, 2) = \mu_1 p(1, 0, 1) \tag{4.85}$$

$$\mu_1 p(1, 1, 0) + \mu_2 p(1, 1, 0) = \mu_2 p(0, 2, 0) + \mu_3 p(1, 0, 1) \tag{4.86}$$

$$\mu_2 p(0, 1, 1) + \mu_3 p(0, 1, 1) = \mu_3 p(0, 0, 2) + \mu_1 p(1, 1, 0) \tag{4.87}$$

$$\mu_1 p(1, 0, 1) + \mu_3 p(1, 0, 1) = \mu_2 p(0, 1, 1) + \mu_1 p(2, 0, 0) \tag{4.88}$$

Each of these global-balance equations is of the form whereby the left-hand side represents the flow out of a state and the right-hand side represents the flow into that state. Equations (4.83)–(4.85) are already local-balance equations as we shall see; Eqs. (4.86)–(4.88) have been written so that the first term on the left-hand side of each equation balances the first term on the right-hand side of the equation, and likewise for the second terms. Thus Eq. (4.86) gives rise to the following local-balance equations:

$$\mu_1 p(1, 1, 0) = \mu_2 p(0, 2, 0) \tag{4.89}$$

$$\mu_2 p(1, 1, 0) = \mu_3 p(1, 0, 1) \tag{4.90}$$

Note, for example, that Eq. (4.89) takes the rate out of state (1, 1, 0) due to a departure from node 1 and equates it to the rate into that state due to arrivals at node 1; similarly, Eq. (4.90) does likewise for departures and arrivals at node 2. This is the principle of local balance and we see therefore that Eqs. (4.83)–(4.85) are already of this form. Thus we generate nine local-balance equations* (four of which must therefore be redundant when we consider the conservation of probability), each of which is extremely simple and therefore permits a straightforward solution to be found. If this set of equations does indeed have a solution, then they certainly guarantee that the global equations are satisfied and therefore that the solution we have found is the unique solution to the original global equations. The reader may easily verify the following solution:

$$p(1, 0, 1) = \frac{\mu_1}{\mu_3} p(2, 0, 0)$$

$$p(1, 1, 0) = \frac{\mu_1}{\mu_2} p(2, 0, 0)$$

$$p(0, 1, 1) = \frac{(\mu_1)^2}{\mu_2 \mu_3} p(2, 0, 0)$$

$$p(0, 0, 2) = \left(\frac{\mu_1}{\mu_3}\right)^2 p(2, 0, 0)$$

$$p(0, 2, 0) = \left(\frac{\mu_1}{\mu_2}\right)^2 p(2, 0, 0)$$

$$p(2, 0, 0) = \left[1 + \frac{\mu_1}{\mu_3} + \frac{\mu_1}{\mu_2} + \frac{(\mu_1)^2}{\mu_2 \mu_3} + \left(\frac{\mu_1}{\mu_3}\right)^2 + \left(\frac{\mu_1}{\mu_2}\right)^2\right]^{-1} \tag{4.91}$$

Had we allowed all possible transitions among nodes (rather than the cyclic behavior in this example) then the state-transition-rate diagram would have

* The reader should write them out directly from Figure 4.17.

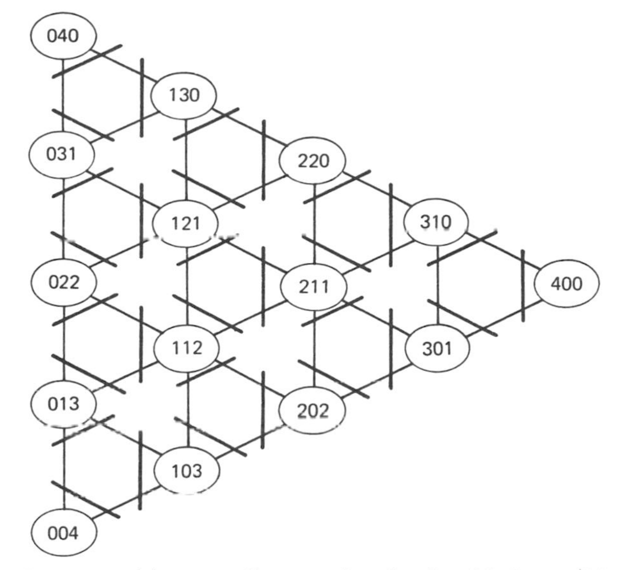

Figure 4.18 State-transition-rate diagram showing local balance ($N = 3$, $K = 4$).

permitted transitions in both directions where now only unidirectional transitions are permitted; however, it will always be true that only transitions to nearest-neighbor states (in this two-dimensional diagram) are permitted so that such a diagram can always be drawn in a planar fashion. For example, had we allowed four customers in an arbitrarily connected three-node network, then the state-transition-rate diagram would have been as shown in Figure 4.18. In this diagram we represent possible transitions between nodes by an undirected branch (representing two one-way branches in opposite directions). Also, we have collected together sets of branches by joining them with a heavy line, and these are meant to represent branches whose contributions appear in the same local-balance equation. These diagrams can be extended to higher dimensions when there are more than three nodes in the system. In particular, with four nodes we get a tetrahedron (that is, a three-dimensional simplex). In general, with N nodes we will get an $(N - 1)$-dimensional simplex with $K + 1$ nodes along each edge (where K = number of customers in the closed system). We note in these diagrams that all nodes lying in a given straight line (parallel to any base of the simplex) maintain one component of the state vector at a constant value and that this value increases or decreases by unity as one moves to a parallel set of nodes. The local-balance equations are identified as balancing flow in that set of branches that connects a given node on one of these constant lines to all other nodes on that constant line adjacent and parallel to this node, and that decreases by unity that component that had been held constant. In summary, then, the local-balance equations are trivial to write down, and if one can succeed in finding a solution that satisfies them, then one has found the solution to the global-balance equations as well!

As we see, most of these Markovian networks lead to rather complex systems of linear equations. Wallace and Rosenberg [WALL 66] propose a numerical solution method for a large class of these equations which is computationally efficient. They discuss a computer program, which is designed to evaluate the equilibrium probability distributions of state variables in very large finite Markovian queueing networks. Specifically, it is designed to solve the equilibrium equations of the form given in Eqs. (2.50) and (2.116), namely, $\boldsymbol{\pi} = \boldsymbol{\pi}\mathbf{P}$ and $\boldsymbol{\pi}\mathbf{Q} = 0$. The procedure is of the "power-iteration type" such that if $\boldsymbol{\pi}(i)$ is the ith iterate then $\boldsymbol{\pi}(i + 1) = \boldsymbol{\pi}(i)\mathbf{R}$ is the $(i + 1)$th iterate; the matrix $\mathbf{R}$ is either equal to the matrix $\alpha\mathbf{P} + (1 - \alpha)\mathbf{I}$ (where α is a scalar) or equal to the matrix $\beta\mathbf{Q} + \mathbf{I}$ (where β is a scalar and $\mathbf{I}$ is the identity matrix), depending upon which of the two above equations is to be solved. The scalars α and β are chosen carefully so as to give an efficient convergence to the solution of these equations. The speed of solution is quite remarkable and the reader is referred to [WALL 66] and its references for further details.

Thus ends our study of purely Markovian systems in equilibrium. The unifying feature throughout Chapters 3 and 4 has been that these systems give rise to product-type solutions; one is therefore urged to look for solutions of this form whenever Markovian queueing systems are encountered. In the next chapter we permit either $A(t)$ or $B(x)$ (but not both) to be of arbitrary form, requiring the other to remain in exponential form.

REFERENCES

BURK 56 Burke, P. J., "The Output of a Queueing System," *Operations Research*, **4,** 699–704 (1966).

CHAN 72 Chandy, K. M., "The Analysis and Solutions for General Queueing Networks," Proc. Sixth Annual Princeton Conference on Information Sciences and Systems, Princeton University, March 1972.

COX 55 Cox, D. R., "A Use of Complex Probabilities in the Theory of Stochastic Processes," *Proceedings Cambridge Philosophical Society*, **51,** 313–319 (1955).

GORD 67 Gordon, W. J. and G. F. Newell, "Closed Queueing Systems with Exponential Servers," *Operations Research*, **15,** 254–265 (1967).

JACK 57 Jackson, J. R., "Networks of Waiting Lines," *Operations Research*, **5,** 518–521 (1957).

JACK 63 Jackson, J. R., "Jobshop-Like Queueing Systems," *Management Science*, **10,** 131–142 (1963).

KING 69 Kingman, J. F. C., "Markov Population Processes," *Journal of Applied Probability*, **6,** 1–18 (1969).

KOEN 58 Koenigsberg, E., "Cyclic Queues," *Operations Research Quarterly*, **9**, 22–35 (1958).

SAAT 65 Saaty, T. L., "Stochastic Network Flows: Advances in Networks of Queues," *Proc. Symp. Congestion Theory*, Univ. of North Carolina Press, (Chapel Hill), 86–107, (1965).

WALL 66 Wallace, V. L. and R. S. Rosenberg, "Markovian Models and Numerical Analysis of Computer System Behavior," *AFIPS Spring Joint Computer Conference Proc.*, 141–148, (1966).

WHIT 44 Whittaker, E. and G. Robinson, *The Calculus of Observations*, 4th ed., Blackie (London), (1944).

EXERCISES

4.1. Consider the Markovian queueing system shown below. Branch labels are birth and death rates. Node labels give the number of customers in the system.

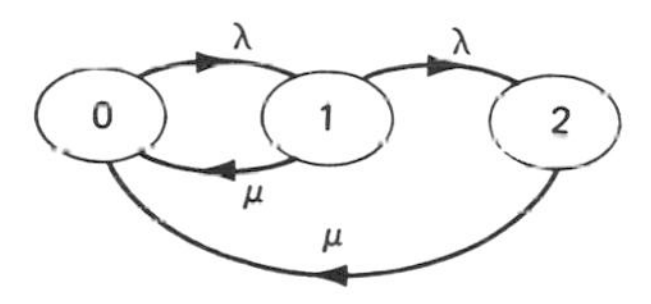

(a) Solve for p_k.

(b) Find the average number in the system.

(c) For $\lambda = \mu$, what values do we get for parts (a) and (b)? Try to interpret these results.

(d) Write down the transition rate matrix **Q** for this problem and give the matrix equation relating **Q** to the probabilities found in part (a).

4.2. Consider an $E_k/E_n/1$ queueing system where *no* queue is permitted to form. A customer who arrives to find the service facility busy is "lost" (he departs with no service). Let E_{ij} be the system state in which the "arriving" customer is in the ith arrival stage and the customer in service is in the jth service stage (note that there is always some customer in the arrival mechanism and that if there is no customer in the service facility, then we let $j = 0$). Let $1/k\lambda$ be the average time spent in any arrival stage and $1/n\mu$ be the average time spent in any service stage.

(a) Draw the state transition diagram showing all the transition *rates*.

(b) Write down the equilibrium equation for E_{ij} where $1 < i < k$, $0 < j < n$.

4.3. Consider an $M/E_r/1$ system in which *no* queue is allowed to form. Let j = the number of stages of service left in the system and let P_j be the equilibrium probability of being in state E_j.

(a) Find $P_j, j = 0, 1, \ldots, r$.

(b) Find the probability of a busy system.

4.4. Consider an $M/H_2/1$ system in which *no* queue is allowed to form. Service is of the hyperexponential type as shown in Figure 4.10 with $\mu_1 = 2\mu\alpha_1$ and $\mu_2 = 2\mu(1 - \alpha_1)$.

(a) Solve for the equilibrium probability of an empty system.

(b) Find the probability that server 1 is occupied.

(c) Find the probability of a busy system.

4.5. Consider an M/M/1 system with parameters λ and μ in which exactly two customers arrive at each arrival instant.

(a) Draw the state-transition-rate diagram.

(b) By inspection, write down the equilibrium equations for p_k $(k = 0, 1, 2, \ldots)$.

(c) Let $\rho = 2\lambda/\mu$. Express $P(z)$ in terms of ρ and z.

(d) Find $P(z)$ by using the bulk arrival results from Section 4.5.

(e) Find the mean and variance of the number of customers in the system from $P(z)$.

(f) Repeat parts (a)–(e) with exactly r customers arriving at each arrival instant (and $\rho = r\lambda/\mu$).

4.6. Consider an M/M/1 queueing system with parameters λ and μ. At each of the arrival instants one new customer will enter the system with probability 1/2 or two new customers will enter simultaneously with probability 1/2.

(a) Draw the state-transition-rate diagram for this system.

(b) Using the method of non-nearest-neighbor systems write down the equilibrium equations for p_k.

(c) Find $P(z)$ and also evaluate any constants in this expression so that $P(z)$ is given in terms only of λ and μ. If possible eliminate any common factors in the numerator and denominator of this expression [this makes life simpler for you in part (d)].

(d) From part (c) find the expected number of customers in the system.

(e) Repeat part (c) using the results obtained in Section 4.5 directly.

4.7. For the bulk arrival system of Section 4.5, assume (for $0 < \alpha < 1$) that

$$g_i = (1 - \alpha)\alpha^i \qquad i = 0, 1, 2, \ldots$$

Find p_k = equilibrium probability of finding k in the system.

4.8. For the bulk arrival system studied in Section 4.5, find the mean $\bar{N}$ and variance σ_N^2 for the number of customers in the system. Express your answers in terms of the moments of the bulk arrival distribution.

4.9. Consider an M/M/1 system with the following variation: Whenever the server becomes free, he accepts *two* customers (if at least two are available) from the queue into service simultaneously. Of these two customers, only one receives service; when the service for this one is completed, both customers depart (and so the other customer got a "free ride").

If only one customer is available in the queue when the server becomes free, then that customer is accepted alone and is serviced; if a new customer happens to arrive when this single customer is being served, then the new customer joins the old one in service and this new customer receives a "free ride."

In all cases, the service time is exponentially distributed with mean $1/\mu$ sec and the average (Poisson) arrival rate is λ customers per second.

(a) Draw the appropriate state diagram.
(b) Write down the appropriate difference equations for p_k = equilibrium probability of finding k customers in the system.
(c) Solve for $P(z)$ in terms of p_0 and p_1.
(d) Express p_1 in terms of p_0.

4.10. We consider the denominator polynomial in Eq. (4.35) for the system $E_r/M/1$. Of the $r + 1$ roots, we know that one occurs at $z = 1$. Use Rouche's theorem (see Appendix I) to show that exactly $r - 1$ of the remaining r roots lie in the unit disk $|z| \leq 1$ and therefore exactly one root, say z_0, lies in the region $|z_0| > 1$.

4.11. Show that the solution to Eq. (4.71) gives a set of variables $\{x_i\}$ which guarantee that Eq. (4.72) is indeed the solution to Eq. (4.69).

4.12. **(a)** Draw the state-transition-rate diagram showing local balance for the case ($N = 3$, $K = 5$) with the following structure:

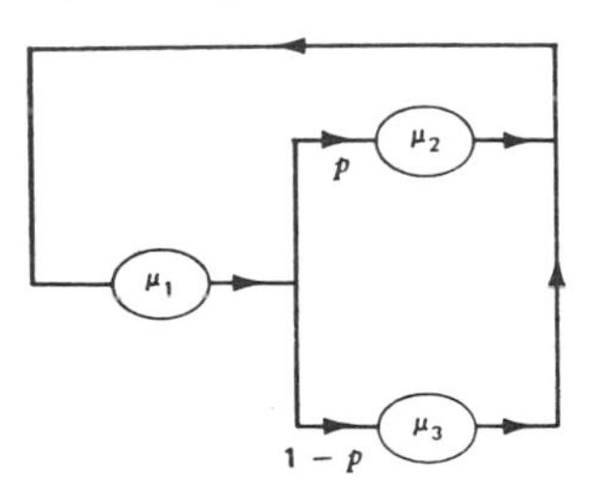

(b) Solve for $p(k_1, k_2, k_3)$.

4.13. Consider a two-node Markovian queueing network (of the more general type considered by Jackson) for which $N = 2$, $m_1 = m_2 = 1$, $\mu_{k_i} = \mu_i$ (constant service rate), and which has transition probabilities (r_{ij}) as described in the following matrix:

$$r_{ij} = \begin{array}{c|cccc} i \backslash j & 0 & 1 & 2 & 3 \\ \hline 0 & 0 & 1 & 0 & 0 \\ 1 & 0 & 0 & 1-\alpha & \alpha \\ 2 & 0 & 1 & 0 & 0 \end{array}$$

where $0 < \alpha < 1$ and nodes 0 and $N + 1$ are the "source" and "sink" nodes, respectively. We also have (for some integer K)

$$\gamma(S(k_1, k_2)) = \begin{cases} \infty & k_1 + k_2 \neq K \\ 0 & k_1 + k_2 = K \end{cases}$$

and assume the system initially contains K customers.

(a) Find e_i $(i = 1, 2)$ as given in Eq. (4.75).

(b) Since $N = 2$, let us denote $p(k_1, k_2) = p(k_1, K - k_1)$ by p_{k_1}. Find the balance equations for p_{k_1}.

(c) Solve these equations for p_{k_1} explicitly.

(d) By considering the fraction of time the first node is busy, find the time between customer departures from the network (via node 1, of course).

PART III

INTERMEDIATE QUEUEING THEORY

We are here concerned with those queueing systems for which we can still apply certain simplifications due to their Markovian nature. We encounter those systems that are representable as *imbedded* Markov chains, namely, the M/G/1 and the G/M/m queues. In Chapter 5 we rapidly develop the basic equilibrium equations for M/G/1 giving the notorious *Pollaczek–Khinchin* equations for queue length and waiting time. We next discuss the busy period and, finally, introduce some moderately advanced techniques for studying these systems, even commenting a bit on the time-dependent solutions. Similarly for the queue G/M/m in Chapter 6, we find that we can make some very specific statements about the equilibrium system behavior and, in fact, find that the conditional distribution of waiting time will always be exponential regardless of the interarrival time distribution! Similarly, the conditional queue-length distribution is shown to be geometric. We note in this part that the methods of solution are quite different from that studied in Part II, but that much of the underlying behavior is similar; in particular the mean queue size, the mean waiting time, and the mean busy period duration all are inversely proportional to $1 - \rho$ as earlier. In Chapter 7 we briefly investigate a rather pleasing interpretation of transforms in terms of probabilities.

The techniques we had used in Chapter 3 [the explicit product solution of Eq. (3.11)] and in Chapter 4 (flow conservation) are replaced by an indirect z-transform approach in Chapter 5. However, in Chapter 6, we return once again to the flow conservation inherent in the $\boldsymbol{\pi} = \boldsymbol{\pi}P$ solution.

5

The Queue M/G/1

That which makes elementary queueing theory so appealing is the simplicity of its state description.* In particular, all that is required in order to summarize the entire past history of the queueing system is a specification of the number of customers† present. All other historical information is irrelevant to the future behavior of pure Markovian systems. Thus the state description is not only one dimensional but also countable (and in some cases finite). It is this latter property (the countability) that simplifies our calculations.

In this chapter and the next we study queueing systems that are driven by non-Markovian stochastic processes. As a consequence we are faced with new problems for which we must find new methods of solution.

In spite of the non-Markovian nature of these two systems there exists an abundance of techniques for handling them. Our approach in this chapter will be the method of the *imbedded Markov chain* due to Palm [PALM 43] and Kendall [KEND 51]. However, we have in reality already seen a second approach to this class of problems, namely, the *method of stages*, in which it was shown that so long as the interarrival time and service time pdf's have Laplace transforms that are rational, then the stage method can be applied (see Section 4.7); the disadvantage of that approach is that it merely gives a procedure for carrying out the solution but does not show the solution as an explicit expression, and therefore properties of the solution cannot be studied for a class of systems. The third approach, to be studied in Chapter 8, is to solve *Lindley's integral equation* [LIND 52]; this approach is suitable for the system G/G/1 and so obviously may be specialized to some of the systems we consider in this chapter. A fourth approach, the *method of supplementary*

* Usually a state description is given in terms of a vector which describes the system's state at time t. A vector $\mathbf{v}(t)$ is a *state vector* if, given $\mathbf{v}(t)$ and all inputs to this system during the interval (t, t_1) (where $t < t_1$), then we are capable of solving for the state vector $\mathbf{v}(t_1)$. Clearly it behooves us to choose a state vector containing that information that permits us to calculate quantities of importance for understanding system behavior.

† We saw in Chapter 4 that occasionally we record the number of stages in the system rather than the number of customers.

variables, is discussed in the exercises at the end of this chapter; more will be said about this method in the next section. We also discuss the *busy period analysis* [GAVE 59], which leads to the waiting-time distribution (see Section 5.10). Beyond these there exist other approaches to non-Markovian queueing systems, among which are the *random-walk* and *combinatorial* approaches [TAKA 67] and the *method of Green's function* [KEIL 65].

5.1. THE M/G/1 SYSTEM

The M/G/1 queue is a single-server system with Poisson arrivals and arbitrary service-time distribution denoted by $B(x)$ [and a service time pdf denoted by $b(x)$]. That is, the interarrival time distribution is given by

$$A(t) = 1 - e^{-\lambda t} \qquad t \geq 0$$

with an average arrival rate of λ customers per second, a mean interarrival time of $1/\lambda$ sec, and a variance $\sigma_a^2 = 1/\lambda^2$. As defined in Chapter 2 we denote the kth moment of service time by

$$x^k \triangleq \int_0^\infty x^k b(x)\, dx$$

and we sometimes express these service time moments by $b_k \triangleq \overline{x^k}$.

Let us discuss the state description (vector) for the M/G/1 system. If at some time t we hope to summarize the complete past history of this system, then it is clear that we must certainly specify $N(t)$, the number of customers present at time t. Moreover, we must specify $X_0(t)$, the service time already received by the customer in service at time t; this is necessary since the service-time distribution is not necessarily of the memoryless type. (Clearly, we need not specify how long it has been since the last arrival entered the system, since the arrival process *is* of the memoryless type.) Thus we see that the random process $N(t)$ is a non-Markovian process. However, the vector $[N(t), X_0(t)]$ is a Markov process and is an appropriate state vector for the M/G/1 system, since it completely summarizes all past history relevant to the future system development.

We have thus gone from a single-component description of state in elementary queueing theory to what appears to be a two-component description here in intermediate queueing theory. Let us examine the inherent difference between these two state descriptions. In elementary queueing theory, it is sufficient to provide $N(t)$, the number in the system at time t, and we then have a Markov process with a discrete-state space, where the states themselves are either finite or countable in number. When we proceed to the current situation where we need a two-dimensional state description, we find that the number in the system $N(t)$ is still denumerable, but now we must also provide $X_0(t)$, the expended service time, which is *continuous.* We have thus evolved from a discrete-state description to a continuous-state description, and this essential difference complicates the analysis.

It is possible to proceed with a general theory based upon the couplet $[N(t), X_0(t)]$ as a state vector and such a method of solution is referred to as the *method of supplementary variables.* For a treatment of this sort the reader is referred to Cox [COX 55] and Kendall [KEND 53]; Henderson [HEND 72] also discusses this method, but chooses the remaining service time instead of the expended service time as the supplementary variable. In this text we choose to use the *method of the imbedded Markov chain* as discussed below. However, before we proceed with the method itself, it is clear that we should understand some properties of the expended service time; this we do in the following section.

5.2. THE PARADOX OF RESIDUAL LIFE: A BIT OF RENEWAL THEORY

We are concerned here with the case where an arriving customer finds a partially served customer in the service facility. Problems of this sort occur repeatedly in our studies, and so we wish to place this situation in a more general context. We begin with an apparent paradox illustrated through the following example. Assume that our hippie from Chapter 2 arrives at a roadside cafe at an arbitrary instant in time and begins hitchhiking. Assume further that automobiles arrive at this cafe according to a Poisson process at an average rate of λ cars per minute. How long must the hippie wait, on the average, until the next car comes along?

There are two apparently logical answers to this question. First, we might argue that since the average time between automobile arrivals is $1/\lambda$ min, and since the hippie arrives at a random point in time, then "obviously" the hippie will wait on the average $1/2\lambda$ min. On the other hand, we observe that since the Poisson process is memoryless, the time until the next arrival is independent of how long it has been since the previous arrival and therefore the hippie will wait on the average $1/\lambda$ min; this second argument can be extended to show that the average time from the last arrival until the hippie begins hitchhiking is also $1/\lambda$ min. The second solution therefore implies that the average time between the last car and the next car to arrive will be $2/\lambda$ min! It appears that this interval is twice as long as it should be for a Poisson process! Nevertheless, the second solution is the correct one, and so we are faced with an apparent paradox!

Let us discuss the solution to this problem in the case of an arbitrary interarrival time distribution. This study properly belongs to renewal theory, and we quote results freely from that field; most of these results can be found

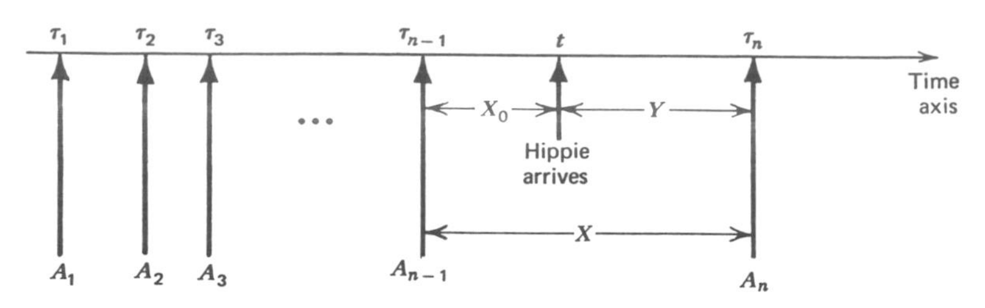

Figure 5.1 Life, age and residual life.

in the excellent monograph by Cox [COX 62] or in the fine expository article by Smith [SMIT 58]; the reader is also encouraged to see Feller [FELL 66]. The basic diagram is that given in Figure 5.1. In this figure we let A_k denote the kth automobile, which we assume arrives at time τ_k. We assume that the intervals $\tau_{k+1} - \tau_k$ are independent and identically distributed random variables with distribution given by

$$F(x) \triangleq P[\tau_{k+1} - \tau_k \leq x] \tag{5.1}$$

We further define the common pdf for these intervals as

$$f(x) \triangleq \frac{dF(x)}{dx} \tag{5.2}$$

Let us now choose a random point in time, say t, when our hippie arrives at the roadside cafe. In this figure, A_{n-1} is the last automobile to arrive prior to t and A_n will be the first automobile to arrive after t. We let X denote this "special" interarrival time and we let Y denote the time that our hippie must wait until the next arrival. Clearly, the sequence of arrival points $\{\tau_k\}$ forms a renewal process; renewal theory discusses the instantaneous replacement of components. In this case, $\{\tau_k\}$ forms the sequence of instants when the old component fails and is replaced by a new component. In the language of renewal theory X is said to be the *lifetime* of the component under consideration, Y is said to be the *residual life* of that component at time t, and $X_0 = X - Y$ is referred to as the *age* of that component at time t. Let us adopt that terminology and proceed to find the pdf for X and Y, the lifetime and residual life of our selected component. We assume that the renewal process has been operating for an arbitrarily long time since we are interested only in limiting distributions.

The amazing result we will find is that X is not distributed according to $F(x)$. In terms of our earlier example this means that the interval which the hippie happens to select by his arrival at the cafe is not a typical interval. In fact, herein lies the solution to our paradox: A long interval is more likely to be "intercepted" by our hippie than a short one. In the case of a Poisson process we shall see that this bias causes the selected interval to be on the average twice as long as a typical interval.

Let the residual life have a distribution

$$\hat{F}(x) \triangleq P[Y \leq x] \tag{5.3}$$

with density

$$\hat{f}(x) = \frac{d\hat{F}(x)}{dx} \tag{5.4}$$

Similarly, let the selected lifetime X have a pdf $f_X(x)$ and PDF $F_X(x)$ where

$$F_X(x) \triangleq P[X \leq x] \tag{5.5}$$

In Exercise 5.2 we direct the reader through a rigorous derivation for the residual lifetime density $\hat{f}(x)$. Rather than proceed through those details, let us give an intuitive derivation for the density that takes advantage of our physical intuition regarding this problem. Our basic observation is that long intervals between renewal points occupy larger segments of the time axis than do shorter intervals, and therefore it is more likely that our random point t will fall in a long interval. If we accept this, then we recognize that the probability that an interval of length x is chosen should be proportional to the length (x) as well as to the relative occurrence of such intervals [which is given by $f(x)\,dx$]. Thus, for the selected interval, we may write

$$f_X(x)\,dx = Kxf(x)\,dx \tag{5.6}$$

where the left-hand side is $P[x < X \leq x + dx]$ and the right-hand side expresses the linear weighting with respect to interval length and includes a constant K, which must be evaluated so as to properly normalize this density. Integrating both sides of Eq. (5.6) we find that $K = 1/m_1$, where

$$m_1 \triangleq E[\tau_k - \tau_{k-1}] \tag{5.7}$$

and is the common average time between renewals (between arrivals of automobiles). Thus we have shown that the density associated with the selected interval is given in terms of the density of typical intervals by

$$f_X(x) = \frac{xf(x)}{m_1} \quad \blacksquare \tag{5.8}$$

This is our first result. Let us proceed now to find the density of residual life $\hat{f}(x)$. If we are told that $X = x$, then the probability that the residual life Y does not exceed the value y is given by

$$P[Y \leq y \mid X = x] = \frac{y}{x}$$

for $0 \leq y \leq x$; this last is true since we have randomly chosen a point within this selected interval, and therefore this point must be uniformly distributed within that interval. Thus we may write down the joint density of X and Y as

$$P[y < Y \leq y + dy, x < X \leq x + dx] = \left(\frac{dy}{x}\right)\left(\frac{xf(x)\,dx}{m_1}\right)$$
$$= \frac{f(x)\,dy\,dx}{m_1} \tag{5.9}$$

for $0 \leq y \leq x$. Integrating over x we obtain $\hat{f}(y)$, which is the unconditional density for Y, namely,

$$\hat{f}(y)\,dy = \int_{x=y}^{\infty} \frac{f(x)\,dy\,dx}{m_1}$$

This immediately gives the final result:

$$\hat{f}(y) = \frac{1 - F(y)}{m_1} \qquad \blacksquare(5.10)$$

This is our second result. It gives the density of residual life in terms of the common distribution of interval length and its mean.*

Let us express this last result in terms of transforms. Using our usual transform notation we have the following correspondences:

$$f(x) \Leftrightarrow F^*(s)$$
$$\hat{f}(x) \Leftrightarrow \hat{F}^*(s)$$

Clearly, all the random variables we have been discussing in this section are nonnegative, and so the relationship in Eq. (5.10) may be transformed directly by use of entry 5 in Table I.4 and entry 13 in Table I.3 to give

$$\hat{F}^*(s) = \frac{1 - F^*(s)}{sm_1} \qquad \blacksquare(5.11)$$

It is now a trivial matter to find the moments of residual life in terms of the moments of the lifetimes themselves. We denote the nth moment of the lifetime by m_n and the nth moment of the residual life by r_n, that is,

$$m_n \triangleq E[(\tau_k - \tau_{k-1})^n] \tag{5.12}$$
$$r_n \triangleq E[Y^n] \tag{5.13}$$

Using our moment formula Eq. (II.26), we may differentiate Eq. (5.11) to obtain the moments of residual life. As $s \to 0$ we obtain indeterminate forms

* It may also be shown that the limiting pdf for age (X_0) is the same as for residual life (Y) given in Eq. (5.10).

which may be evaluated by means of L'Hospital's rule; this computation gives the moments of residual life as

$$r_n = \frac{m_{n+1}}{(n+1)m_1} \qquad (5.14)$$

This important formula is most often used to evaluate r_1, the mean residual life, which is found equal to

$$r_1 = \frac{m_2}{2m_1} \qquad (5.15)$$

and may also be expressed in terms of the lifetime variance (denoted by $\sigma^2 \triangleq m_2 - m_1^2$) to give

$$r_1 = \frac{m_1}{2} + \frac{\sigma^2}{2m_1} \qquad (5.16)$$

This last form shows that the correct answer to the hippie paradox is $m_1/2$, half the mean interarrival time, *only* if the variance is zero (regularly spaced arrivals); however, for the Poisson arrivals, $m_1 = 1/\lambda$ and $\sigma^2 = 1/\lambda^2$, giving $r_1 = 1/\lambda = m_1$, which confirms our earlier solution to the hippie paradox of residual life. Note that $m_1/2 \le r_1$ and r_1 will grow without bound as $\sigma^2 \to \infty$. The result for the mean residual life (r_1) is a rather counterintuitive result; we will see it appear again and again.

Before leaving renewal theory we take this opportunity to quote some other useful results. In the language of renewal theory the age-dependent failure rate $r(x)$ is defined as the instantaneous rate at which a component will fail given that it has already attained an age of x; that is, $r(x)\,dx \triangleq P[x <$ lifetime of component $\le x + dx \mid$ lifetime $> x]$. From first principles, we see that this conditional density is

$$r(x) = \frac{f(x)}{1 - F(x)} \qquad (5.17)$$

where once again $f(x)$ and $F(x)$ refer to the common distribution of component lifetime. The *renewal function* $H(x)$ is defined to be

$$H(x) \triangleq E[\text{number of renewals in an interval of length } x] \qquad (5.18)$$

and the *renewal density* $h(x)$ is merely the renewal rate at time x defined by

$$h(x) \triangleq \frac{dH(x)}{dx} \qquad (5.19)$$

Renewal theory seems to be obsessed with limit theorems, and one of the important results is the *renewal theorem*, which states that

$$\lim_{x\to\infty} h(x) = \frac{1}{m_1} \tag{5.20}$$

This merely says that in the limit one cannot identify when the renewal process began, and so the rate at which components are renewed is equal to the inverse of the average time between renewals (m_1). We note that $h(x)$ is not a pdf; in fact, its integral diverges in the typical case. Nevertheless, it does possess a Laplace transform which we denote by $H^*(s)$. It is easy to show that the following relationship exists between this transform and the transform of the underlying pdf for renewals, namely:

$$H^*(s) = \frac{F^*(s)}{1 - F^*(s)} \tag{5.21}$$

This last is merely the transform expression of the *integral equation of renewal theory*, which may be written as

$$h(x) = f(x) + \int_0^x h(x - t)f(t)\,dt \tag{5.22}$$

More will not be said about renewal theory at this point. Again the reader is urged to consult the references mentioned above.

5.3. THE IMBEDDED MARKOV CHAIN

We now consider the method of the imbedded Markov chain and apply it to the M/G/1 queue. The fundamental idea behind this method is that we wish to simplify the description of state from the two-dimensional description $[N(t), X_0(t)]$ into a one-dimensional description $N(t)$. If indeed we are to be successful in calculating future values for our state variable we must also implicitly give, along with this one-dimensional description of the number in system, the time expended on service for the customer in service. Furthermore (and here is the crucial point), we agree that we may gain this simplification by looking not at all points in time but rather at a select set of points in time. Clearly, these special epochs must have the property that, if we specify the number in the system at one such point and also provide future inputs to the system, then at the next suitable point in time we can again calculate the number in system; thus somehow we must implicitly be specifying the expended service for the man in service. How are we to identify a set of points with this property? There are many such sets. An extremely convenient set of points with this property is the set of *departure* instants from service. It is clear if we specify the number of customers left behind by a departing customer that we can calculate this same quantity at some point in the future given only the additional inputs to the system. Certainly, we have specified the expended service time at these instants: it is in fact zero for the customer (if any) currently in service since he has just at that instant entered service!* (There are other sets of points with this property, for example, the set of points that occur exactly 1 sec after customers enter service; if we specify the number in the system at these instants, then we are capable of solving for the number of customers in the system at such future instants of time. Such a set as just described is not as useful as the departure instants since we must worry about the case where a customer in service does not remain for a duration exceeding 1 sec.)

The reader should recognize that what we are describing is, in fact, a semi-Markov process in which the state transitions occur at customer departure instants. At these instants we define the imbedded Markov chain to be the number of customers present in the system immediately following the departure. The transitions take place only at the imbedded points and form a discrete-state space. The distribution of time between state transitions is equal to the service time distribution $B(x)$ whenever a departure leaves behind at least one customer, whereas it equals the convolution of the interarrival-time distribution (exponentially distributed) with $b(x)$ in the case that the departure leaves behind an empty system. In any case, the behavior of the chain at these imbedded points is completely describable as a Markov process, and the results we have discussed in Chapter 2 are applicable.

Our approach then is to focus attention upon departure instants from service and to specify as our state variable the *number of customers left behind* by such a departing customer. We will proceed to solve for the system behavior at these instants in time. Fortunately, the solution at these imbedded Markov points happens also to provide the solution for *all* points in time.† In Exercise 5.7 the reader is asked to rederive some M/G/1 results using the method of supplementary variables; this method is good at all points in time and (as it must) turns out to be identical to the results we get here by using the imbedded Markov chain approach. This proves once again that our solution

* Moreover, we assume that no service has been expended on any other customer in the queue.

† This happy circumstance is due to the fact that we have a Poisson input and therefore (as shown in Section 4.1) an arriving customer takes what amounts to a "random" look at the system. Furthermore, in Exercise 5.6 we assist the reader in proving that the limiting distribution for the number of customers left behind by a departure is the same as the limiting distribution of customers found by a new arrival for any system that changes state by unit step values (positive or negative); this result is true for arbitrary arrival- and arbitrary service-time distributions! Thus, for M/G/1, arrivals, departures, and random observers all see the *same* distribution of number in the system.

is good for all time. In the following pages we establish results for the queue-length distribution, the waiting-time distribution, and the busy-period distribution (all in terms of transforms); the waiting-time and busy-period duration results are in no way restricted by the imbedding we have described. So even if the other methods were not available, these results would still hold and would be unconstrained due to the imbedding process. As a final reassurance to the reader we now offer an intuitive justification for the equivalence between the limiting distributions seen by departures and arrivals. Taking the state of the system as the number of customers therein, we may observe the changes in system state as time evolves; if we follow the system state in continuous time, then we observe that these changes are of the nearest-neighbor type. In particular, if we let E_k be the system state when k customers are in the system, then we see that the only transitions from this state are $E_k \rightarrow E_{k+1}$ and $E_k \rightarrow E_{k-1}$ (where this last can only occur if $k > 0$). This is denoted in Figure 5.2. We now make the observation that the number of transitions of the type $E_k \rightarrow E_{k+1}$ can differ by at most one from the number of transitions of the type $E_{k+1} \rightarrow E_k$. The former correspond to customer arrivals and occur at the arrival instants; the latter refer to customer departures and occur at the departure instants. After the system has been in operation for an arbitrarily long time, the number of such transitions upward must essentially equal the number of transitions downward. Since this up-and-down motion with respect to E_k occurs with essentially the same frequency, we may therefore conclude that the system states found by arrivals must have the same limiting distribution (r_k) as the system states left behind by departures (which we denote by d_k). Thus, if we let $N(t)$ be the number in the system at time t, we may summarize our two conclusions as follows:

1. For *Poisson* arrivals, it is always true that [see Eq. (4.6)]

$$P[N(t) = k] = P[\text{arrival at time } t \text{ finds } k \text{ in system}]$$

that is,

$$P_k(t) = R_k(t) \tag{5.23}$$

2. If in any (perhaps non-Markovian) system $N(t)$ makes only discontinuous changes of size (plus or minus) one, then if either one of the following limiting distributions exists, so does the other and they are equal (see Exercise 5.6):

$$r_k \triangleq \lim_{t \to \infty} P[\text{arrival at } t \text{ finds } k \text{ customers in system}]$$

$$d_k \triangleq \lim_{t \to \infty} P[\text{departure at } t \text{ leaves } k \text{ customers behind}]$$

$$r_k = d_k \tag{5.24}$$ ■

Thus, for M/G/1,

$$r_k = p_k = d_k$$ ■

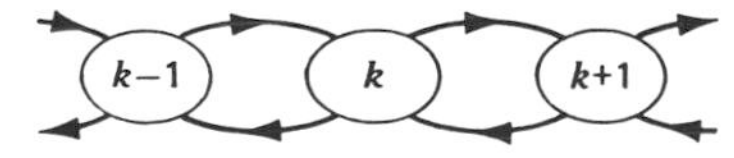

Figure 5.2 State transitions for unit step-change systems.

Our approach for the balance of this chapter is first to find the mean number in system, a result referred to as the Pollaczek–Khinchin mean-value formula.* Following that we obtain the generating function for the distribution of number of customers in the system and then the transform for both the waiting-time and total system-time distributions. These last transform results we shall refer to as Pollaczek–Khinchin transform equations.* Furthermore, we solve for the transform of the busy-period duration and for the number served in the busy period; we then show how to derive waiting-time results from the busy-period analysis. Lastly, we derive the Takács integrodifferential equation for the unfinished work in the system. We begin by defining some notation and identifying the transition probabilities associated with our imbedded Markov chain.

5.4. THE TRANSITION PROBABILITIES

We have already discussed the use of customer departure instants as a set of imbedded points in the time axis; at these instants we define the imbedded Markov chain as the number of customers left behind by these departures (this forms our imbedded Markov chain). It should be clear to the reader that this is a complete state description since we know for sure that zero service has so far been expended on the customer in service and that the time since the last arrival is irrelevant to the future development of the process, since the interarrival-time distribution is memoryless. Early in Chapter 2 we introduced some symbolical and graphical notation; we ask that the reader refresh his understanding of Figure 2.2 and that he recall the following definitions:

C_n represents the nth customer to enter the system

τ_n = arrival time of C_n

$t_n = \tau_n - \tau_{n-1}$ = interarrival time between C_{n-1} and C_n

x_n = service time for C_n

In addition, we introduce two new random variables of considerable interest:

q_n = number of customers left behind by departure of C_n from service

v_n = number of customers arriving during the service of C_n

* There is considerable disagreement within the queueing theory literature regarding the names for the mean-value and transform equations. Some authors refer to the mean-value expression as the Pollaczek–Khinchin formula, whereas others reserve that term for the transform equations. We attempt to relieve that confusion by adding the appropriate adjectives to these names.

We are interested in solving for the distribution of q_n, namely, $P[q_n = k]$, which is, in fact, a time-dependent probability; its limiting distribution (as $n \to \infty$) corresponds to d_k, which we know is equal to p_k, the basic distribution discussed in Chapters 3 and 4 previously. In carrying out that solution we will find that the number of arriving customers v_n plays a crucial role.

As in Chapter 2, we find that the transition probabilities describe our Markov chain; thus we define the one-step transition probabilities

$$p_{ij} \triangleq P[q_{n+1} = j \mid q_n = i] \tag{5.25}$$

Since these transitions are observed only at departures, it is clear that $q_{n+1} < q_n - 1$ is an impossible situation; on the other hand, $q_{n+1} \geq q_n - 1$ is possible for all values due to the arrivals v_{n+1}. It is easy to see that the matrix of transition probabilities $\mathbf{P} = [p_{ij}]$ $(i, j = 0, 1, 2, \ldots)$ takes the following form:

$$\mathbf{P} = \begin{bmatrix} \alpha_0 & \alpha_1 & \alpha_2 & \alpha_3 & \cdots \\ \alpha_0 & \alpha_1 & \alpha_2 & \alpha_3 & \cdots \\ 0 & \alpha_0 & \alpha_1 & \alpha_2 & \cdots \\ 0 & 0 & \alpha_0 & \alpha_1 & \cdots \\ 0 & 0 & 0 & \alpha_0 & \cdots \\ \cdot & \cdot & \cdot & \cdot & \cdot \\ \cdot & \cdot & \cdot & \cdot & \cdot \\ \cdot & \cdot & \cdot & \cdot & \cdot \end{bmatrix}$$

where

$$\alpha_k \triangleq P[v_{n+1} = k] \tag{5.26}$$

For example, the jth component of the first row of this matrix gives the probability that the previous customer left behind an empty system and that during the service of C_{n+1} exactly j customers arrived (all of whom were left behind by the departure of C_{n+1}); similarly, for other than the first row, the entry p_{ij} for $j \geq i - 1$ gives the probability that exactly $j - i + 1$ customers arrived during the service period for C_{n+1}, given that C_n left behind exactly i customers; of these i customers one was indeed C_{n+1} and this accounts for the $+1$ term in this last computation. The state-transition-probability diagram for this Markov chain is shown in Figure 5.3, in which we show only transitions out of E_i.

Let us now calculate α_k. We observe first of all that the arrival process (a Poisson process at a rate of λ customers per second) is independent of the state of the queueing system. Similarly, x_n, the service time for C_n, is independent

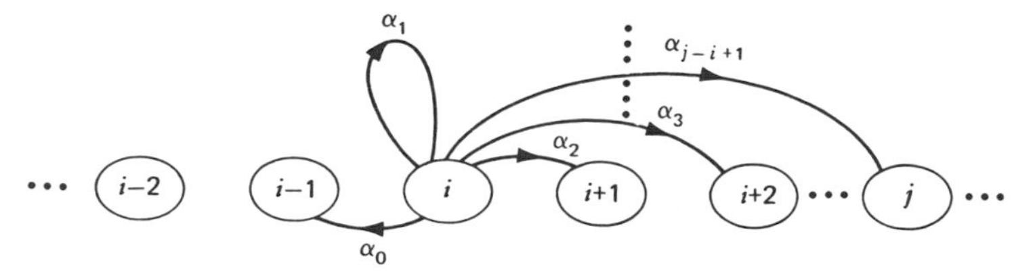

Figure 5.3 State-transition-probability diagram for the M/G/1 imbedded Markov Chain.

of n and is distributed according to $B(x)$. Therefore, v_n, the number of arrivals during the service time x_n depends only upon the duration of x_n and not upon n at all. We may therefore dispense with the subscripts on v_n and x_n, replacing them with the random variables $\tilde{v}$ and $\tilde{x}$ so that we may write $P[x_n \le x] = P[\tilde{x} \le x] = B(x)$ and $P[v_n = k] = P[\tilde{v} = k] = \alpha_k$. We may now proceed with the calculation of α_k. We have by the law of total probability

$$\alpha_k = P[\tilde{v} = k] = \int_0^\infty P[\tilde{v} = k, x < \tilde{x} \le x + dx]\, dx$$

By conditional probabilities we further have

$$\alpha_k = \int_0^\infty P[\tilde{v} = k \mid \tilde{x} = x] b(x)\, dx \tag{5.27}$$

where again $b(x) = dB(x)/dx$ is the pdf for service time. Since we have a Poisson arrival process, we may replace the probability beneath the integral by the expression given in Eq. (2.131), that is,

$$\alpha_k = \int_0^\infty \frac{(\lambda x)^k}{k!} e^{-\lambda x} b(x)\, dx \tag{5.28}$$

This then completely specifies the transition probability matrix **P**.

We note that since $\alpha_k > 0$ for all $k \ge 0$ it is possible to reach all other states from any given state; thus our Markov chain is irreducible (and aperiodic). Moreover, let us make our usual definition:

$$\rho = \lambda \bar{x}$$

and point out that this Markov chain is ergodic if $\rho < 1$ (unless specified otherwise, we shall assume $\rho < 1$ below).

The stationary probabilities may be obtained from the vector equation $\mathbf{p} = \mathbf{pP}$ where $\mathbf{p} = [p_0, p_1, p_2, \ldots]$ whose kth component p_k $(= d_k)$ is merely the limiting probability that a departing customer will leave behind k customers, namely,

$$p_k = P[\tilde{q} = k] \tag{5.29}$$

In the following section we find the mean value $E[\tilde{q}]$ and in the section following that we find the z-transform for p_k.

5.5. THE MEAN QUEUE LENGTH

In this section we derive the Pollaczek–Khinchin formula for the mean value of the limiting queue length. In particular, we define

$$\tilde{q} = \lim_{n\to\infty} q_n \tag{5.30}$$

which certainly will exist in the case where our imbedded chain is ergodic.

Our first step is to find an equation relating the random variable q_{n+1} to the random variable q_n by considering two cases. The first is shown in Figure 5.4 (using our time-diagram notation) and corresponds to the case where C_n leaves behind a nonempty system (i.e., $q_n > 0$). Note that we are assuming a first-come-first-served queueing discipline, although this assumption only affects waiting times and not queue lengths or busy periods. We see from Figure 5.4 that q_n is clearly greater than zero since C_{n+1} is already in the system when C_n departs. We purposely do not show when customer C_{n+2} arrives since that is unimportant to our developing argument. We wish now to find an expression for q_{n+1}, the number of customers left behind when C_{n+1} departs. This is clearly given as equal to q_n the number of customers present when C_n departed less 1 (since customer C_{n+1} departs himself) plus the number of customers that arrive during the service interval x_{n+1}. This last term is clearly equal to v_{n+1} by definition and is shown as a "set" of arrivals

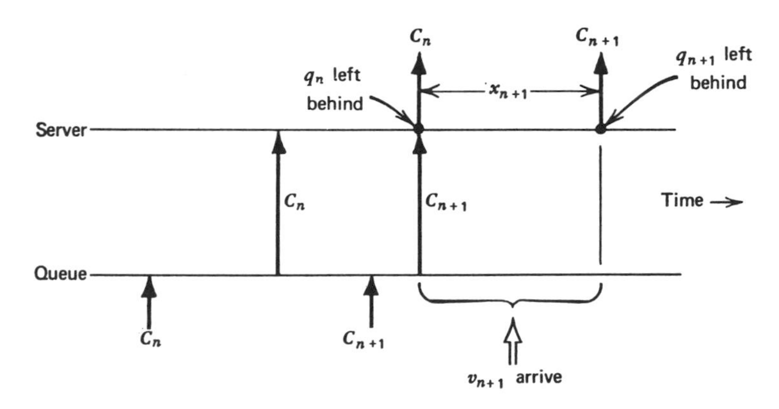

Figure 5.4 Case where $q_n > 0$.

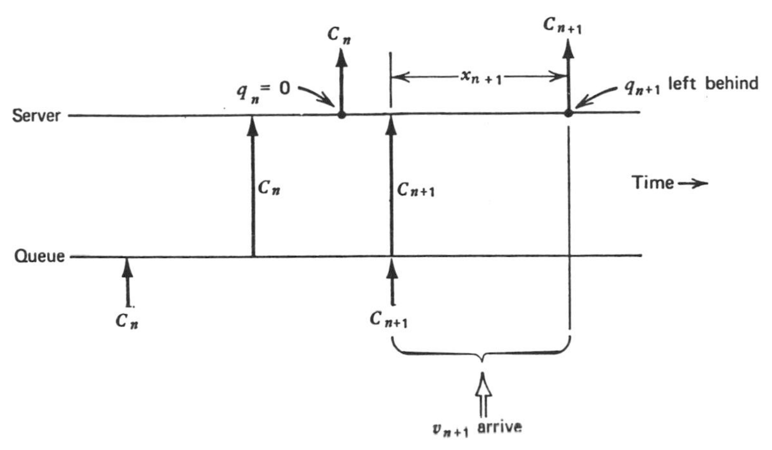

Figure 5.5 Case where $q_n = 0$.

in the diagram. Thus we have

$$q_{n+1} = q_n - 1 + v_{n+1} \qquad q_n > 0 \tag{5.31}$$

Now consider the second case where $q_n = 0$, that is, our departing customer leaves behind an empty system; this is illustrated in Figure 5.5. In this case we see that q_n is clearly zero since C_{n+1} has not yet arrived by the time C_n departs. Thus q_{n+1}, the number of customers left behind by the departure of C_{n+1}, is merely equal to the number of arrivals during his service time. Thus

$$q_{n+1} = v_{n+1} \qquad q_n = 0 \tag{5.32}$$

Collecting together Eq. (5.31) and Eq. (5.32) we have

$$q_{n+1} = \begin{cases} q_n - 1 + v_{n+1} & q_n > 0 \\ v_{n+1} & q_n = 0 \end{cases} \tag{5.33}$$

It is convenient at this point to introduce Δ_k, the shifted discrete step function

$$\Delta_k = \begin{cases} 1 & k = 1, 2, \ldots \\ 0 & k \leq 0 \end{cases} \tag{5.34}$$

which is related to the discrete step function δ_k [defined in Eq. (4.70)] through $\Delta_k = \delta_{k-1}$. Applying this definition to Eq. (5.33) we may now write the single defining equation for q_{n+1} as

$$q_{n+1} = q_n - \Delta_{q_n} + v_{n+1} \tag{5.35}$$

Equation (5.35) is the key equation for the study of M/G/1 systems. It remains for us to extract from Eq. (5.35) the mean value* for q_n. As usual, we concern ourselves not with the time-dependent behavior (which is inferred by the subscript n) but rather with the limiting distribution for the random variable q_n, which we denote by $\tilde{q}$. Accordingly we assume that the jth moment of q_n exists in the limit as n goes to infinity independent of n, namely,

$$\lim_{n\to\infty} E[q_n{}^j] = E[\tilde{q}^j] \tag{5.36}$$

(We are in fact requiring ergodicity here.)

As a first attempt let us hope that forming the expectation of both sides of Eq. (5.35) and then taking the limit as $n \to \infty$ will yield the average value we are seeking. Proceeding as described we have

$$E[q_{n+1}] = E[q_n] - E[\Delta_{q_n}] + E[v_{n+1}]$$

Using Eq. (5.36) we have, in the limit as $n \to \infty$,

$$E[\tilde{q}] = E[\tilde{q}] - E[\Delta_{\tilde{q}}] + E[\tilde{v}]$$

Alas, the expectation we were seeking drops out of this equation, which yields instead

$$E[\Delta_{\tilde{q}}] = E[\tilde{v}] \tag{5.37}$$

What insight does this last equation provide us? (Note that since $\tilde{v}$ is the number of arrivals during a customer's service time, which is independent of n, the index on v_n could have been dropped even before we went to the limit.) We have by definition that

$$E[\tilde{v}] = \text{average number of arrivals in a service time}$$

Let us now interpret the left-hand side of Eq. (5.37). By definition we may calculate this directly as

$$\begin{aligned} E[\Delta_{\tilde{q}}] &= \sum_{k=0}^{\infty} \Delta_k P[\tilde{q} = k] \\ &= \Delta_0 P[\tilde{q} = 0] + \Delta_1 P[\tilde{q} = 1] + \cdots \end{aligned}$$

* We could at this point proceed to the next section to obtain the (z-transform of the) limiting distribution for number in system and from that expression evaluate the average number in system. Instead, let us calculate the average number in system directly from Eq. (5.35) following the method of Kendall [KEND 51]; we choose to carry out this extra work to demonstrate to the student the simplicity of the argument.

But, from the definition in Eq. (5.34) we may rewrite this as

$$E[\Delta_{\tilde{q}}] = 0\{P[\tilde{q} = 0]\} + 1\{P[\tilde{q} > 0]\}$$

or

$$E[\Delta_{\tilde{q}}] = P[\tilde{q} > 0] \tag{5.38}$$

Since we are dealing with a single-server system, Eq. (5.38) may also be written as

$$E[\Delta_{\tilde{q}}] = P[\text{busy system}] \tag{5.39}$$

And from our definition of the utilization factor we further have

$$P[\text{busy system}] = \rho \tag{5.40}$$

as we had observed* in Eq. (2.32). Thus from Eqs. (5.37), (5.39), and (5.40) we conclude that

$$E[\tilde{v}] = \rho \qquad \blacksquare (5.41)$$

We thus have the perfectly reasonable conclusion that the expected number of arrivals per service interval is equal to ρ $(= \lambda\bar{x})$. For stability we of course require $\rho < 1$, and so Eq. (5.41) indicates that customers must arrive more slowly than they can be served (on the average).

We now return to the task of solving for the expected value of $\tilde{q}$. Forming the *first* moment of Eq. (5.35) yielded interesting results but failed to give the desired expectation. Let us now attempt to find this average value by *first squaring* Eq. (5.35) and *then* taking expectations as follows:

$$q_{n+1}^2 = q_n^{\,2} + \Delta_{q_n}^{\,2} + v_{n+1}^2 - 2q_n\Delta_{q_n} + 2q_n v_{n+1} - 2\Delta_{q_n} v_{n+1} \tag{5.42}$$

From our definition in Eq. (5.34) we have $(\Delta_{q_n})^2 = \Delta_{q_n}$ and also $q_n\Delta_{q_n} = q_n$. Applying this to Eq. (5.42) and taking expectations,we have

$$E[q_{n+1}^2] = E[q_n^{\,2}] + E[\Delta_{q_n}] + E[v_{n+1}^2] - 2E[q_n] + 2E[q_n v_{n+1}] - 2E[\Delta_{q_n} v_{n+1}]$$

In this equation, we have the expectation of the product of two random variables in the last two terms. However, we observe that v_{n+1} [the number of arrivals during the $(n + 1)$th service interval] is independent of q_n (the number of customers left behind by C_n). Consequently, the last two expectations may each be written as a product of the expectations. Taking the limit as n goes to infinity, and using our limit assumptions in Eq. (5.36), we have

$$0 = E[\Delta_{\tilde{q}}] + E[\tilde{v}^2] - 2E[\tilde{q}] + 2E[\tilde{q}]E[\tilde{v}] - 2E[\Delta_{\tilde{q}}]E[\tilde{v}]$$

* For any M/G/1 system, we see that $P[\tilde{q} = 0] = 1 - P[\tilde{q} > 0] = 1 - \rho$ and so P[new customer need *not* queue] $= 1 - \rho$. This agrees with our earlier observation for G/G/1.

We now make use of Eqs. (5.37) and (5.41) to obtain, as an intermediate result for the expectation of $\tilde{q}$,

$$E[\tilde{q}] = \rho + \frac{E[\tilde{v}^2] - E[\tilde{v}]}{2(1-\rho)} \tag{5.43}$$

The only unknown here is $E[\tilde{v}^2]$.

Let us solve not only for the second moment of $\tilde{v}$ but, in fact, let us describe a method for obtaining *all* the moments. Equation (5.28) gives an expression for $\alpha_k = P[\tilde{v} = k]$. From this expression we should be able to calculate the moments. However, we find it expedient first to define the z-transform for the random variable $\tilde{v}$ as

$$V(z) \triangleq E[z^{\tilde{v}}] \triangleq \sum_{k=0}^{\infty} P[\tilde{v} = k] z^k \tag{5.44}$$

Forming $V(z)$ from Eqs. (5.28) and (5.44) we have

$$V(z) = \sum_{k=0}^{\infty} \int_0^{\infty} \frac{(\lambda x)^k}{k!} e^{-\lambda x} b(x)\, dx\, z^k$$

Our summation and integral are well behaved, and we may interchange the order of these two operations to obtain

$$\begin{aligned} V(z) &= \int_0^{\infty} e^{-\lambda x} \left(\sum_{k=0}^{\infty} \frac{(\lambda x z)^k}{k!} \right) b(x)\, dx \\ &= \int_0^{\infty} e^{-\lambda x} e^{\lambda x z} b(x)\, dx \\ &= \int_0^{\infty} e^{-(\lambda - \lambda z)x} b(x)\, dx \end{aligned} \tag{5.45}$$

At this point we define (as usual) the Laplace transform $B^*(s)$ for the service time pdf as

$$B^*(s) \triangleq \int_0^{\infty} e^{-sx} b(x)\, dx$$

We note that Eq. (5.45) is of this form, with the complex variable s replaced by $\lambda - \lambda z$, and so we recognize the important result that

$$V(z) = B^*(\lambda - \lambda z) \tag{5.46}$$

This last equation is extremely useful and represents a relationship between the z-transform of the probability distribution of the random variable $\tilde{v}$ and the Laplace transform of the pdf of the random variable $\tilde{x}$ when the Laplace transform is evaluated at the critical point $\lambda - \lambda z$. These two random variables are such that $\tilde{v}$ represents the number of arrivals occurring during the interval $\tilde{x}$ where the arrival process is Poisson at an average rate of λ arrivals per second. We will shortly have occasion to incorporate this interpretation of Eq. (5.46) in our further results.

From Appendix II we note that various derivatives of z-transforms evaluated for $z = 1$ give the various moments of the random variable under consideration. Similarly, the appropriate derivative of the Laplace transform evaluated at its argument $s = 0$ also gives rise to moments. In particular, from that appendix we recall that

$$B^{*(k)}(0) \triangleq \left. \frac{d^k B^*(s)}{ds^k} \right|_{s=0} = (-1)^k E[\tilde{x}^k] \tag{5.47}$$

$$V^{(1)}(1) \triangleq \left. \frac{dV(z)}{dz} \right|_{z=1} = E[\tilde{v}] \tag{5.48}$$

$$V^{(2)}(1) \triangleq \left. \frac{d^2 V(z)}{dz^2} \right|_{z=1} = E[\tilde{v}^2] - E[\tilde{v}] \tag{5.49}$$

In order to simplify the notation for these limiting derivative operations, we have used the more usual superscript notation with the argument replaced by its limit. Furthermore, we now resort to the overbar notation to denote expected value of the random variable below that bar.† Thus Eqs. (5.47)–(5.49) become

$$B^{*(k)}(0) = (-1)^k \overline{x^k} \tag{5.50}$$

$$V^{(1)}(1) = \bar{v} \tag{5.51}$$

$$V^{(2)}(1) = \overline{v^2} - \bar{v} \tag{5.52}$$

Of course, we must also have the conservation of probability given by

$$B^*(0) = V(1) = 1 \tag{5.53}$$

We now wish to exploit the relationship given in Eq. (5.46) so as to be able to obtain the moments of the random variable $\tilde{v}$ from the expressions given in Eqs. (5.50)–(5.53). Thus from Eq. (5.46) we have

$$\frac{dV(z)}{dz} = \frac{dB^*(\lambda - \lambda z)}{dz} \tag{5.54}$$

† Recall from Eq. (2.19) that $E[x_n{}^k] \to \overline{x^k} = b_k$ (rather than the more cumbersome notation $\overline{(\tilde{x})^k}$ which one might expect). We take the same liberties with $\tilde{v}$ and $\tilde{q}$, namely, $\overline{(\tilde{v})^k} = \overline{v^k}$ and $\overline{(\tilde{q})^k} = \overline{q^k}$.

This last may be calculated as

$$\frac{dB^*(\lambda - \lambda z)}{dz} = \left(\frac{dB^*(\lambda - \lambda z)}{d(\lambda - \lambda z)}\right)\left(\frac{d(\lambda - \lambda z)}{dz}\right)$$

$$= -\lambda \frac{dB^*(y)}{dy} \tag{5.55}$$

where

$$y = \lambda - \lambda z \tag{5.56}$$

Setting $z = 1$ in Eq. (5.54) we have

$$V^{(1)}(1) = -\lambda \frac{dB^*(y)}{dy}\bigg|_{z=1}$$

But from Eq. (5.56) the case $z = 1$ is the case $y = 0$, and so we have

$$V^{(1)}(1) = -\lambda B^{*(1)}(0) \tag{5.57}$$

From Eqs. (5.50), (5.51), and (5.57), we finally have

$$\bar{v} = \lambda \bar{x} \tag{5.58}$$

But $\lambda\bar{x}$ is just ρ and we have once again established that which we knew from Eq. (5.41), namely, $\bar{v} = \rho$. (This certainly is encouraging.) We may continue to pick up higher moments by differentiating Eq. (5.54) once again to obtain

$$\frac{d^2V(z)}{dz^2} = \frac{d^2B^*(\lambda - \lambda z)}{dz^2} \tag{5.59}$$

Using the first derivative of $B^*(y)$ we now form its second derivative as follows:

$$\frac{d^2B^*(\lambda - \lambda z)}{dz^2} = \frac{d}{dz}\left[-\lambda \frac{dB^*(y)}{dy}\right]$$

$$= -\lambda\left(\frac{d^2B^*(y)}{dy^2}\right)\left(\frac{dy}{dz}\right)$$

or

$$\frac{d^2B^*(\lambda - \lambda z)}{dz^2} = \lambda^2 \frac{d^2B^*(y)}{dy^2} \tag{5.60}$$

Setting z equal to 1 in Eq. (5.59) and using Eq. (5.60) we have

$$V^{(2)}(1) = \lambda^2 B^{*(2)}(0)$$

Thus, from earlier results in Eqs. (5.50) and (5.52), we obtain

$$\overline{v^2} - \bar{v} = \lambda^2\overline{x^2} \tag{5.61}$$

We have thus finally solved for $\overline{v^2}$. This clearly is the quantity required in order to evaluate Eq. (5.43). If we so desired (and with suitable energy) we could continue this differentiation game and extract additional moments of $\tilde{v}$ in terms of the moments of $\tilde{x}$; we prefer not to yield to that temptation here.

Returning to Eq. (5.43) we apply Eq. (5.61) to obtain

$$\bar{q} = \rho + \frac{\lambda^2\overline{x^2}}{2(1-\rho)} \tag{5.62}$$

This is the result we were after! It expresses the average queue size at customer departure instants in terms of known quantities, namely, the utilization factor ($\rho = \lambda\bar{x}$), λ, and $\overline{x^2}$ (the second moment of the service-time distribution). Let us rewrite this result in terms of $C_b^2 = \sigma_b^2/(\bar{x})^2$, the squared coefficient of variation for service time:

$$\bar{q} = \rho + \rho^2\frac{(1 + C_b^2)}{2(1-\rho)} \tag{5.63}$$

This last is the extremely well-known formula for the average number of customers in an M/G/1 system and is commonly* referred to as the *Pollaczek-Khinchin (P–K) mean-value formula*. Note with emphasis that this average depends only upon the *first two* moments ($\bar{x}$ and $\overline{x^2}$) of the service-time distribution. Moreover, observe that $\bar{q}$ grows *linearly* with the variance of the service-time distribution (or, if you will, linearly with its squared coefficient of variation).

The P–K mean-value formula provides an expression for $\bar{q}$ that represents the average number of customers in the system at departure instants; however, we already know that this also represents the average number at the arrival instants and, in fact, at all points in time. We already have a notation for the average number of customers in the system, namely $\bar{N}$, which we introduced in Chapter 2 and have used in previous chapters; we will continue to use the $\bar{N}$ notation outside of this chapter. Furthermore, we have defined $\bar{N}_q$ to be the average number of customers in the queue (not counting the customer in service). Let us take a moment to develop a relationship between these two quantities. By definition we have

$$\bar{N} \overset{\Delta}{=} \sum_{k=0}^{\infty} kP[\tilde{q} = k] \tag{5.64}$$

* See footnote on p. 177.

Similarly we may calculate the average queue size by subtracting unity from this previous calculation so long as there is at least one customer in the system, that is (note the lower limit),

$$\bar{N}_q = \sum_{k=1}^{\infty}(k - 1)P[\tilde{q} = k]$$

This easily gives us

$$\bar{N}_q = \sum_{k=0}^{\infty} kP[\tilde{q} = k] - \sum_{k=1}^{\infty} P[\tilde{q} = k]$$

But the second sum is merely ρ and so we have the result

$$\bar{N}_q = \bar{N} - \rho \qquad \blacksquare (5.65)$$

This simple formula gives the general relationship we were seeking.

As an example of the P–K mean-value formula, in the case of an M/M/1 system, we have that the coefficient of variation for the exponential distribution is unity [see Eq. (2.145)]. Thus for this system we have

$$\bar{q} = \rho + \rho^2 \frac{(2)}{2(1 - \rho)}$$

or

$$\bar{q} = \frac{\rho}{1 - \rho} \qquad \text{M/M/1} \qquad (5.66)$$

Equation (5.66) gives the expected number of customers left behind by a departing customer. Compare this to the expression for the average number of customers in an M/M/1 system as given in Eq. (3.24). They are identical and lend validity to our earlier statements that the method of the imbedded Markov chain in the M/G/1 case gives rise to a solution that is good at all points in time. As a second example, let us consider the service-time distribution in which service time is a constant and equal to $\bar{x}$. Such systems are described by the notation M/D/1, as we mentioned earlier. In this case clearly $C_b^2 = 0$ and so we have

$$\bar{q} = \rho + \rho^2 \frac{1}{2(1 - \rho)}$$

$$\bar{q} = \frac{\rho}{1 - \rho} - \frac{\rho^2}{2(1 - \rho)} \qquad \text{M/D/1} \qquad \blacksquare (5.67)$$

Thus the M/D/1 system has $\rho^2/2(1 - \rho)$ fewer customers on the average than the M/M/1 system, demonstrating the earlier statement that $\bar{q}$ increases with the variance of the service-time distribution.

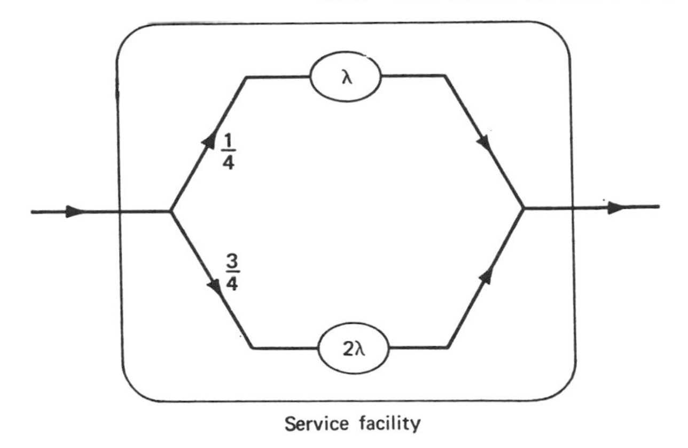

Figure 5.6 The M/H$_2$/1 example.

For a third example, we consider an M/H$_2$/1 system in which

$$b(x) = \tfrac{1}{4}\lambda e^{-\lambda x} + \tfrac{3}{4}(2\lambda)e^{-2\lambda x} \qquad x \geq 0 \tag{5.68}$$

That is, the service facility consists of two parallel service stages, as shown in Figure 5.6. Note that λ is also the arrival rate, as usual. We may immediately calculate $\bar{x} = 5/(8\lambda)$ and $\sigma_b^2 = 31/(64\lambda^2)$, which yields $C_b^2 = 31/25$. Thus

$$\begin{aligned}\bar{q} &= \rho + \frac{\rho^2(2.24)}{2(1-\rho)} \\ &= \frac{\rho}{1-\rho} + \frac{0.12\rho^2}{1-\rho}\end{aligned}$$

Thus we see the (small) increase in $\bar{q}$ for the (small) increase in C_b^2 over the value of unity for M/M/1. We note in this example that ρ is fixed at $\rho = \lambda\bar{x} = 5/8$; therefore, $\bar{q} = 1.79$, whereas for M/M/1 at this value of ρ we get $\bar{q} = 1.66$. We have introduced this M/H$_2$/1 example here since we intend to carry it (and the M/M/1 example) through our M/G/1 discussion.

The main result of this section is the Pollaczek–Khinchin formula for the mean number in system, as given in Eq. (5.63). This result becomes a special case of our results in the next section, but we feel that its development has been useful as a pedagogical device. Moreover, in obtaining this result we established the basic equation for M/G/1 given in Eq. (5.35). We also obtained the general relationship between $V(z)$ and $B^*(s)$, as given in Eq. (5.46); from this we are able to obtain the moments for the number of arrivals during a service interval.

We have not as yet derived any results regarding *time* spent in the system; we are now in a position to do so. We recall Little's result:

$$\bar{N} = \lambda T$$

This result relates the expected number of customers $\bar{N}$ in a system to λ, the arrival rate of customers and to T, their average time in the system. For M/G/1 we have derived Eq. (5.63), which is the expected number in the system at customer departure instants. We may therefore apply Little's result to this expected number in order to obtain the average time spent in the system (queue + service). We know that $\bar{q}$ also represents the average number of customers found at random, and so we may equate $\bar{q} = \bar{N}$. Thus we have

$$\bar{N} = \rho + \rho^2 \frac{(1 + C_b^{\,2})}{2(1 - \rho)} = \lambda T$$

Solving for T we have

$$T = \bar{x} + \frac{\rho\bar{x}(1 + C_b^{\,2})}{2(1 - \rho)} \tag{5.69}$$

This last is easily interpreted. The average total time spent in system is clearly the average time spent in service plus the average time spent in the queue. The first term above is merely the average service time and thus the second term must represent the average queueing time (which we denote by W). Thus we have that the average queueing time is

$$W = \frac{\rho\bar{x}(1 + C_b^{\,2})}{2(1 - \rho)}$$

or

$$W = \frac{W_0}{1 - \rho} \tag{5.70}$$

where $W_0 \triangleq \lambda\overline{x^2}/2$; W_0 is the average remaining service time for the customer (if any) found in service by a new arrival (work it out using the mean residual life formula). A particularly nice normalization factor is now apparent. Consider T, the average time spent in system. It is natural to compare this time to $\bar{x}$, the average service time required of the system by a customer. Thus the ratio $T/\bar{x}$ expresses the ratio of time spent in system to time required of the system and represents the *factor* by which the system inconveniences customers due to the fact that they are sharing the system with other customers. If we use this normalization in Eqs. (5.69) and (5.70), we arrive at the following, where now time is expressed in units of average service intervals:

$$\frac{T}{\bar{x}} = 1 + \rho \frac{(1 + C_b^2)}{2(1 - \rho)} \tag{5.71}$$

$$\frac{W}{\bar{x}} = \rho \frac{(1 + C_b^2)}{2(1 - \rho)} \tag{5.72}$$

Each of these last two equations is also referred to as the P–K mean-value formula [along with Eq. (5.63)]. Here we see the linear fashion in which the statistical fluctuations of the input processes create delays (i.e., $1 + C_b^2$ is the sum of the squared interarrival-time and service-time coefficients of variation). Further, we see the highly nonlinear dependence of delays upon the average load ρ.

Let us now compare the mean normalized queueing time for the systems* M/M/1 and M/D/1; these have a squared coefficient of variation C_b^2 equal to 1 and 0, respectively. Applying this to Eq. (5.72) we have

$$\frac{W}{\bar{x}} = \frac{\rho}{(1 - \rho)} \qquad \text{M/M/1} \tag{5.73}$$

$$\frac{W}{\bar{x}} = \frac{\rho}{2(1 - \rho)} \qquad \text{M/D/1} \tag{5.74}$$

Note that the system with constant service time (M/D/1) has *half* the average waiting time of the system with exponentially distributed service time (M/M/1). Thus, as we commented earlier, the time in the system and the number in the system both grow in proportion to the variance of the service-time distribution.

Let us now proceed to find the *distribution* of the number in the system.

5.6. DISTRIBUTION OF NUMBER IN SYSTEM

In the previous sections we characterized the M/G/1 queueing system as an imbedded Markov chain and then established the fundamental equation (5.35) repeated here:

$$q_{n+1} = q_n - \Delta_{q_n} + v_{n+1} \tag{5.75}$$

By forming the average of this last equation we obtained a result regarding the utilization factor ρ [see Eq. (5.41)]. By first *squaring* Eq. (5.75) and then

* Of less interest is our highly specialized $M/H_2/1$ example for which we obtain $W/\bar{x} = 1.12\rho/(1 - \rho)$.

taking expectations we were able to obtain P–K formulas that gave the expected number in the system [Eq. (5.63)] and the normalized expected time in the system [Eq. (5.71)]. If we were now to seek the *second* moment of the number in the system we could obtain this quantity by first *cubing* Eq. (5.75) and then taking expectations. In this operation it is clear that the expectation $E[\tilde{q}^3]$ would cancel on both sides of the equation once the limit on n was taken; this would then leave an expression for the second moment of $\tilde{q}$. Similarly, all higher moments can be obtained by raising Eq. (5.75) to successively higher powers and then forming expectations.* In this section, however, we choose to go after the *distribution* for q_n itself (actually we consider the limiting random variable $\tilde{q}$). As it turns out, we will obtain a result which gives the z-transform for this distribution rather than the distribution itself. In principle, these last two are completely equivalent; in practice, we sometimes face great difficulty in inverting from the z-transform back to the distribution. Nevertheless, we can pick off the moments of the distribution of $\tilde{q}$ from the z-transform in extremely simple fashion by making use of the usual properties of transforms and their derivatives.

Let us now proceed to calculate the z-transform for the probability of finding k customers in the system immediately following the departure of a customer. We begin by defining the z-transform for the random variable q_n as

$$Q_n(z) \triangleq \sum_{k=0}^{\infty} P[q_n = k]z^k \tag{5.76}$$

From Appendix II (and from the definition of expected value) we have that this z-transform (or probability generating function) is also given by

$$Q_n(z) \triangleq E[z^{q_n}] \tag{5.77}$$

Of interest is the z-transform for our limiting random variable $\tilde{q}$:

$$Q(z) = \lim_{n\to\infty} Q_n(z) = \sum_{k=0}^{\infty} P[\tilde{q} = k]z^k = E[z^{\tilde{q}}] \tag{5.78}$$

As is usual in these definitions for transforms, the sum on the right-hand side of Eq. (5.76) converges to Eq. (5.77) only within some circle of convergence in the z-plane which defines a maximum value for $|z|$ (certainly $|z| \le 1$ is allowed).

The system M/G/1 is characterized by Eq. (5.75). We therefore use both sides of this equation as an exponent for z as follows:

$$z^{q_{n+1}} = z^{q_n - \Delta_{q_n} + v_{n+1}}$$

* Specifically, the kth power leads to an expression for $E[\tilde{q}^{k-1}]$ that involves the first k moments of service time.

Let us now take expectations:

$$E[z^{q_{n+1}}] = E[z^{q_n - \Delta_{q_n} + v_{n+1}}]$$

Using Eq. (5.77) we recognize the left-hand side of this last as $Q_{n+1}(z)$. Similarly, we may write the right-hand side of this equation as the expectation of the product of two factors, giving us

$$Q_{n+1}(z) = E[z^{q_n - \Delta_{q_n}} z^{v_{n+1}}] \tag{5.79}$$

We now observe, as earlier, that the random variable v_{n+1} (which represents the number of arrivals during the service of C_{n+1}) is independent of the random variable q_n (which is the number of customers left behind upon the departure of C_n). Since this is true, then the two factors within the expectation on the right-hand side of Eq. (5.79) must themselves be independent (since functions of independent random variables are also independent). We may thus write the expectation of the product in that equation as the product of the expectations:

$$Q_{n+1}(z) = E[z^{q_n - \Delta_{q_n}}]E[z^{v_{n+1}}] \tag{5.80}$$

The second of these two expectations we again recognize as being independent of the subscript $n + 1$; we thus remove the subscript and consider the random variable $\tilde{v}$ again. From Eq. (5.44) we then recognize that the second expectation on the right-hand side of Eq. (5.80) is merely

$$E[z^{v_{n+1}}] = E[z^{\tilde{v}}] = V(z)$$

We thus have

$$Q_{n+1}(z) = V(z)E[z^{q_n - \Delta_{q_n}}] \tag{5.81}$$

The only complicating factor in this last equation is the expectation. Let us examine this term separately; from the definition of expectation we have

$$E[z^{q_n - \Delta_{q_n}}] = \sum_{k=0}^{\infty} P[q_n = k]z^{k-\Delta_k}$$

The difficult part of this summation is that the exponent on z contains Δ_k, which takes on one of two values according to the value of k. In order to simplify this special behavior we write the summation by exposing the first term separately:

$$E[z^{q_n - \Delta_{q_n}}] = P[q_n = 0]z^{0-0} + \sum_{k=1}^{\infty} P[q_n = k]z^{k-1} \tag{5.82}$$

Regarding the sum in this last equation we see that it is almost of the form given in Eq. (5.76); the differences are that we have one fewer powers of z and also that we are missing the first term in the sum. Both these deficiencies may be corrected as follows:

$$\sum_{k=1}^{\infty} P[q_n = k]z^{k-1} = \frac{1}{z}\sum_{k=0}^{\infty} P[q_n = k]z^k - \frac{1}{z} P[q_n = 0]z^0 \tag{5.83}$$

Applying this to Eq. (5.82) and recognizing that the sum on the right-hand side of Eq. (5.83) is merely $Q_n(z)$, we have

$$E[z^{q_n - \Delta_{q_n}}] = P[q_n = 0] + \frac{Q_n(z) - P[q_n = 0]}{z}$$

We may now substitute this last in Eq. (5.81) to obtain

$$Q_{n+1}(z) = V(z)\left(P[q_n = 0] + \frac{Q_n(z) - P[q_n = 0]}{z}\right)$$

We now take the limit as n goes to infinity and recognize the limiting value expressed in Eq. (5.36). We thus have

$$Q(z) = V(z)\left(P[\tilde{q} = 0] + \frac{Q(z) - P[\tilde{q} = 0]}{z}\right) \tag{5.84}$$

Using $P[\tilde{q} = 0] = 1 - \rho$, and solving Eq. (5.84) for $Q(z)$ we find

$$Q(z) = V(z)\frac{(1 - \rho)(1 - 1/z)}{1 - V(z)/z} \tag{5.85}$$

Finally we multiply numerator and denominator of this last by $(-z)$ and use our result in Eq. (5.46) to arrive at the well-known equation that gives the z-transform for the number of customers in the system,

$$Q(z) = B^*(\lambda - \lambda z)\frac{(1 - \rho)(1 - z)}{B^*(\lambda - \lambda z) - z} \quad ■ \tag{5.86}$$

We shall refer to this as one form of the *Pollaczek–Khinchin (P–K) transform equation.*†

The P–K transform equation readily yields the moments for the distribution of the number of customers in the system. Using the moment-generating properties of our transform expressed in Eqs. (5.50)–(5.52) we see that certainly $Q(1) = 1$; when we attempt to set $z = 1$ in Eq. (5.86), we obtain an indeterminant form‡ and so we are required to use L'Hospital's rule. In carrying out this operation we find that we must evaluate lim $dB^*(\lambda - \lambda z)/dz$ as $z \to 1$, which was carried out in the previous section and shown to be equal to ρ. This computation verifies that $Q(1) = 1$. In Exercise 5.5, the reader is asked to show that $Q^{(1)}(1) = \bar{q}$.

† This formula was found in 1932 by A. Y. Khinchin [KHIN 32]. Shortly we will derive two other equations (each of which follow from and imply this equation), which we also refer to as P–K transform equations; these were studied by F. Pollaczek [POLL 30] in 1930 and Khinchin in 1932. See also the footnote on p. 177.

‡ We note that the denominator of the P–K transform equation must always contain the factor $(1 - z)$ since $B^*(0) = 1$.

Usually, the inversion of the P–K transform equation is difficult, and therefore one settles for moments. However, the system M/M/1 yields very nicely to inversion (and to almost everything else). Thus, by way of example, we shall find its distribution. We have

$$B^*(s) = \frac{\mu}{s + \mu} \qquad \text{M/M/1} \tag{5.87}$$

Clearly, the region of convergence for this last form is Re $(s) > -\mu$. Applying this to the P–K transform equation we find

$$Q(z) = \left(\frac{\mu}{\lambda - \lambda z + \mu}\right)\frac{(1 - \rho)(1 - z)}{[\mu/(\lambda - \lambda z + \mu)] - z}$$

Noting that $\rho = \lambda/\mu$, we have

$$Q(z) = \frac{1 - \rho}{1 - \rho z} \tag{5.88}$$

Equation (5.88) is the solution for the z-transform of the distribution of the number of people in the system. We can reach a point such as this with many service-time distributions $B(x)$; for the exponential distribution we can evaluate the inverse transform (by inspection!). We find immediately that

$$P[\tilde{q} = k] = (1 - \rho)\rho^k \qquad \text{M/M/1} \tag{5.89}$$

This then is the familiar solution for M/M/1. If the reader refers back to Eq. (3.23), he will find the same function for the probability of k customers in the M/M/1 system. However, Eq. (3.23) gives the solution for *all* points in time whereas Eq. (5.89) gives the solution only at the imbedded Markov points (namely, at the departure instants for customers). The fact that these two answers are identical is no surprise for two reasons: first, because we told you so (we said that the imbedded Markov points give solutions that are good at all points); and second, because we recognize that the M/M/1 system forms a continuous-time Markov chain.

As a second example, we consider the system M/H_2/1 whose pdf for service time was given in Eq. (5.68). By inspection we may find $B^*(s)$, which gives

$$B^*(s) = \left(\frac{1}{4}\right)\frac{\lambda}{s + \lambda} + \left(\frac{3}{4}\right)\frac{2\lambda}{s + 2\lambda}$$

$$= \frac{7\lambda s + 8\lambda^2}{4(s + \lambda)(s + 2\lambda)} \tag{5.90}$$

where the plane of convergence is Re $(s) > -\lambda$. From the P–K transform equation we then have

$$Q(z) = \frac{(1-\rho)(1-z)[8+7(1-z)]}{8+7(1-z)-4z(2-z)(3-z)}$$

Factoring the denominator and canceling the common term $(1-z)$ we have

$$Q(z) = \frac{(1-\rho)[1-(7/15)z]}{[1-(2/5)z][1-(2/3)z]}$$

We now expand $Q(z)$ in partial fractions, which gives

$$Q(z) = (1-\rho)\left(\frac{1/4}{1-(2/5)z} + \frac{3/4}{1-(2/3)z}\right)$$

This last may be inverted by inspection (by now the reader should recognize the sixth entry in Table I.2) to give

$$p_k = P[\tilde{q} = k] = (1-\rho)\left[\frac{1}{4}\left(\frac{2}{5}\right)^k + \frac{3}{4}\left(\frac{2}{3}\right)^k\right] \tag{5.91}$$

Lastly, we note that the value for ρ has already been calculated at 5/8, and so for a final solution we have

$$p_k = \frac{3}{32}\left(\frac{2}{5}\right)^k + \frac{9}{32}\left(\frac{2}{3}\right)^k \qquad k = 0, 1, 2, \ldots \tag{5.92}$$

It should not surprise us to find this sum of geometric terms for our solution.

Further examples will be found in the exercises. For now we terminate the discussion of how *many* customers are in the system and proceed with the calculation of how *long* a customer spends in the system.

5.7. DISTRIBUTION OF WAITING TIME

Let us now set out to find the distribution of time spent in the system and in the queue. These particular quantities are rather easy to obtain from our earlier principal result, namely, the P–K transform equation (and as we have said, lead to expressions which share that name). Note that the order in which customers receive service has so far not affected our results. Now, however, we must use our assumption that the order of service is first-come–first-served.

In order to proceed in the simplest possible fashion, let us re-examine the derivation of the following equation:

$$V(z) = B^*(\lambda - \lambda z) \tag{5.93}$$

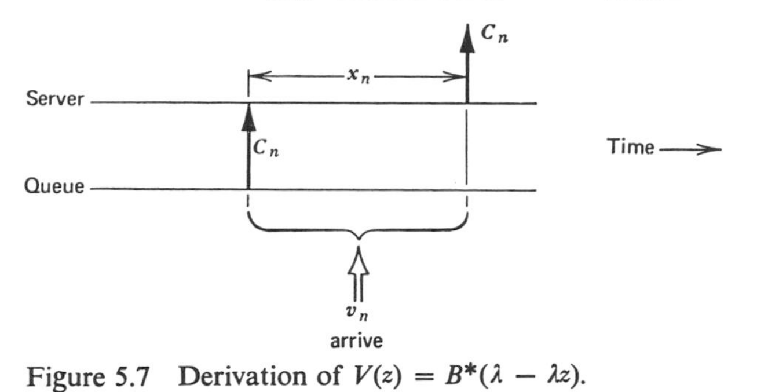

Figure 5.7 Derivation of $V(z) = B^*(\lambda - \lambda z)$.

In Figure 5.7, the reader is reminded of the structure from which we obtained this equation. Recall that $V(z)$ is the z-transform of the number of customer arrivals in a particular interval, where the arrival process is Poisson at a rate λ customers per second. The particular time interval involved happens to be the service interval for C_n; this interval has distribution $B(x)$ with Laplace transform $B^*(s)$. The derived relation between $V(z)$ and $B^*(s)$ is given in Eq. (5.93). The important observation to make now is that a relationship of this form must exist between *any* two random variables where the one identifies the number of customer arrivals from a Poisson process and the other describes the time interval over which we are counting these customer arrivals. It clearly makes no difference what the interpretation of this time interval is, only that we give the distribution of its length; in Eq. (5.93) it just so happens that the interval involved is a service interval. Let us now direct our attention to Figure 5.8, which concentrates on the *time spent in the system* for C_n. In this figure we have traced the history of C_n. The interval labeled w_n identifies the time from when C_n enters the queue until that customer leaves the queue and enters service; it is clearly the *waiting time in queue* for C_n. We have also identified the service time x_n for C_n. We may thus

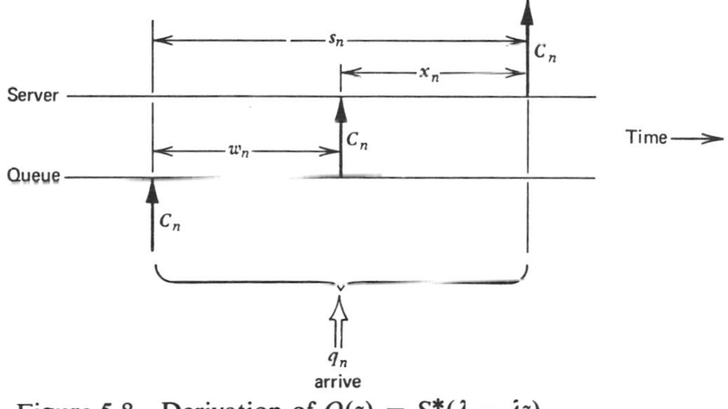

Figure 5.8 Derivation of $Q(z) = S^*(\lambda - \lambda z)$.

identify the *total time spent in system* s_n for C_n,

$$s_n = w_n + x_n \tag{5.94}$$

We have earlier defined q_n as the number of customers left behind upon the departure of C_n. In considering a first-come–first-served system it is clear that all those customers present upon the arrival of C_n must depart before he does; consequently, those customers that C_n leaves behind him (a total of q_n) must be precisely those who arrive during his stay in the system. Thus, referring to Figure 5.8, we may identify those customers who arrive during the time interval s_n as being our previously defined random variable q_n. The reader is now asked to compare Figures 5.7 and 5.8. In both cases we have a Poisson arrival process at rate λ customers per second. In Figure 5.7 we inquire into the number of arrivals (v_n) during the interval whose duration is given by x_n; in Figure 5.8 we inquire into the number of arrivals (q_n) during an interval whose duration is given by s_n. We now define the distribution for the total time spent in system for C_n as

$$S_n(y) \triangleq P[s_n \leq y] \tag{5.95}$$

Since we are assuming ergodicity, we recognize immediately that the limit of this distribution (as n goes to infinity) must be independent of n. We denote this limit by $S(y)$ and the limiting random variable by $\tilde{s}$ [i.e., $S_n(y) \to S(y)$ and $s_n \to \tilde{s}$]. Thus

$$S(y) \triangleq P[\tilde{s} \leq y] \tag{5.96}$$

Finally, we define the Laplace transform of the pdf for total time in system as

$$S^*(s) \triangleq \int_0^\infty e^{-sy}\, dS(y) = E[e^{-s\tilde{s}}] \tag{5.97}$$

With these definitions we go back to the analogy between Figures 5.7 and 5.8. Clearly, since v_n is analogous to q_n, then $V(z)$ must be analogous to $Q(z)$, since each describes the generating function for the respective number distribution. Similarly, since x_n is analogous to s_n, then $B^*(s)$ must be analogous to $S^*(s)$. We have therefore by direct analogy from Eq. (5.93) that†

$$Q(z) = S^*(\lambda - \lambda z) \tag{5.98}$$

Since we already have an explicit expression for $Q(z)$ as given in the P–K transform equation, we may therefore use that with Eq. (5.98) to give an explicit expression for $S^*(s)$ as

$$S^*(\lambda - \lambda z) = B^*(\lambda - \lambda z)\frac{(1-\rho)(1-z)}{B^*(\lambda - \lambda z) - z} \tag{5.99}$$

† This can be derived directly by the unconvinced reader in a fashion similar to that which led to Eqs. (5.28) and (5.46).

This last equation is just crying for the obvious change of variable

$$s = \lambda - \lambda z$$

which gives

$$z = 1 - \frac{s}{\lambda}$$

Making this change of variable in Eq. (5.99) we then have

$$S^*(s) = B^*(s) \frac{s(1 - \rho)}{s - \lambda + \lambda B^*(s)} \tag{5.100}$$

Equation (5.100) is the desired explicit expression for the Laplace transform of the distribution of total time spent in the M/G/1 system. It is given in terms of known quantities derivable from the initial statement of the problem [namely, the specification of the service-time distribution $B(x)$ and the parameters λ and $\bar{x}$]. This is the second of the three equations that we refer to as the P–K transform equation.

From Eq. (5.100) it is trivial to derive the Laplace transform of the distribution of waiting time, which we shall denote by $W^*(s)$. We define the PDF for C_n's waiting time (in queue) to be $W_n(y)$, that is,

$$W_n(y) \triangleq P[w_n \le y] \tag{5.101}$$

Furthermore, we define the limiting quantities (as $n \to \infty$), $W_n(y) \to W(y)$ and $w_n \to \tilde{w}$, so that

$$W(y) \triangleq P[\tilde{w} \le y] \tag{5.102}$$

The corresponding Laplace transform is

$$W^*(s) \triangleq \int_0^\infty e^{-sy}\, dW(y) = E[e^{-s\tilde{w}}] \tag{5.103}$$

From Eq. (5.94) we may derive the distribution of $\tilde{w}$ from the distribution of $\tilde{s}$ and $\tilde{x}$ (we drop subscript notation now since we are considering equilibrium behavior). Since a customer's service time is independent of his queueing time, we have that $\tilde{s}$, the time spent in system for some customer, is the sum of two independent random variables: $\tilde{w}$ (his queueing time) and $\tilde{x}$ (his service time). That is, Eq. (5.94) has the limiting form

$$\tilde{s} = \tilde{w} + \tilde{x} \tag{5.104}$$

As derived in Appendix II the Laplace transform of the pdf of a random variable that is itself the sum of two independent random variables is equal to the product of the Laplace transforms for the pdf of each. Consequently, we have

$$S^*(s) = W^*(s)B^*(s)$$

Thus from Eq. (5.100) we obtain immediately that

$$W^*(s) = \frac{s(1-\rho)}{s - \lambda + \lambda B^*(s)} \tag{5.105}$$

This is the desired expression for the Laplace transform of the queueing (waiting)-time distribution. Here we have the third equation that will be referred to as the P–K transform equation.

Let us rewrite the P–K transform equation for waiting time as follows:

$$W^*(s) = \frac{1-\rho}{1 - \rho\left[\dfrac{1 - B^*(s)}{s\bar{x}}\right]} \tag{5.106}$$

We recognize the bracketed term in the denominator of this equation to be exactly the Laplace transform associated with the density of residual service time from Eq. (5.11). Using our special notation for residual densities and their transforms, we define

$$\hat{B}^*(s) \triangleq \frac{1 - B^*(s)}{s\bar{x}} \tag{5.107}$$

and are therefore permitted to write

$$W^*(s) = \frac{1-\rho}{1 - \rho\hat{B}^*(s)} \tag{5.108}$$

This observation is truly amazing since we recognized at the outset that the problem with the M/G/1 analysis was to take account of the expended service time for the man in service. From that investigation we found that the residual service time remaining for the customer in service had a pdf given by $\hat{b}(x)$, whose Laplace transform is given in Eq. (5.107). In a sense there is a poetic justice in its appearance at this point in the final solution. Let us follow Beneš [BENE 56] in inverting this transform in terms of these residual service time densities. Equation (5.108) may be expanded as the following power series:

$$W^*(s) = (1-\rho)\sum_{k=0}^{\infty} \rho^k [\hat{B}^*(s)]^k \tag{5.109}$$

From Appendix I we know that the kth power of a Laplace transform corresponds to the k-fold convolution of the inverse transform with itself. As in Appendix I the symbol $\circledast$ is used to denote the convolution operator, and we now choose to denote the k-fold convolution of a function $f(x)$ with itself by the use of a parenthetical subscript as follows:

$$f_{(k)}(x) \triangleq \underbrace{f(x) \circledast f(x) \circledast \cdots \circledast f(x)}_{k\text{-fold convolution}} \tag{5.110}$$

Using this notation we may by inspection invert Eq. (5.109) to obtain the waiting-time pdf, which we denote by $w(y) \triangleq dW(y)/dy$; it is given by

$$w(y) = \sum_{k=0}^{\infty}(1 - \rho)\rho^k \hat{b}_{(k)}(y) \tag{5.111}$$

This is a most intriguing result! It states that the waiting time pdf is given by a weighted sum of convolved residual service time pdf's. The interesting observation is that the weighting factor is simply $(1 - \rho)\rho^k$, which we now recognize to be the probability distribution for the number of customers in an M/M/1 system. Tempting as it is to try to give a physical explanation for the simplicity of this result and its relation to M/M/1, no satisfactory, intuitive explanation has been found to explain this dramatic form. We note that the contribution to the waiting-time density decreases geometrically with ρ in this series. Thus, for ρ not especially close to unity, we expect the high-order terms to be of less and less significance, and one practical application of this equation is to provide a rapidly converging approximation to the density of waiting time.

So far in this section we have established two principle results, namely, the P–K transform equations for time in system and time in queue given in Eqs. (5.100) and (5.105), respectively. In the previous section we have already given the first moment of these two random variables [see Eqs. (5.69) and (5.70)]. We wish now to give a recurrence formula for the moments of the waiting time. We denote the kth moment of the waiting time $E[\tilde{w}^k]$, as usual, by $\overline{w^k}$. Takács [TAKA 62b] has shown that if $\overline{x^{i+1}}$ is finite, then so also are $\bar{w}, \overline{w^2}, \ldots, \overline{w^i}$; we now adopt our slightly simplified notation for the ith moment of service time as follows: $b_i \triangleq \overline{x^i}$. The Takács recurrence formula is

$$\overline{w^k} = \frac{\lambda}{1 - \rho} \sum_{i=1}^{k} \binom{k}{i} \frac{b_{i+1}}{(i + 1)} \overline{w^{k-i}} \tag{5.112}$$

where $\overline{w^0} \triangleq 1$. From this formula we may write down the first couple of moments for waiting time (and note that the first moment of waiting time agrees with the P–K formula):

$$\bar{w}\,(= W) = \frac{\lambda b_2}{2(1 - \rho)} \tag{5.113}$$

$$\overline{w^2} = 2(\bar{w})^2 + \frac{\lambda b_3}{3(1 - \rho)} \tag{5.114}$$

In order to obtain similar moments for the total time in system, that is, $E[\tilde{s}^k]$, which we denote by $\overline{s^k}$, we need merely take advantage of Eq. (5.104); from this equation we find

$$\overline{s^k} = \overline{(\tilde{w} + \tilde{x})^k} \tag{5.115}$$

Using the binomial expansion and the independence between waiting time and service time for a given customer, we find

$$\overline{s^k} = \sum_{i=0}^{k} \binom{k}{i} \overline{w^{k-i}} b_i \tag{5.116}$$

Thus calculating the moments of the waiting time from Eq. (5.112) also permits us to calculate the moments of time in system from this last equation. In Exercise 5.25, we drive a relationship between $\overline{s^k}$ and the moments of the number in system; the simplest of these is Little's result, and the others are useful generalizations.

At the end of Section 3.2, we promised the reader that we would develop the pdf for the time spent in the system for an M/M/1 queueing system. We are now in a position to fulfill that promise. Let us in fact find both the distribution of waiting time and distribution of system time for customers in M/M/1. Using Eq. (5.87) for the system M/M/1 we may calculate $S^*(s)$ from Eq. (5.100) as follows:

$$S^*(s) = \frac{\mu}{(s + \mu)} \left[\frac{s(1 - \rho)}{s - \lambda + \lambda\mu/(s + \mu)} \right]$$

$$S^*(s) = \frac{\mu(1 - \rho)}{s + \mu(1 - \rho)} \qquad \text{M/M/1} \tag{5.117}$$

This equation gives the Laplace transform of the pdf for time in the system which we denote, as usual, by $s(y) \triangleq dS(y)/dy$. Fortunately (as is usual with the case M/M/1), we recognize the inverse of this transform by inspection. Thus we have immediately that

$$s(y) = \mu(1 - \rho)e^{-\mu(1-\rho)y} \qquad y \geq 0 \qquad \text{M/M/1} \tag{5.118}$$

The corresponding PDF is given by

$$S(y) = 1 - e^{-\mu(1-\rho)y} \qquad y \geq 0 \qquad \text{M/M/1} \tag{5.119}$$

Similarly, from Eq. (5.105) we may obtain $W^*(s)$ as

$$W^*(s) = \frac{s(1 - \rho)}{s - \lambda + \lambda\mu/(s + \mu)} = \frac{(s + \mu)(1 - \rho)}{s + (\mu - \lambda)} \tag{5.120}$$

Before we can invert this we must place the right-hand side in proper form, namely, where the numerator polynomial is of lower degree than the denominator. We do this by dividing out the constant term and obtain

$$W^*(s) = (1 - \rho) + \frac{\lambda(1 - \rho)}{s + \mu(1 - \rho)} \tag{5.121}$$

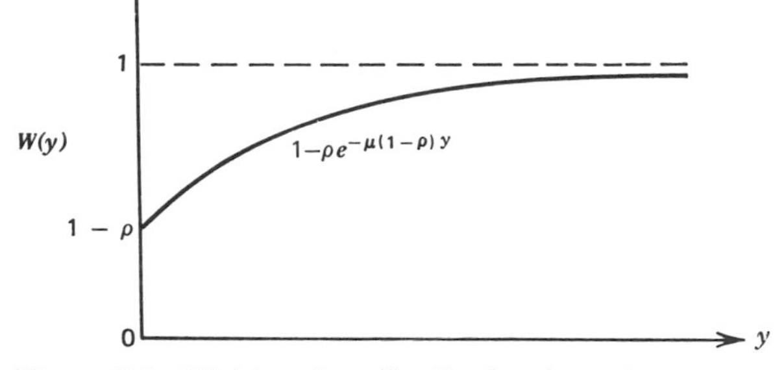

Figure 5.9 Waiting-time distribution for M/M/1.

This expression gives the Laplace transform for the pdf of waiting time which we denote, as usual, by $w(y) \triangleq dW(y)/dy$. From entry 2 in Table I.4 of Appendix I, we recognize that the inverse transform of $(1 - \rho)$ must be an impulse at the origin; thus by inspection we have

$$w(y) = (1 - \rho)u_0(y) + \lambda(1 - \rho)e^{-\mu(1-\rho)y} \qquad y \geq 0 \qquad \text{M/M/1} \quad (5.122)$$

From this we find the PDF of waiting time simply as

$$W(y) = 1 - \rho e^{-\mu(1-\rho)y} \qquad y \geq 0 \qquad \text{M/M/1} \quad (5.123)$$

This distribution is shown in Figure 5.9.

Observe that the probability of not queueing is merely $1 - \rho$; compare this to Eq. (5.89) for the probability that $\tilde{q} = 0$. Clearly, they are the same; both represent the probability of not queueing. This also was found in Eq. (5.40). Recall further that the mean normalized queueing time was given in Eq. (5.73); we obtain the same answer, of course, if we calculate this mean value from (5.123). It is interesting to note for M/M/1 that all of the interesting distributions are memoryless: this applies not only to the given interarrival time and service time processes, but also to the distribution of the number in the system given by Eq. (5.89), the pdf of time in the system given by Eq. (5.119), and the pdf of waiting time* given by Eq. (5.122).

It turns out that it is possible to find the density given in Eq. (5.118) by a more *direct* calculation, and we display this method here to indicate its simplicity. Our point of departure is our early result given in Eq. (3.23) for the probability of finding k customers in system upon arrival, namely,

$$p_k = (1 - \rho)\rho^k \qquad (5.124)$$

* A simple exponential form for the tail of the waiting-time distribution (that is, the probabilities associated with long waits) can be derived for the system M/G/1. We postpone a discussion of this asymptotic result until Chapter 2, Volume II, in which we establish this result for the more general system G/G/1.

We repeat again that this is the same expression we found in Eq. (5.89) and we know by now that this result applies for all points in time. We wish to form the Laplace transform of the pdf of total time in the system by considering this Laplace transform *conditioned* on the number of customers found in the system upon arrival of a new customer. We begin as generally as possible and first consider the system M/G/1. In particular, we define the conditional distribution

$$S(y \mid k) = P[\text{customer's total time in system} \le y \mid \text{he finds } k \text{ in system upon his arrival}]$$

We now define the Laplace transform of this conditional density

$$S^*(s \mid k) \triangleq \int_0^\infty e^{-sy}\, dS(y \mid k) \tag{5.125}$$

Now it is clear that if a customer finds no one in system upon his arrival, then he must spend an amount of time in the system exactly equal to his own service time, and so we have

$$S^*(s \mid 0) = B^*(s)$$

On the other hand, if our arriving customer finds exactly one customer ahead of him, then he remains in the system for a time equal to the time to finish the man in service, plus his own service time; since these two intervals are independent, then the Laplace transform of the density of this sum must be the product of the Laplace transform of each density, giving

$$S^*(s \mid 1) = \hat{B}^*(s)B^*(s)$$

where $\hat{B}^*(s)$ is, again, the transform for the pdf for residual service time. Similarly, if our arriving customer finds k in front of him, then his total system time is the sum of the k service times associated with each of these customers plus his own service time. These $k + 1$ random variables are all independent, and k of them are drawn from the same distribution $B(x)$. Thus we have the k-fold product of $B^*(s)$ with $\hat{B}^*(s)$ giving

$$S^*(s \mid k) = [B^*(s)]^k \hat{B}^*(s) \tag{5.126}$$

Equation (5.126) holds for M/G/1. Now for our M/M/1 problem, we have that $B^*(s) = \mu/(s + \mu)$ and, similarly, for $\hat{B}^*(s)$ (memoryless); thus we have

$$S^*(s \mid k) = \left(\frac{\mu}{s + \mu}\right)^{k+1} \tag{5.127}$$

In order to obtain $S^*(s)$ we need merely weight the transform $S^*(s \mid k)$ with the probability p_k of our customer finding k in the system upon his arrival, namely,

$$S^*(s) = \sum_{k=0}^{\infty} S^*(s \mid k) p_k$$

Substituting Eqs. (5.127) and (5.124) into this last we have

$$S^*(s) = \sum_{k=0}^{\infty} \left(\frac{\mu}{s + \mu}\right)^{k+1} (1 - \rho)\rho^k$$

$$= \frac{\mu(1 - \rho)}{s + \mu(1 - \rho)} \tag{5.128}$$

We recognize that Eq. (5.128) is identical to Eq. (5.117) and so the remaining steps leading to Eq. (5.118) follow immediately. This demonstration of a simpler method for calculating the distribution of system time in the M/M/1 queue demonstrates the following important fact: In the development of Eq. (5.128) we were required to consider a sum of random variables, each distributed by the same exponential distribution; the number of terms in that sum was itself a random variable distributed geometrically. What we found was that this geometrical weighting on a sum of identically distributed exponential random variables was itself exponential [see Eq. (5.118)]. This result is true in general, namely, that a geometric sum of exponential random variables is itself exponentially distributed.

Let us now carry out the calculations for our $M/H_2/1$ example. Using the expression for $B^*(s)$ given in Eq. (5.90), and applying this to the P–K transform equation for waiting-time density, we have

$$W^*(s) = \frac{4s(1 - \rho)(s + \lambda)(s + 2\lambda)}{4(s - \lambda)(s + \lambda)(s + 2\lambda) + 8\lambda^3 + 7\lambda^2 s}$$

This simplifies upon factoring the denominator, to give

$$W^*(s) = \frac{(1 - \rho)(s + \lambda)(s + 2\lambda)}{[s + (3/2)\lambda][s + (1/2)\lambda]}$$

Once again, we must divide numerator by denominator to reduce the degree of the numerator by one, giving

$$W^*(s) = (1 - \rho) + \frac{\lambda(1 - \rho)[s + (5/4)\lambda]}{[s + (3/2)\lambda][s + (1/2)\lambda]}$$

We may now carry out our partial-fraction expansion:

$$W^*(s) = (1 - \rho)\left[1 + \frac{\lambda/4}{s + (3/2)\lambda} + \frac{3\lambda/4}{s + (1/2)\lambda}\right]$$

This we may now invert by inspection to obtain the pdf for waiting time (and recalling that $\rho = 5/8$):

$$w(y) = \frac{3}{8} u_0(y) + \frac{3\lambda}{32} e^{-(3/2)\lambda y} + \frac{9\lambda}{32} e^{-(1/2)\lambda y} \qquad y \geq 0 \qquad (5.129)$$

This completes our discussion of the waiting-time and system-time distributions for M/G/1. We now introduce the busy period, an important stochastic process in queueing systems.

5.8. THE BUSY PERIOD AND ITS DURATION

We now choose to study queueing systems from a different point of view. We make the observation that the system passes through alternating cycles of busy period, idle period, busy period, idle period, and so on. Our purpose in this section is to derive the distribution for the length of the idle period and the length of the busy period for the M/G/1 queue.

As we already understand, the pertinent sequences of random variables that drive a queueing system are the instants of arrival and the sequence of service times. As usual let

C_n = the nth customer
τ_n = arrival time of C_n
$t_n = \tau_n - \tau_{n-1}$ = interarrival time between C_{n-1} and C_n
x_n = service time for C_n

We now recall the important stochastic process $U(t)$ as defined in Eq. (2.3):

$U(t) \triangleq$ the unfinished work in the system at time t
$\triangleq$ the remaining time required to empty the system of all customers present at time t

This function $U(t)$ is appropriately referred to as the unfinished work at time t since it represents the interval of time that is required to empty the system completely if no new customers are allowed to enter after the instant t. This function is sometimes referred to as the "virtual" waiting time at time t since, for a first-come–first-served system it represents how long a (virtual) customer would wait in queue *if* he entered at time t; however, this waiting-time interpretation is good only for first-come–first-served disciplines, whereas the unfinished work interpretation applies for all disciplines. Behavior of this function is extremely important in understanding queueing systems when one studies them from the point of view of the busy period.

Let us refer to Figure 5.10*a*, which shows the fashion in which busy periods alternate with idle periods. The busy-period durations are denoted by $Y_1, Y_2, Y_3, \ldots$ and the idle period durations by $I_1, I_2, \ldots$. Customer C_1

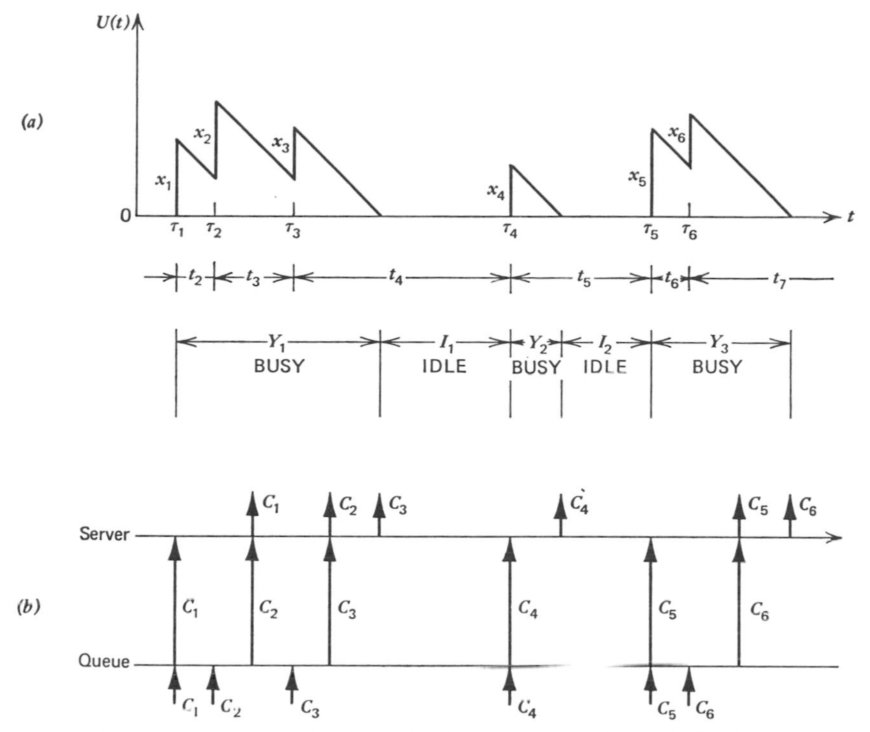

Figure 5.10 (*a*) The unfinished work, the busy period, and (*b*) the customer history.

enters the system at time τ_1 and brings with him an amount of work (that is, a required service time) of size x_1. This customer finds the system idle and therefore his arrival terminates the previous idle period and initiates a new busy period. Prior to his arrival we assumed the system to be empty and therefore the unfinished work was clearly zero. At the instant of the arrival of C_1 the system backlog or unfinished work jumps to the size x_1, since it would take this long to empty the system if we allowed no further entries beyond this instant. As time progresses from τ_1 and the server works on C_1, this unfinished work reduces at the rate of 1 sec/sec and so $U(t)$ decreases with slope equal to -1. t_2 sec later at time τ_2 we observe that C_2 enters the system and forces the unfinished work $U(t)$ to make another vertical jump of magnitude x_2 equal to the service time for C_2. The function then decreases again at a rate of 1 sec/sec until customer C_3 enters at time τ_3 forcing a vertical jump again of size x_3. $U(t)$ continues to decrease as the server works on the customers in the system until it reaches the instant $\tau_1 + Y_1$, at which time he has successfully emptied the system of all customers and of all work. This then terminates the busy period and initiates a new idle period. The idle period is terminated at time τ_4 when C_4 enters. This second busy period serves only one customer before the system goes idle again. The third busy period serves two customers. And so it continues. For reference we show in Figure 5.10*b* our usual double-time-axis representation for the same sequence of customer arrivals and service times drawn to the same scale as Figure 5.10*a* and under an assumed first-come–first-served discipline. Thus we can say that $U(t)$ is a function which has vertical jumps at the customer-arrival instants (these jumps equaling the service times for those customers) and decreases at a rate of 1 sec/sec so long as it is positive; when it reaches a value of zero, it remains there until the next customer arrival. This stochastic process is a continuous-state Markov process subject to discontinuous jumps; we have not seen such as this before.

Observe for Figure 5.10*a* that the departure instants may be obtained by extrapolating the linearly decreasing portion of $U(t)$ down to the horizontal axis; at these intercepts, a customer departure occurs and a new customer service begins. Again we emphasize that the last observation is good only for the first-come–first-served system. What is important, however, is to observe that the function $U(t)$ itself is *independent of the order of service*! The only requirement for this last statement to hold is that the server remain busy as long as some customer is in the system and that no customers depart before they are completely served; such a system is said to be "work conserving" (see Chapter 3, Volume II). The truth of this independence is evident when one considers the definition of $U(t)$.

Now for the idle-period and busy-period distributions. Recall

$$A(t) = P[t_n \le t] = 1 - e^{-\lambda t} \qquad t \ge 0 \tag{5.130}$$
$$B(x) = P[x_n \le x]$$

where $A(t)$ and $B(x)$ are each independent of n. Our interest lies in the two following distributions:

$$F(y) \triangleq P[I_n \le y] \triangleq \text{idle-period distribution} \tag{5.131}$$

$$G(y) \triangleq P[Y_n \le y] \triangleq \text{busy-period distribution} \tag{5.132}$$

The calculation of the idle-period distribution is trivial for the system M/G/1. Observe that when the system terminates a busy period, a new idle period must begin, and this idle period will terminate immediately upon the arrival of the next customer. Since we have a memoryless distribution, the time until the next customer arrival is distributed according to Eq. (5.130), and therefore we have

$$F(y) = 1 - e^{-\lambda y} \qquad y \ge 0 \tag{5.133}$$

So much for the idle-time distribution in M/G/1.

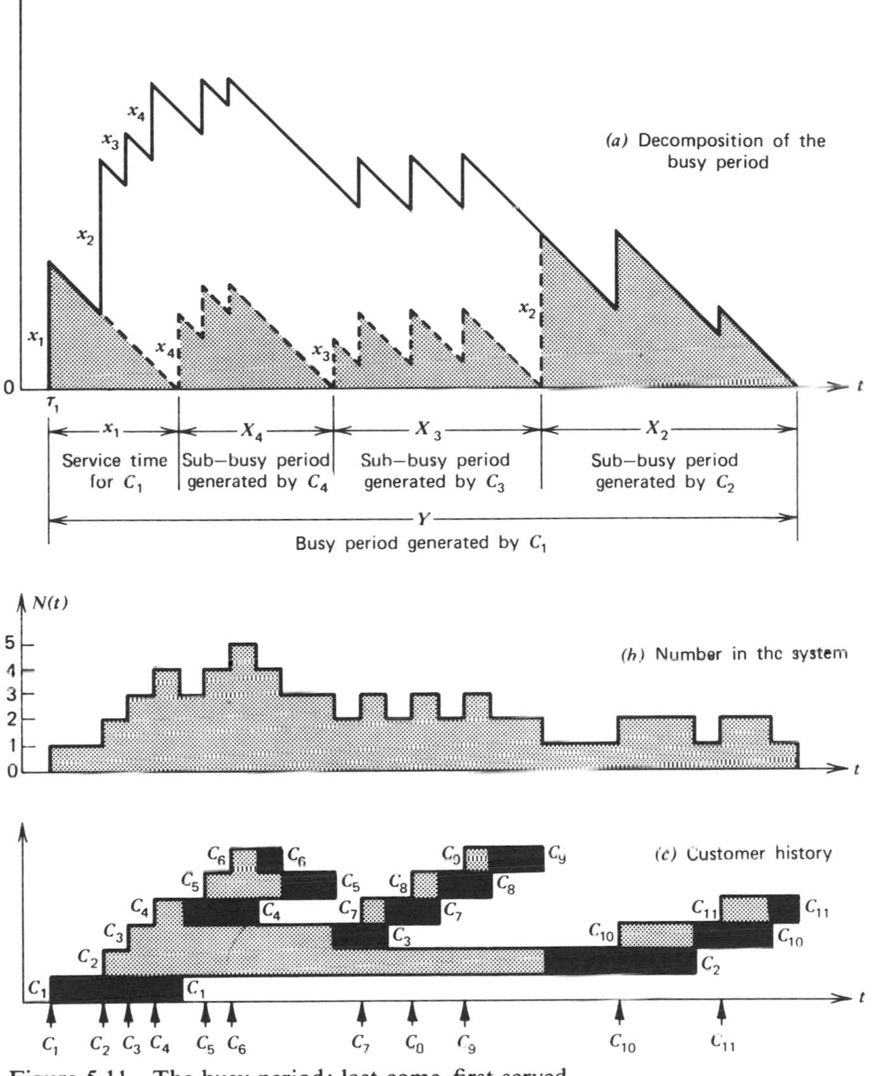

Figure 5.11 The busy period: last-come-first-served

Now for the busy-period distribution; this is not quite so simple. The reader is referred to Figure 5.11. In part (*a*) of this figure we once again observe the unfinished work $U(t)$. We assume that the system is empty just prior to the instant τ_1, at which time customer C_1 initiates a busy period of duration Y. His service time is equal to x_1. It is clear that this customer will depart from the system at a time $\tau_1 + x_1$. During his service other customers may arrive to the system and it is they who will continue the busy period. For the function shown, three other customers (C_2, C_3, and C_4) arrive during the interval of C_1's service. We now make use of a brilliant device due to Takács [TAKA 62*a*]. In particular, we choose to permute the order in which customers are served so as to create a last-come–first-served (LCFS) queueing discipline* (recall that the duration of a busy period is *independent* of the order in which customers are served). The motivation for the reordering of customers will soon be apparent. At the departure of C_1 we then take into service the *newest* customer, which in our example is C_4. In addition, since all future arrivals during this busy period must be served before (LCFS!) any customers (besides C_4) who arrived during C_1's service (in this case C_2 and C_3), then we may as well consider them to be (temporarily) out of the system. Thus, when C_4 enters service, it is *as if he initiated a new busy period*, which we will refer to as a "sub-busy period"; the sub-busy period generated by C_4 will have a duration X_4 exactly as long as it takes to service C_4 and all those who enter into the system to find it busy (remember that C_2 and C_3 are not considered to be in the system at this time). Thus in Figure 5.11*a* we show the sub-busy period generated by C_4 during which customers C_4, C_6, and C_5 get serviced in that order. At time $\tau_1 + x_1 + X_4$ this sub-busy period ends and we now continue the last-come–first-served order of service by bringing C_3 back into the system. It is clear that he may be considered as generating his own sub-busy period, of duration X_3, during which all of his "descendents" receive service in the last-come–first-served order (namely, C_3, C_7, C_8, and C_9). Finally, then, the system empties again, we reintroduce C_2, and permit his sub-busy period (of length X_2) to run its course (and complete the major busy period) in which customers get serviced in the order C_2, C_{10}, and finally C_{11}.

Figure 5.11*a* shows that the contour of any sub-busy period is identical with the contour of the main busy period over the same time interval and is merely shifted down by a constant amount; this shift, in fact, is equal to the summed service time of all those customers who arrived during C_1's service time and who have not yet been allowed to generate their own sub-busy periods. The details of customer history are shown in Figure 5.11*c* and the total number in the system at any time under this discipline is shown in Figure 5.11*b*. Thus, as far as the queueing system is concerned, it is strictly a last-come–first-served system from start to finish. However, our analysis is simplified if we focus upon the sub-busy periods and observe that *each behaves statistically in a fashion identical to the major busy period* generated by C_1. This is clear since all the sub-busy periods as well as the major busy period

* This is a "push-down" stack. This is only one of many permutations that "work"; it happens that LCFS is convenient for pedagogical purposes.

are each initiated by a single customer whose service times are all drawn from the same distribution independently; each sub-busy period continues until the system catches up to the work load, in the sense that the unfinished work function $U(t)$ drops to zero. Thus we recognize that the random variables $\{X_k\}$ are each independent and identically distributed and have the same distribution as Y, the duration of the major busy period.

In Figure 5.11*c* the reader may follow the customer history in detail; the solid black region in this figure identifies the customer being served during that time interval. At each customer departure the server "floats up" to the top of the customer contour to engage the most recent arrival at that time; occasionally the server "floats down" to the customer directly below him such as at the departure of C_6. The server may truly be thought of as floating up to the highest customer there to be held by him until his departure, and so on. Occasionally, however, we see that our server "falls down" through a gap in order to pick up the most recent arrival to the system, for example, at the departure of C_5. It is at such instants that new sub-busy periods begin and only when the server falls down to hit the horizontal axis does the major busy period terminate.

Our point of view is now clear: the duration of a busy period Y is the sum of $1 + \tilde{v}$ random variables, the first of which is the service time for C_1 and the remainder of which are each random variables describing the duration of the sub-busy periods, each of which is distributed as a busy period itself. $\tilde{v}$ is a random variable equal to the number of customer arrivals during C_1's service interval. Thus we have the important relation

$$Y = x_1 + X_{\tilde{v}+1} + X_{\tilde{v}} + \cdots + X_3 + X_2 \tag{5.134}$$

We define the busy-period distribution as $G(y)$:

$$G(y) \overset{\Delta}{=} P[Y \le y] \tag{5.135}$$

We also know that x_1 is distributed according to $B(x)$ and that X_k is distributed as $G(y)$ from our earlier comments. We next derive the Laplace transform for the pdf associated with Y, which we define, as usual, by

$$G^*(s) \overset{\Delta}{=} \int_0^\infty e^{-sy}\, dG(y) \tag{5.136}$$

Once again we remind the reader that these transforms may also be expressed as expectation operators, namely:

$$G^*(s) \overset{\Delta}{=} E[e^{-sY}]$$

Let us now take advantage of the powerful technique of conditioning used so often in probability theory; this technique permits one to write down the probability associated with a complex event by conditioning that event on enough given conditions, so that the conditional probability may be written down by inspection. The unconditional probability is then obtained by multiplying by the probability of each condition and summing over all mutually exclusive and exhaustive conditions. In our case we choose to condition Y on two events: the duration of C_1's service and the number of customer arrivals during his service. With this point of view we then calculate the following conditional transform:

$$
\begin{aligned}
E[e^{-sY} \mid x_1 = x, \tilde{v} = k] &= E[e^{-s(x+X_{k+1}+\cdots+X_2)}] \\
&= E[e^{-sx}e^{-sX_{k+1}} \ldots e^{-sX_2}]
\end{aligned}
$$

Since the sub-busy periods have durations that are independent of each other, we may write this last as

$$E[e^{-sY} \mid x_1 = x, \tilde{v} = k] = E[e^{-sx}]E[e^{-sX_{k+1}}] \cdots E[e^{-sX_2}]$$

Since x is a given constant we have $E[e^{-sx}] = e^{-sx}$, and further, since the sub-busy periods are identically distributed with corresponding transforms $G^*(s)$, we have

$$E[e^{-sY} \mid x_1 = x, \tilde{v} = k] = e^{-sx}[G^*(s)]^k$$

Since $\tilde{v}$ represents the number of arrivals during an interval of length x, then $\tilde{v}$ must have a Poisson distribution whose mean is λx. We may therefore remove the condition on $\tilde{v}$ as follows:

$$
\begin{aligned}
E[e^{-sY} \mid x_1 = x] &= \sum_{k=0}^{\infty} E[e^{-sY} \mid x_1 = x, \tilde{v} = k]P[\tilde{v} = k] \\
&= \sum_{k=0}^{\infty} e^{-sx}[G^*(s)]^k \frac{(\lambda x)^k}{k!} e^{-\lambda x} \\
&= e^{-x[s+\lambda-\lambda G^*(s)]}
\end{aligned}
$$

Similarly, we may remove the condition on x_1 by integrating with respect to $B(x)$, finally to obtain $G^*(s)$ thusly

$$G^*(s) = \int_0^\infty e^{-x[s+\lambda-\lambda G^*(s)]}\, dB(x)$$

This last we recognize as the transform of the pdf for service time evaluated at a value equal to the bracketed term in the exponent, that is,

$$G^*(s) = B^*[s + \lambda - \lambda G^*(s)] \qquad \blacksquare(5.137)$$

This major result gives the transform for the M/G/1 busy-period distribution (for any order of service) expressed as a functional equation (which is usually impossible to invert). It was obtained by identifying sub-busy periods within the busy period all of which had the same distribution as the busy period itself.

Later in this chapter we give an explicit expression for the busy period PDF $G(y)$, but unfortunately it is not in closed form [see Eq. (5.169)]. We point out, however, that it is possible to solve Eq. (5.137) numerically for $G^*(s)$ at any given value of s through the following iterative equation:

$$G_{n+1}^*(s) = B^*[s + \lambda - \lambda G_n^*(s)] \tag{5.138}$$

in which we choose $0 \leq G_0^*(s) \leq 1$; for $\rho = \lambda\bar{x} < 1$ the limit of this iterative scheme will converge to $G^*(s)$ and so one may attempt a numerical inversion of these calculated values if so desired.

In view of these inversion difficulties we obtain what we can from our functional equation, and one calculation we can make is for the *moments* of the busy period. We define

$$g_k \triangleq E[Y^k] \tag{5.139}$$

as the kth moment of the busy-period distribution, and we intend to express the first few moments in terms of the moments of the service-time distribution, namely, $\overline{x^k}$. As usual we have

$$\begin{aligned} g_k &= (-1)^k G^{*(k)}(0) \\ \overline{x^k} &= (-1)^k B^{*(k)}(0) \end{aligned} \tag{5.140}$$

From Eq. (5.137) we then obtain directly

$$\begin{aligned} g_1 = -G^{*(1)}(0) &= -B^{*(1)}(0) \frac{d}{ds}[s + \lambda - \lambda G^*(s)]\bigg|_{s=0} \\ &= -B^{*(1)}(0)[1 - \lambda G^{*(1)}(0)] \end{aligned}$$

[note, for $s = 0$, that $s + \lambda - \lambda G^*(s) = 0$] and so

$$g_1 = \bar{x}(1 + \lambda g_1)$$

Solving for g_1 and recalling that $\rho = \lambda\bar{x}$, we then have

$$g_1 = \frac{\bar{x}}{1 - \rho} \tag{5.141}$$

If we compare this last result with Eq. (3.26), we find that the *average length of a busy period for the system M/G/1 is equal to the average time a customer spends in an M/M/1 system and depends only on λ and $\bar{x}$*

Let us now chase down the second moment of the busy period. Proceeding from Eq. (5.140) and (5.137) we obtain

$$\begin{aligned} g_2 = G^{*(2)}(s) &= \frac{d}{ds}[B^{*(1)}(s + \lambda - \lambda G^*(s))][1 - \lambda G^{*(1)}(s)]\bigg|_{s=0} \\ &= B^{*(2)}(0)[1 - \lambda G^{*(1)}(0)]^2 + B^{*(1)}(0)[-\lambda G^{*(2)}(0)] \end{aligned}$$

and so

$$g_2 = \overline{x^2}(1 + \lambda g_1)^2 + \bar{x}\lambda g_2$$

Solving for g_2 and using our result for g_1, we have

$$g_2 = \frac{\overline{x^2}(1 + \lambda g_1)^2}{1 - \lambda\bar{x}}$$

$$= \frac{\overline{x^2}[1 + \lambda\bar{x}/(1 - \rho)]^2}{1 - \rho}$$

and so finally

$$g_2 = \frac{\overline{x^2}}{(1 - \rho)^3} \qquad ■(5.142)$$

This last result gives the second moment of the busy period and it is interesting to note the cube in the denominator; this effect does not occur when one calculates the second moment of the wait in the system where only a square power appears [see Eq. (5.114)]. We may now easily calculate the variance of the busy period, denoted by σ_g^2, as follows:

$$\sigma_g^2 = g_2 - g_1^2$$

$$= \frac{\overline{x^2}}{(1 - \rho)^3} - \frac{(\bar{x})^2}{(1 - \rho)^2}$$

and so

$$\sigma_g^2 = \frac{\sigma_b^2 + \rho(\bar{x})^2}{(1 - \rho)^3} \qquad ■(5.143)$$

where σ_b^2 is the variance of the service-time distribution.

Proceeding as above we find that

$$g_3 = \frac{\overline{x^3}}{(1 - \rho)^4} + \frac{3\lambda(\overline{x^2})^2}{(1 - \rho)^5} \qquad ■$$

$$g_4 = \frac{\overline{x^4}}{(1 - \rho)^5} + \frac{10\lambda\overline{x^2}\,\overline{x^3}}{(1 - \rho)^6} + \frac{15\lambda^2(\overline{x^2})^3}{(1 - \rho)^7} \qquad ■$$

We observe that the factor $(1 - \rho)$ goes up in powers of 2 for the dominant term of each succeeding moment of the busy period and this determines the behavior as $\rho \to 1$.

We now consider some examples of inverting Eq. (5.137). We begin with the M/M/1 queueing system. We have

$$B^*(s) = \frac{\mu}{s + \mu}$$

which we apply to Eq. (5.137) to obtain

$$G^*(s) = \frac{\mu}{s + \lambda - \lambda G^*(s) + \mu}$$

or

$$\lambda[G^*(s)]^2 - (\mu + \lambda + s)G^*(s) + \mu = 0$$

Solving for $G^*(s)$ and restricting our solution to the required (stable) case for which $|G^*(s)| \leq 1$ for Re $(s) \geq 0$, gives

$$G^*(s) = \frac{\mu + \lambda + s - [(\mu + \lambda + s)^2 - 4\mu\lambda]^{1/2}}{2\lambda} \tag{5.144}$$

This equation may be inverted (by referring to transform tables) to obtain the pdf for the busy period, namely,

$$g(y) \triangleq \frac{dG(y)}{dy} = \frac{1}{y(\rho)^{1/2}} e^{-(\lambda+\mu)y} I_1[2y(\lambda\mu)^{1/2}] \tag{5.145}$$

where I_1 is the modified Bessel function of the first kind of order one.

Consider the limit

$$\lim_{0<s\to 0} G^*(s) = \lim_{0<s\to 0} \int_0^\infty e^{-sy}\, dG(y) \tag{5.146}$$

Examining the right side of this equation we observe that this limit is merely the probability that the busy period is finite, which is equivalent to the probability of the busy period ending. Clearly, for $\rho < 1$ the busy period ends with probability one, but Eq. (5.146) provides information in the case $\rho > 1$. We have

$$P[\text{busy period ends}] = G^*(0)$$

Let us examine this computation in the case of the system M/M/1. We have directly from Eq. (5.144)

$$G^*(0) = \frac{\mu + \lambda - [(\mu + \lambda)^2 - 4\mu\lambda]^{1/2}}{2\lambda}$$

$$= \frac{\mu}{\lambda}$$

and so

$$G^*(0) = \frac{1}{\rho}$$

Thus

$$P[\text{busy period ends in M/M/1}] = \begin{cases} 1 & \rho < 1 \\ \dfrac{1}{\rho} & \rho > 1 \end{cases} \tag{5.147}$$

The busy period pdf given in Eq. (5.145) is much more complex than we would have wished for this simplest of interesting queueing systems! It is indicative of the fact that Eq. (5.137) is usually uninvertible for more general service-time distributions.

As a second example, let's see how well we can do with our $M/H_2/1$ example. Using the expression for $B^*(s)$ in our functional equation for the busy period we get

$$G^*(s) = \frac{8\lambda^2 + 7\lambda[s + \lambda - \lambda G^*(s)]}{4[s + \lambda - \lambda G^*(s) + \lambda][s + \lambda - \lambda G^*(s) + 2\lambda]}$$

which leads directly to the cubic equation

$$4[G^*(s)]^3 - 4(2s + 5)[G^*(s)]^2 + (4s^2 + 20s + 31)G^*(s) - (15 + 7s) = 0$$

This last is not easily solved and so we stall at this point in our attempt to invert $G^*(s)$. We will return to the functional equation for the busy period when we discuss priority queueing in Chapter 3, Volume II. This will lead us to the concept of a *delay cycle*, which is a slight generalization of the busy-period analysis we have just carried out and greatly simplifies priority queueing calculations.

5.9. THE NUMBER SERVED IN A BUSY PERIOD

In this section we discuss the distribution of the number of customers served in a busy period. The development parallels that of the previous section very closely, both in the spirit of the derivation and in the nature of the result we will obtain.

Let N_{bp} be the number of customers served in a busy period. We are interested in its probability distribution f_n defined as

$$f_n = P[N_{bp} = n] \tag{5.148}$$

The best we can do is to obtain a functional equation for its z-transform defined as

$$F(z) \triangleq E[z^{N_{bp}}] \triangleq \sum_{n=1}^{\infty} f_n z^n \tag{5.149}$$

The term for $n = 0$ is omitted from this definition since at least one customer must be served in a busy period. We recall that the random variable $\tilde{v}$ represents the number of arrivals during a service period and its z-transform $V(z)$ obeys the equation derived earlier, namely,

$$V(z) = B^*(\lambda - \lambda z) \tag{5.150}$$

Proceeding as we did for the duration of the busy period, we condition our argument on the fact that $\tilde{v} = k$, that is, we assume that k customers arrive during the service of C_1. Moreover, we recognize immediately that each of these arrivals will generate a sub-busy period and the number of customers served in each of these sub-busy periods will have a distribution given by f_n. Let the random variable M_i denote the number of customers served in the ith sub-busy period. We may then write down immediately

$$E[z^{N_{bp}} \mid \tilde{v} = k] = E[z^{1+M_1+M_2+\cdots+M_k}]$$

and since the M_i are independent and identically distributed we have

$$E[z^{N_{bp}} \mid \tilde{v} = k] = z \prod_{i=1}^{k} E[z^{M_i}]$$

But each of the M_i is distributed exactly the same as N_{bp} and, therefore,

$$E[z^{N_{bp}} \mid \tilde{v} = k] = z[F(z)]^k$$

Removing the condition on the number of arrivals we have

$$F(z) = \sum_{k=0}^{\infty} E[z^{N_{bp}} \mid \tilde{v} = k] P[\tilde{v} = k]$$

$$= z \sum_{k=0}^{\infty} P[\tilde{v} = k][F(z)]^k$$

From Eq. (5.44) we recognize this last summation as $V(z)$ (the z-transform associated with $\tilde{v}$) with transform variable $F(z)$; thus we have

$$F(z) = zV[F(z)] \tag{5.151}$$

But from Eq. (5.150) we may finally write

$$F(z) = zB^*[\lambda - \lambda F(z)] \quad \blacksquare(5.152)$$

This functional equation for the z-transform of the number served in a busy period is not unlike the equation given earlier in Eq. (5.137).

From this fundamental equation we may easily pick off the moments for the number served in a busy period. We define the kth moment of the number served in a busy period as h_k. We recognize then

$$h_1 = F^{(1)}(1)$$

$$= B^{*(1)}(0)[-\lambda F^{(1)}(1)] + B^*(0)$$

Thus

$$h_1 = \lambda \bar{x} h_1 + 1$$

which immediately gives us

$$h_1 = \frac{1}{1 - \rho} \quad \blacksquare(5.153)$$

We further recognize

$$F^{(2)}(1) = h_2 - h_1$$

Carrying out this computation in the usual way, we obtain the second moment and variance of the number served in the busy period:

$$h_2 = \frac{2\rho(1-\rho) + \lambda^2\overline{x^2}}{(1-\rho)^3} + \frac{1}{1-\rho} \qquad \blacksquare(5.154)$$

$$\sigma_h^{\ 2} = \frac{\rho(1-\rho) + \lambda^2\overline{x^2}}{(1-\rho)^3} \qquad \blacksquare(5.155)$$

As an example we again use the simple case of the M/M/1 system to solve for $F(z)$ from Eq. (5.152). Carrying this out we find

$$F(z) = z\frac{\mu}{\mu + \lambda - \lambda F(z)}$$

$$\lambda F^2(z) - (\mu + \lambda)F(z) + \mu z = 0$$

Solving,

$$F(z) = \frac{1+\rho}{2\rho}\left[1 - \left(1 - \frac{4\rho z}{(1+\rho)^2}\right)^{1/2}\right] \qquad (5.156)$$

Fortunately, it turns out that the equation (5.156) can be inverted to obtain f_n, the probability of having n served in the busy period:

$$f_n = \frac{1}{n}\binom{2n-2}{n-1}\rho^{n-1}(1+\rho)^{1-2n} \qquad \blacksquare(5.157)$$

As a second example we consider the system M/D/1. For this system we have $b(x) = u_0(x - \bar{x})$ and from entry three in Table I.4 we have immediately that

$$B^*(s) = e^{-s\bar{x}}$$

Using this in our functional equation we obtain

$$F(z) = ze^{-\rho}e^{\rho F(z)} \qquad (5.158)$$

where as usual $\rho = \lambda\bar{x}$. It is convenient to make the substitution $u = z\rho e^{-\rho}$ and $H(u) = \rho F(z)$, which then permits us to rewrite Eq. (5.158) as

$$u = H(u)e^{-H(u)}$$

The solution to this equation may be obtained [RIOR 62] and then our original function may be evaluated to give

$$F(z) = \sum_{n=1}^{\infty}\frac{(n\rho)^{n-1}}{n!}e^{-n\rho}z^n$$

From this power series we recognize immediately that the distribution for the number served in the M/D/1 busy period is given explicitly by

$$f_n = \frac{(n\rho)^{n-1}}{n!} e^{-n\rho} \qquad \blacksquare(5.159)$$

For the case of a constant service time we know that if the busy period serves n customers then it must be of duration $n\bar{x}$, and therefore we may immediately write down the solution for the M/D/1 busy-period distribution as

$$G(y) = \sum_{n=1}^{[y/\bar{x}]} \frac{(n\rho)^{n-1}}{n!} e^{-n\rho} \qquad \blacksquare(5.160)$$

where $[y/\bar{x}]$ is the largest integer not exceeding $y/\bar{x}$.

5.10. FROM BUSY PERIODS TO WAITING TIMES

We had mentioned in the opening paragraphs of this chapter that waiting times could be obtained from the busy-period analysis. We are now in a position to fulfill that claim. As the reader may be aware (and as we shall show in Chapter 3, Volume II), whereas the distribution of the busy-period duration is independent of the queueing discipline, the distribution of waiting time is strongly dependent upon order of service. Therefore, in this section we consider only first-come first-served M/G/1 systems. Since we restrict ourselves to this discipline, the reordering of customers used in Section 5.8 is no longer permitted. Instead, we must now decompose the busy period into a sequence of intervals whose lengths are *dependent* random variables as follows. Consider Figure 5.12 in which we show a single busy period for the first-come–first-served system [in terms of the unfinished work $U(t)$]. Here we see that customer C_1 initiates the busy period upon his arrival at time τ_1. The first interval we consider is his service time x_1, which we denote by X_0; during this interval more customers arrive (in this case C_2 and C_3). All those customers who arrive during X_0 are served during the next interval, whose duration is X_1 and which equals the sum of the service times of all arrivals during X_0 (in this case C_2 and C_3). At the expiration of X_1, we then create a new interval of duration X_2 in which all customers arriving during X_1 are served, and so on. Thus X_i is the length of time required to service all those customers who arrive during the previous interval whose duration is X_{i-1}. If we let n_i denote the number of customer arrivals during the interval X_i, then n_i customers are served during the interval X_{i+1}. We let n_0 equal the number of customers who arrive during X_0 (the first customer's service time).

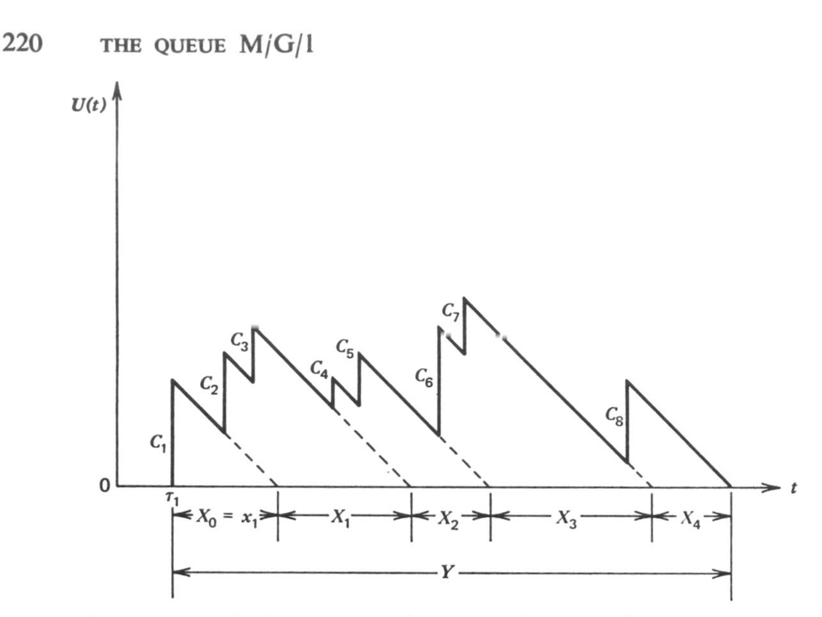

Figure 5.12 The busy period: first-come–first-served.

Thus we see that Y, the duration of the total busy period, is given by

$$Y = \sum_{i=0}^{\infty} X_i$$

where we permit the possibility of an infinite sequence of such intervals. Clearly, we define $X_i = 0$ for those intervals that fall beyond the termination of this busy period; for $\rho < 1$ we know that with probability 1 there will be a finite i_0 for which X_{i_0} (and all its successors) will be 0. Furthermore, we know that X_{i+1} will be the sum of n_i service intervals [each of which is distributed as $B(x)$].

We now define $X_i(y)$ to be the PDF for X_i, that is,

$$X_i(y) \triangleq P[X_i \leq y]$$

and the corresponding Laplace transform of the associated pdf to be

$$\begin{aligned} X_i^*(s) &\triangleq \int_0^\infty e^{-sy}\, dX_i(y) \\ &= E[e^{-sX_i}] \end{aligned}$$

We wish to derive a recurrence relation among the $X_i^*(s)$. This derivation is much like that in Section 5.8, which led up to Eq. (5.137). That is, we first condition our transform sufficiently so that we may write it down by inspection; the conditions are on the interval length X_{i-1} and on the number of arrivals n_{i-1} during that interval, that is, we may write

$$E[e^{-sX_i} \mid X_{i-1} = y, n_{i-1} = n] = [B^*(s)]^n$$

This last follows from our convolution property leading to the multiplication of transforms in the case when the variables are independent; here we have n independent service times, all with identical distributions. We may uncondition first on n:

$$E[e^{-sX_i} \mid X_{i-1} = y] = \sum_{n=0}^{\infty} \frac{(\lambda y)^n}{n!} e^{-\lambda y}[B^*(s)]^n$$

and next on y:

$$E[e^{-sX_i}] = \int_{y=0}^{\infty} \sum_{n=0}^{\infty} \frac{(\lambda y)^n}{n!} e^{-\lambda y}[B^*(s)]^n \, dX_{i-1}(y)$$

Clearly, the left-hand side is $X_i^*(s)$; evaluating the sum on the right-hand side leads us to

$$X_i^*(s) = \int_{y=0}^{\infty} e^{-[\lambda - \lambda B^*(s)]y} \, dX_{i-1}(y)$$

This integral is recognized as the transform of the pdf for X_{i-1}, namely,

$$X_i^*(s) = X_{i-1}^*[\lambda - \lambda B^*(s)] \tag{5.161}$$

This is the first step.

We now condition our calculations on the event that a new ("tagged") arrival occurs during the busy period and, in particular, while the busy period is in its ith interval (of duration X_i). From our observations in Section 4.1, we know that Poisson arrivals find the system in a given state with a probability equal to the equilibrium probability of the system being in that state. Now we know that if the system is in a busy period, then the fraction of time it spends in the interval of duration X_i is given by $E[X_i]/E[Y]$ (this can be made rigorous by renewal theory arguments). Consider a customer who arrives during an interval of duration X_i. Let his waiting time in system be denoted by $\tilde{w}$; it is clear that this waiting time will equal the sum of the remaining time (residual life) of the ith interval plus the sum of the service times of all jobs who arrived before he did during the ith interval. We wish to calculate $E[e^{-s\tilde{w}} \mid i]$, which is the transform of the waiting time pdf for an arrival during the ith interval; again, we perform this calculation by conditioning on the three variables X_i, Y_i (defined to be the residual life of this ith interval) and on N_i (defined to be the number of arrivals during the ith interval but prior to our customer's arrival—that is, in the interval $X_i - Y_i$). Thus, using our convolution property as before, we may write

$$E[e^{-s\tilde{w}} \mid i, X_i = y, Y_i = y', N_i = n] = e^{-sy'}[B^*(s)]^n$$

Now since we assume that n customers have arrived during an interval of duration $y - y'$ we uncondition on N_i as follows:

$$E[e^{-s\tilde{w}} \mid i, X_i = y, Y_i = y'] = e^{-sy'} \sum_{n=0}^{\infty} \frac{[\lambda(y - y')]^n}{n!} e^{-\lambda(y-y')}[B^*(s)]^n$$
$$= e^{-sy'-\lambda(y-y')+\lambda(y-y')B^*(s)} \tag{5.162}$$

We have already observed that Y_i is the residual life of the lifetime X_i. Equation (5.9) gives the joint density for the residual life Y and lifetime X; in that equation Y and X play the roles of Y_i and X_i in our problem. Therefore, replacing $f(x)\,dx$ in Eq. (5.9) by $dX_i(y)$ and noting that y and y' have replaced x and y in that development, we see that the joint density for X_i and Y_i is given by $dX_i(y)\,dy'/E[X_i]$ for $0 \le y' \le y \le \infty$. By means of this joint density we may remove the condition on X_i and Y_i in Eq. (5.162) to obtain

$$E[e^{-s\tilde{w}} \mid i] = \int_{y=0}^{\infty} \int_{y'=0}^{y} e^{-[s-\lambda+\lambda B^*(s)]y'} e^{-[\lambda-\lambda B^*(s)]y}\, dX_i(y)\, dy'/E[X_i]$$
$$= \int_{y=0}^{\infty} \frac{[e^{-sy} - e^{-[\lambda-\lambda B^*(s)]y}]}{[-s + \lambda - \lambda B^*(s)]E[X_i]}\, dX_i(y)$$

These last integrals we recognize as transforms and so

$$E[e^{-s\tilde{w}} \mid i] = \frac{X_i^*(s) - X_i^*(\lambda - \lambda B^*(s))}{[-s + \lambda - \lambda B^*(s)]E[X_i]}$$

But now Eq. (5.161) permits us to rewrite the second of these transforms to obtain

$$E[e^{-s\tilde{w}} \mid i] = \frac{X_{i+1}^*(s) - X_i^*(s)}{[s - \lambda + \lambda B^*(s)]E[X_i]}$$

Now we may remove the condition on our arrival entering during the ith interval by weighting this last expression by the probability that we have formerly expressed for the occurrence of this event (still conditioned on our arrival entering during a busy period), and so we have

$$E[e^{-s\tilde{w}} \mid \text{enter in busy period}]$$
$$= \sum_{i=0}^{\infty} E[e^{-s\tilde{w}} \mid i] \frac{E[X_i]}{E[Y]}$$
$$= \frac{1}{[s - \lambda + \lambda B^*(s)]E[Y]} \sum_{i=0}^{\infty} [X_{i+1}^*(s) - X_i^*(s)]$$

This last sum nicely collapses to yield $1 - X_0^*(s)$ since $X_i^*(s) = 1$ for those intervals beyond the busy period (recall $X_i = 0$ for $i \geq i_0$); also, since $X_0 = x_1$, a service time, then $X_0^*(s) = B^*(s)$, and so we arrive at

$$E[e^{-s\tilde{w}} \mid \text{enter in busy period}] = \frac{1 - B^*(s)}{[s - \lambda + \lambda B^*(s)]E[Y]}$$

From previous considerations we know that the probability of an arrival entering during a busy period is merely $\rho = \lambda\bar{x}$ (and for sure he must wait for service in such a case); further, we may evaluate the average length of the busy period $E[Y]$ either from our previous calculation in Eq. (5.141) or from elementary considerations* to give $E[Y] = \bar{x}/(1 - \rho)$. Thus, unconditioning on an arrival finding the system busy, we finally have

$$\begin{aligned} E[e^{-s\tilde{w}}] &= (1 - \rho)E[e^{-s\tilde{w}} \mid \text{enter in idle period}] + \rho E[e^{-s\tilde{w}} \mid \text{enter in busy period}] \\ &= (1 - \rho) + \rho \frac{[1 - B^*(s)](1 - \rho)}{[s - \lambda + \lambda B^*(s)]\bar{x}} \\ &= \frac{s(1 - \rho)}{s - \lambda + \lambda B^*(s)} \end{aligned} \tag{5.163}$$

Voilà! This is exactly the P–K transform equation for waiting time, namely, $W^*(s) \triangleq E[e^{-s\tilde{w}}]$ given in Eq. (5.105).

Thus we have shown how to go from a busy-period analysis to the calculation of waiting time in the system. This method is reported upon in [CONW 67] and we will have occasion to return to it in Chapter 3, Volume II.

5.11. COMBINATORIAL METHODS

We had mentioned in the opening remarks of this chapter that consideration of random walks and combinatorial methods was applicable to the study of the M/G/1 queue. We take this opportunity to indicate some aspects of those methods. In Figure 5.13 we have reproduced $U(t)$ from Figure 5.10*a*. In addition, we have indicated the "random walk" $R(t)$, which is the same as $U(t)$ except that it does not saturate at zero but rather continues to decline at a rate of 1 sec/sec below the horizontal axis; of course, it too takes vertical jumps at the customer-arrival instants. We introduce this diagram in order to define what are known as *ladder indices*. The kth (descending) ladder index

* The following simple argument enables us to calculate $E[Y]$. In a long interval (say, t) the server is busy a fraction ρ of the time. Each idle period in M/G/1 is of average length $1/\lambda$ sec and therefore we expect to have $(1 - \rho)t/(1/\lambda)$ idle periods. This will also be the number of busy periods, approximately; therefore, since the time spent in busy periods is ρt, the average duration of each must be $\rho t/\lambda t(1 - \rho) = \bar{x}/(1 - \rho)$. As $t \to \infty$, this argument becomes exact.

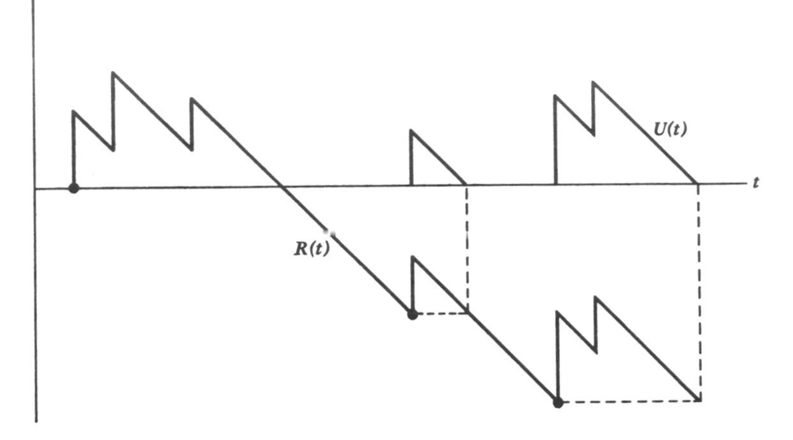

Figure 5.13 The descending ladder indices.

is defined as the instant when the random walk $R(t)$ rises from its kth new minimum (and the value of this minimum is referred to as the ladder height). In Figure 5.13 the first three ladder indices are indicated by heavy dots. Fluctuation theory concerns itself with the distribution of such ladder indices and is amply discussed both in Feller [FELL 66] and in Prabhu [PRAB 65] in which they consider the applications of that theory to queueing processes. Here we merely make the observation that each ladder index identifies the arrival instants for *those customers who begin new busy periods* and it is this observation that makes them interesting for queueing theory. Moreover, whenever $R(t)$ drops below its previous ladder height then a busy period terminates as shown in Figure 5.13. Thus, between the occurrence of a ladder index and the first time $R(t)$ drops below the corresponding ladder height, a busy period ensues and both $R(t)$ and $U(t)$ have exactly the same shape, where the former is shifted down from the latter by an amount exactly equal to the accumulated idle time since the end of the first busy period. One sees that we are quickly led into methods from combinatorial theory when we deal with such indices.

In a similar vein, Takács has successfully applied combinatorial theory to the study of the busy period. He considers this subject in depth in his book [TAKA 67] on combinatorial methods as applied to queueing theory and develops, as his cornerstone, a generalization of the *classical ballot theorem.* The classical ballot theorem concerns itself with the counting of votes in a two-way contest involving candidate A and candidate B. If we assume that A scores a votes and B scores b votes and that $a \geq mb$, where m is a non-negative integer and if we let P be the probability that throughout the counting of votes A continually leads B by a factor greater than m and further, if all possible sequences of voting records are equally likely, then the classical ballot theorem states that

$$P = \frac{a - mb}{a + b} \tag{5.164}$$

This theorem originated in 1887 (see [TAKA 67] for its history). Takács generalized this theorem and phrased it in terms of cards drawn from an urn in the following way. Consider an urn with n cards, where the cards are marked with the nonnegative integers $k_1, k_2, \ldots, k_n$ and where

$$\sum_{i=1}^{n} k_i = k \le n$$

(that is, the ith card in the set is marked with the integer k_i). Assume that all n cards are drawn without replacement from the urn. Let v_r ($r = 1, \ldots, n$) be the number on the card drawn at the rth drawing. Let

$$\tilde{N}_r = v_1 + v_2 + \cdots + v_r \qquad r = 1, 2, \ldots, n$$

$\tilde{N}_r$ is thus the sum of the numbers on all cards drawn up through the rth draw. Takács' generalization of the classical ballot theorem states that

$$P[\tilde{N}_r < r \text{ for } all\ r = 1, 2, \ldots, n] = \frac{n - k}{n} \tag{5.165}$$

The proof of this theorem is not especially difficult but will not be reproduced here. Note the simplicity of the theorem and, in particular, that the probability expressed is independent of the particular set of integers k_i and depends only upon their sum k. We may identify v_r as the number of customer arrivals during the service of the rth customer in a busy period of an M/G/1 queueing system. Thus $\tilde{N}_r + 1$ is the cumulative number of arrivals up to the conclusion of the rth customer's service during a busy period. We are thus involved in a race between $\tilde{N}_r + 1$ and r: As soon as r equals $\tilde{N}_r + 1$ then the busy period must terminate since, at this point, we have served exactly as many as have arrived (including the customer who initiated the busy period) and so the system empties. If we now let N_{bp} be the number of customers served in a busy period it is possible to apply Eq. (5.165) and obtain the following result [TAKA 67]:

$$P[N_{\text{bp}} = n] = \frac{1}{n} P[\tilde{N}_n = n - 1] \tag{5.166}$$

It is easy to calculate the probability on the right-hand side of this equation since we have Poisson arrivals: All we need do is condition this number of arrivals on the duration of the busy period, multiply by the probability that n service intervals will, in fact, sum to this length and then integrate over all possible lengths. Thus

$$P[\tilde{N}_n = n - 1] = \int_0^\infty \frac{(\lambda y)^{n-1}}{(n-1)!} e^{-\lambda y} b_{(n)}(y)\, dy \tag{5.167}$$

where $b_{(n)}(y)$ is the n-fold convolution of $b(y)$ with itself [see Eq. (5.110)] and represents the pdf for the sum of n independent random variables, where each is drawn from the common density $b(y)$. Thus we arrive at an explicit expression for the probability distribution for the number served in a busy period:

$$P[N_{\text{bp}} = n] = \int_0^\infty \frac{(\lambda y)^{n-1}}{n!} e^{-\lambda y} b_{(n)}(y)\, dy \tag{5.168}$$

We may go further and calculate $G(y)$, the distribution of the busy period, by integrating in Eq. (5.168) only up to some point y (rather than ∞) and then summing over all possible numbers served in the busy period, that is,

$$G(y) = \int_0^y \sum_{n=1}^{\infty} P[N_{\text{bp}} = n \mid Y = x] b_{(n)}(x)\, dx$$

and so,

$$G(y) = \int_0^y \sum_{n=1}^{\infty} e^{-\lambda x} \frac{(\lambda x)^{n-1}}{n!} b_{(n)}(x)\, dx \tag{5.169}$$

Thus Eq. (5.169) is an explicit expression in terms of known quantities for the distribution of the busy period and in fact may be used in place of the expression given in Eq. (5.137), the Laplace transform of $dG(y)/dy$. This is the expression we had promised earlier, although we have expressed it as an infinite summation; nevertheless, it does provide the ability to approximate the busy-period distribution numerically in any given situation. Similarly, Eq. (5.168) gives an explicit expression for the number served in the busy period.

The reader may have observed that our study of the busy period has really been the study of a transient phenomenon and this is one of the reasons that the development bogged down. In the next section we consider certain aspects of the transient solution for M/G/1 a bit further.

5.12. THE TAKACS INTEGRODIFFERENTIAL EQUATION

In this section we take a closer look at the unfinished work and derive the forward Kolmogorov equation for its time-dependent behavior. A moment's reflection will reveal the fact that the unfinished work $U(t)$ is a continuous-time continuous-state Markov process that is subject to discontinuous changes. It is a Markov process since the entire past history of its motion is summarized in its current value as far as its future behavior is concerned. That is, its vertical discontinuities occur at instants of customer arrivals and for M/G/1 these arrivals form a Poisson process (therefore, we need not know how long it has been since the last arrival), and the current value for $U(t)$ tells us exactly how much work remains in the system at each instant.

We wish to derive the probability distribution function for $U(t)$, given its initial value at time $t = 0$. Accordingly we define

$$F(w, t; w_0) \triangleq P[U(t) \leq w \mid U(0) = w_0] \tag{5.170}$$

This notation is a bit cumbersome and so we choose to suppress the initial value of the unfinished work and use the shorthand notation $F(w, t) \triangleq F(w, t; w_0)$ with the understanding that the initial value is w_0. We wish to relate the probability $F(w, t + \Delta t)$ to its possible values at time t. We observe that we can reach this state from t if, on the one hand, there had been no arrivals during this increment in time [which occurs with probability $1 - \lambda \Delta t + o(\Delta t)$] and the unfinished work was no larger than $w + \Delta t$ at time t; or if, on the other hand, there had been an arrival in this interval [with probability $\lambda \Delta t + o(\Delta t)$] such that the unfinished work at time t, plus the new increment of work brought in by this customer, together do not exceed w. These observations lead us to the following equation:

$$\begin{aligned} F(w, t + \Delta t) &\\ = (1 - \lambda \Delta t)F(w + \Delta t, t) &+ \lambda \Delta t \int_{x=0}^{w} B(w - x) \frac{\partial F(x, t)}{\partial x} dx + o(\Delta t) \end{aligned} \tag{5.171}$$

Clearly, $(\partial F(x, t)/\partial x)\, dx \triangleq d_x F(x, t)$ is the probability that at time t we have $x < U(t) \leq x + dx$. Expanding our distribution function on its first variable we have

$$F(w + \Delta t, t) = F(w, t) + \frac{\partial F(w, t)}{\partial w} \Delta t + o(\Delta t)$$

Using this expansion for the first term on the right-hand side of Eq. (5.171) we obtain

$$\begin{aligned} F(w, t + \Delta t) = F(w, t) + \frac{\partial F(w, t)}{\partial w} \Delta t - \lambda \Delta t \left[F(w, t) + \frac{\partial F(w, t)}{\partial w} \Delta t \right] \\ + \lambda \Delta t \int_{x=0}^{w} B(w - x)\, d_x F(x, t) + o(\Delta t) \end{aligned}$$

Subtracting $F(w, t)$, dividing by Δt, and passing to the limit as $\Delta t \to 0$ we finally obtain the *Takács integrodifferential equation* for $U(t)$:

$$\frac{\partial F(w, t)}{\partial t} = \frac{\partial F(w, t)}{\partial w} - \lambda F(w, t) + \lambda \int_{x=0}^{w} B(w - x)\, d_x F(x, t) \quad \blacksquare \tag{5.172}$$

Takács [TAKA 55] derived this equation for the more general case of a nonhomogeneous Poisson process, namely, where the arrival rate $\lambda(t)$ depends upon t. He showed that this equation is good for *almost* all $w \geq 0$ and $t \geq 0$; it does *not* hold at those w and t for which $\partial F(w, t)/\partial w$ has an accumulation of probability (namely, an impulse). This occurs, in particular, at $w = 0$ and would give rise to the term $F(0, t)u_0(w)$ in $\partial F(w, t)/\partial w$, whereas no other term in the equation contains such an impulse.

We may gain more information from the Takács integrodifferential equation if we transform it on the variable w (and not on t); thus using the transform variable r we define

$$W^{*\cdot}(r, t) \triangleq \int_{0^-}^{\infty} e^{-rw}\, dF_w(w, t) \tag{5.173}$$

We use the notation $(*\cdot)$ to denote transformation on the first, but not the second argument. The symbol W is chosen since, as we shall see, $\lim W^{*\cdot}(r, t) = W^*(r)$ as $t \to \infty$, which is our former transform for the waiting-time pdf [see, for example, Eq. (5.103)].

Let us examine the transform of each term in Eq. (5.172) separately. First we note that since $F(w, t) = \int_{-\infty}^{w} d_x F(x, t)$, then from entry 13 in Table I.3 of Appendix I (and its footnote) we must have

$$\int_{0^-}^{\infty} F(w, t) e^{-rw}\, dw = \frac{W^{*\cdot}(r, t) + F(0^-, t)}{r}$$

and, similarly, we have

$$\int_{0^-}^{\infty} B(w) e^{-rw}\, dw = \frac{B^*(r) + B(0^-)}{r}$$

However, since the unfinished work and the service time are both nonnegative random variables, it must be that $F(0^-, t) = B(0^-) = 0$ always. We recognize that the last term in the Takács integrodifferential equation is a convolution between $B(w)$ and $\partial F(w, t)/\partial w$, and therefore the transform of this convolution (including the constant multiplier λ) must be (by properties 10 and 13 in that same table) $\lambda W^{*\cdot}(r, t)[B^*(r) - B(0^-)]/r = \lambda W^{*\cdot}(r, t)B^*(r)/r$. Now it is clear that the transform for the term $\partial F(w, t)/\partial w$ will be $W^{*\cdot}(r, t)$; but this transform includes $F(0^+, t)$, the transform of the impulse located at the origin for this partial derivative, and since we know that the Takács integrodifferential equation does not contain that impulse it must be subtracted out. Thus, we have from Eq. (5.172),

$$\left(\frac{1}{r}\right)\frac{\partial W^{*\cdot}(r, t)}{\partial t} = W^{*\cdot}(r, t) - F(0^+, t) - \frac{\lambda W^{*\cdot}(r, t)}{r} + \lambda \frac{W^{*\cdot}(r, t)B^*(r)}{r} \tag{5.174}$$

which may be rewritten as

$$\frac{\partial W^{*\cdot}(r, t)}{\partial t} = [r - \lambda + \lambda B^*(r)]W^{*\cdot}(r, t) - rF(0^+, t) \tag{5.175}$$

Takács gives the solution to this equation {p. 51, Eq. (8) in [TAKA 62*b*]}.

We may now transform on our second variable t by first defining the double transform

$$F^{**}(r, s) \triangleq \int_0^\infty e^{-st} W^{*\cdot}(r, t)\, dt \tag{5.176}$$

We also need the definition

$$F_0^*(s) \triangleq \int_0^\infty e^{-st} F(0^+, t)\, dt \tag{5.177}$$

We may now transform Eq. (5.175) using the transform property given as entry 11 in Table I.3 (and its footnote) to obtain

$$sF^{**}(r, s) - W^{*\cdot}(r, 0^-) = [r - \lambda + \lambda B^*(r)]F^{**}(r, s) - rF_0^*(s)$$

From this we obtain

$$F^{**}(r, s) = \frac{W^{*\cdot}(r, 0^-) - rF_0^*(s)}{s - r + \lambda - \lambda B^*(r)} \tag{5.178}$$

The unknown function $F_0^*(s)$ may be determined by insisting that the transform $F^{**}(r, s)$ be analytic in the region Re $(s) > 0$, Re $(r) > 0$. This implies that the zeroes of the numerator and denominator must coincide in this region; Beneš [BENE 56] has shown that in this region $\eta = \eta(s)$ is the unique root of the denominator in Eq. (5.178). Thus $W^{*\cdot}(\eta, 0^-) = \eta F_0^*(s)$ and so (writing 0^- as 0), we have

$$F^{**}(r, s) = \frac{W^{*\cdot}(r, 0) - (r/\eta)W^{*\cdot}(\eta, 0)}{s - r + \lambda - \lambda B^*(r)} \tag{5.179}$$

Now we recall that $U(0) = w_0$ with probability one, and so from Eq. (5.173) we have $W^{*\cdot}(r, 0) = e^{-rw_0}$. Thus $F^{**}(r, s)$ takes the final form

$$F^{**}(r, s) = \frac{(r/\eta)e^{-\eta w_0} - e^{-rw_0}}{\lambda B^*(r) - \lambda + r - s} \tag{5.180}$$

We will return to this equation later in Chapter 2, Volume II, when we discuss the diffusion approximation.

For now it behooves us to investigate the steady-state value of these functions; in particular, it can be shown that $F(w, t)$ has a limit as $t \to \infty$ so long as $\rho < 1$, and this limit will be independent of the initial condition

$F(0, w)$; we denote this limit by $F(w) = \lim F(w, t)$ as $t \to \infty$, and from Eq. (5.172) we find that it must satisfy the following equation:

$$\frac{dF(w)}{dw} = \lambda F(w) - \lambda \int_{x=0}^{w} B(w - x)\, dF(x) \tag{5.181}$$

Furthermore, for $\rho < 1$ then $W^*(r) \triangleq \lim W^*(r, t)$ as $t \to \infty$ will exist and be independent of the initial distribution. Taking the transform of Eq. (5.181) we find as we did in deriving Eq. (5.174)

$$W^*(r) - F(0^+) = \frac{\lambda W^*(r)}{r} - \frac{\lambda B^*(r) W^*(r)}{r}$$

where $F(0^+) = \lim F(0^+, t)$ as $t \to \infty$ and equals the probability that the unfinished work is zero. This last may be rewritten to give

$$W^*(r) = \frac{rF(0^+)}{r - \lambda + \lambda B^*(r)}$$

However, we require $W^*(0) = 1$, which requires that the unknown constant $F(0^+)$ have a value $F(0^+) = 1 - \rho$. Finally we have

$$W^*(r) = \frac{r(1 - \rho)}{r - \lambda + \lambda B^*(r)} \tag{5.182}$$

which is exactly the Pollaczek–Khinchin transform equation for waiting time as we promised!

This completes our discussion of the system M/G/1 (for the time being). Next we consider the "companion" system, G/M/m.

REFERENCES

BENE 56 Beneš, V. E., "On Queues with Poisson Arrivals," *Annals of Mathematical Statistics*, **28,** 670–677 (1956).

CONW 67 Conway, R. W., W. L. Maxwell, and L. W. Miller, *Theory of Scheduling*, Addison-Wesley (Reading, Mass.) 1967.

COX 55 Cox, D. R., "The Analysis of Non-Markovian Stochastic Processes by the Inclusion of Supplementary Variables," *Proc. Camb. Phil. Soc. (Math. and Phys. Sci.)*, **51,** 433–441 (1955).

COX 62 Cox, D. R., *Renewal Theory*, Methuen (London) 1962.

FELL 66 Feller, W., *Probability Theory and its Applications* Vol. II, Wiley (New York), 1966.

GAVE 59 Gaver, D. P., Jr., "Imbedded Markov Chain Analysis of a Waiting-Line Process in Continuous Time," *Annals of Mathematical Statistics* **30,** 698–720 (1959).

HEND 72 Henderson, W., "Alternative Approaches to the Analysis of the M/G/1 and G/M/1 Queues," *Operations Research*, **15,** 92–101 (1972).

KEIL 65 Keilson, J., "The Role of Green's Functions in Congestion Theory," *Proc. Symposium on Congestion Theory*, Univ. of North Carolina Press, 43–71 (1965).

KEND 51 Kendall, D. G., "Some Problems in the Theory of Queues," *Journal of the Royal Statistical Society, Ser. B*, **13,** 151–185 (1951).

KEND 53 Kendall, D. G., "Stochastic Processes Occurring in the Theory of Queues and their Analysis by the Method of the Imbedded Markov Chain," *Annals of Mathematical Statistics*, **24,** 338–354 (1953).

KHIN 32 Khinchin, A. Y., "Mathematical Theory of Stationary Queues," *Mat. Sbornik*, **39,** 73–84 (1932).

LIND 52 Lindley, D. V., "The Theory of Queues with a Single Server," *Proc. Cambridge Philosophical Society*, **48,** 277–289 (1952).

PALM 43 Palm, C., "Intensitatschwankungen im Fernsprechverkehr," *Ericsson Technics*, **6,** 1–189 (1943).

POLL 30 Pollaczek, F., "Über eine Aufgabe dev Wahrscheinlichkeitstheorie," I-II *Math. Zeitschrift.*, **32,** 64–100, 729–750 (1930).

PRAB 65 Prabhu, N. U., *Queues and Inventories*, Wiley (New York) 1965.

RIOR 62 Riordan, J., *Stochastic Service Systems*, Wiley (New York) 1962.

SMIT 58 Smith, W. L., "Renewal Theory and its Ramifications," *Journal of the Royal Statistical Society, Ser. B*, **20,** 243–302 (1958).

TAKA 55 Takács, L., "Investigation of Waiting Time Problems by Reduction to Markov Processes," *Acta Math Acad. Sci. Hung.*, **6,** 101–129 (1955).

TAKA 62a Takács, L., *Introduction to the Theory of Queues*, Oxford University Press (New York) 1962.

TAKA 62b Takács, L., "A Single-Server Queue with Poisson Input," *Operations Research*, **10,** 388–397 (1962).

TAKA 67 Takács, L., *Combinatorial Methods in the Theory of Stochastic Processes*, Wiley (New York) 1967.

EXERCISES

5.1. Prove Eq. (5.14) from Eq. (5.11).

5.2. Here we derive the residual lifetime density $\hat{f}(x)$ discussed in Section 5.2. We use the notation of Figure 5.1.

(a) Observing that the event $\{Y \le y\}$ can occur if and only if $t < \tau_k \le t + y < \tau_{k+1}$ for some k, show that

$$\hat{F}_t(y) \triangleq P[Y \le y \mid t]$$
$$= \sum_{k=1}^{\infty} \int_t^{t+y} [1 - F(t + y - x)]\, dP[\tau_k \le x]$$

(b) Observing that $\tau_k \le x$ if and only if $\alpha(x)$, the number of "arrivals" in $(0, x)$, is at least k, that is, $P[\tau_k \le x] = P[\alpha(x) \ge k]$, show that

$$\sum_{k=1}^{\infty} P[\tau_k \le x] = \sum_{k=1}^{\infty} kP[\alpha(x) = k]$$

(c) For large x, the mean-value expression in (b) is x/m_1. Let $\hat{F}(y) = \lim \hat{F}_t(y)$ as $t \to \infty$ with corresponding pdf $\hat{f}(y)$. Show that we now have

$$\hat{f}(y) = \frac{1 - F(y)}{m_1}$$

5.3. Let us rederive the P–K mean-value formula (5.72).

(a) Recognizing that a new arrival is delayed by one service time for each queued customer plus the residual service time of the customer in service, write an expression for W in terms of $\bar{N}_q$, ρ, $\bar{x}$, σ_b^2 and $P[\tilde{w} > 0]$.

(b) Use Little's result in (a) to obtain Eq. (5.72).

5.4. Replace $1 - \rho$ in Eq. (5.85) by an unknown constant and show that $Q(1) = V(1) = 1$ easily gives us the correct value of $1 - \rho$ for this constant.

5.5. **(a)** From Eq. (5.86) form $Q^{(1)}(1)$ and show that it gives the expression for $\bar{q}$ in Eq. (5.63). Note that L'Hospital's rule will be required twice to remove the indeterminacies in the expression for $Q^{(1)}(1)$.

(b) From Eq. (5.105), find the first two moments of the waiting time and compare with Eqs. (5.113) and (5.114).

5.6. We wish to prove that the limiting probability r_k for the number of customers found by an arrival is equal to the limiting probability d_k for the number of customers left behind by a departure, in any queueing system in which the state changes by unit step values only (positive or negative). Beginning at $t = 0$, let x_n be those instants when $N(t)$ (the number in system) increases by one and y_n be those instants when $N(t)$ decreases by unity, $n = 1, 2, \ldots$. Let $N(x_n^-)$ be denoted by α_n and $N(y_n^+)$ by β_n. Let $N(0) = i$.

(a) Show that if $\beta_{n+1} \le k$, then $\alpha_{n+k+1} \le k$.

(b) Show that if $\alpha_{n+k+1} \le k$, then $\beta_{n+i} \le k$.

(c) Show that (a) and (b) must therefore give, for any k,

$$\lim_{n\to\infty} P[\beta_n \le k] = \lim_{n\to\infty} P[\alpha_n \le k]$$

which establishes that $r_k = d_k$.

5.7. In this exercise, we explore the method of supplementary variables as applied to the M/G/1 queue. As usual, let $P_k(t) = P[N(t) = k]$. Moreover, let $P_k(t, x_0)\,dx_0 = P[N(t) = k,\ \ x_0 < X_0(t) \le x_0 + dx_0]$ where $X_0(t)$ is the service already received by the customer in service at time t.

(a) Show that

$$\frac{\partial P_0(t)}{\partial t} = -\lambda P_0(t) + \int_0^\infty P_1(t, x_0) r(x_0)\, dx_0$$

where

$$r(x_0) = \frac{b(x_0)}{1 - B(x_0)}$$

(b) Let $p_k = \lim P_k(t)$ as $t \to \infty$ and $p_k(x_0) = \lim P_k(t, x_0)$ as $t \to \infty$. From (a) we have the equilibrium result

$$\lambda p_0 = \int_0^\infty p_1(x_0) r(x_0)\, dx_0$$

Show the following equilibrium results [where $p_0(x_0) \triangleq 0$]:

(i) $$\frac{\partial p_k(x_0)}{\partial x_0} = -[\lambda + r(x_0)] p_k(x_0) + \lambda p_{k-1}(x_0) \qquad k \ge 1$$

(ii) $$p_k(0) = \int_0^\infty p_{k+1}(x_0) r(x_0)\, dx_0 \qquad k > 1$$

(iii) $$p_1(0) = \int_0^\infty p_2(x_0) r(x_0)\, dx_0 + \lambda p_0$$

(c) The four equations in (b) determine the equilibrium probabilities when combined with an appropriate normalization equation. In terms of p_0 and $p_k(x_0)$ $(k = 1, 2, \ldots)$ give this normalization equation.

(d) Let $R(z, x_0) = \sum_{k=1}^{\infty} p_k(x_0) z^k$. Show that

$$\frac{\partial R(z, x_0)}{\partial x_0} = [\lambda z - \lambda - r(x_0)] R(z, x_0)$$

and

$$zR(z, 0) = \int_0^\infty r(x_0) R(z, x_0)\, dx_0 + \lambda z(z - 1) p_0$$

(e) Show that the solution for $R(z, x_0)$ from (d) must be

$$R(z, x_0) = R(z, 0) e^{-\lambda x_0 (1-z) - \int_0^{x_0} r(y)\, dy}$$

$$R(z, 0) = \frac{\lambda z(z - 1) p_0}{z - B^*(\lambda - \lambda z)}$$

(f) Defining $R(z) \triangleq \int_0^\infty R(z, x_0)\, dx_0$, show that

$$R(z) = R(z, 0)\,\frac{1 - B^*(\lambda - \lambda z)}{\lambda(1 - z)}$$

(g) From the normalization equation of (c), now show that

$$p_0 = 1 - \rho \qquad (\rho = \lambda\bar{x})$$

(h) Consistent with Eq. (5.78) we now define

$$Q(z) = p_0 + R(z)$$

Show that $Q(z)$ expressed this way is identical to the P–K transform equation (5.86). (See [COX 55] for additional details of this method.)

5.8. Consider the M/G/∞ queue in which each customer always finds a free server; thus

$$s(y) = b(y) \text{ and } T = \bar{x}. \text{ Let } P_k(t) = P[N(t) = k]$$

and assume $P_0(0) = 1$.

(a) Show that

$$P_k(t) = \sum_{n=k}^{\infty} e^{-\lambda t}\frac{(\lambda t)^n}{n!}\binom{n}{k}\left[\frac{1}{t}\int_0^t [1 - B(x)]\, dx\right]^k\left[\frac{1}{t}\int_0^t B(x)\, dx\right]^{n-k}$$

[HINT: $(1/t)\int_0^t B(x)\, dx$ is the probability that a customer's service terminates by time t, given that his arrival time was uniformly distributed over the interval $(0, t)$. See Eq. (2.137) also.]

(b) Show that $p_k \triangleq \lim P_k(t)$ as $t \to \infty$ is

$$p_k = \frac{(\lambda\bar{x})^k}{k!}\, e^{-\lambda\bar{x}}$$

regardless of the form of $B(x)$!

5.9. Consider M/E$_2$/1.

(a) Find the polynomial for $G^*(s)$.

(b) Solve for $S(y) = P[\text{time in system} \le y]$.

5.10. Consider an M/D/1 system for which $\bar{x} = 2$ sec.

(a) Show that the residual service time pdf $\hat{b}(x)$ is a rectangular distribution.

(b) For $\rho = 0.25$, show that the result of Eq. (5.111) with four terms may be used as a good approximation to the distribution of queueing time.

5.11. Consider an M/G/1 queue in which bulk arrivals occur at rate λ and with a probability g_r that r customers arrive together at an arrival instant.

(a) Show that the z-transform of the number of customers arriving in an interval of length t is $e^{-\lambda t[1-G(z)]}$ where $G(z) = \sum g_r z^r$.

(b) Show that the z-transform of the random variables v_n, the number of arrivals during the service of a customer, is $B^*[\lambda - \lambda G(z)]$.

5.12. Consider the M/G/1 bulk arrival system in the previous problem. Using the method of imbedded Markov chains:

(a) Find the expected queue size. [HINT: show that $\bar{v} = \rho$ and

$$\overline{v^2} - \bar{v} = \left.\frac{d^2V(z)}{dz^2}\right|_{z=1} = \rho^2(C_b^{\ 2} + 1) + \frac{\lambda}{\mu}\left(C_g^{\ 2} + 1 - \frac{1}{\bar{g}}\right)(\bar{g})^2$$

where C_g is the coefficient of variation of the bulk group size and $\bar{g}$ is the mean group size.]

(b) Show that the generating function for queue size is

$$Q(z) = \frac{(1 - \rho)(1 - z)B^*[\lambda - \lambda G(z)]}{B^*[\lambda - \lambda G(z)] - z}$$

Using Little's result, find the ratio $W/\bar{x}$ of the expected wait on queue to the average service time.

(c) Using the same method (imbedded Markov chain) find the expected number of groups in the queue (averaged over departure times). [HINTS: Show that $D(z) = \beta^*(\lambda - \lambda z)$, where $D(z)$ is the generating function for the number of *groups* arriving during the service time for an entire *group* and where $\beta^*(s)$ is the Laplace transform of the service-time density for an entire group. Also note that $\beta^*(s) = G[B^*(s)]$, which allows us to show that $\overline{\tau^2} = (\bar{x})^2(\overline{g^2} - \bar{g}) + \overline{x^2}\bar{g}$, where $\overline{\tau^2}$ is the second moment of the *group* service time.]

(d) Using Little's result, find W_g, the expected wait on queue for a group (measured from the arrival time of the group until the start of service of the *first* member of the group) and show that

$$W_g = \frac{\rho\bar{x}\bar{g}}{2(1 - \rho)}\left[1 + \frac{C_b^{\ 2}}{\bar{g}} + C_g^{\ 2}\right]$$

(e) If the customers within a group arriving together are served in random order, show that the ratio of the mean waiting time for a single customer to the average service time for a single customer is $W_g/\bar{x}$ from (d) increased by $(1/2)\bar{g}(1 + C_g^{\ 2}) - 1/2$.

5.13. Consider an M/G/1 system in which service is instantaneous but is only available at "service instants," the intervals between successive service instants being independently distributed with PDF $F(x)$. The maximum number of customers that can be served at any service instant is m. Note that this is a bulk service system.

(a) Show that if q_n is the number of customers in the system just before the nth service instant, then

$$q_{n+1} = \begin{cases} q_n + v_n - m & q_n \geq m \\ v_n & q_n < m \end{cases}$$

where v_n is the number of arrivals in the interval between the nth and $(n + 1)$th service instants.

(b) Prove that the probability generating function of v_n is $F^*(\lambda - \lambda z)$. Hence show that $Q(z)$ is

$$Q(z) = \frac{\sum_{k=0}^{m-1} p_k(z^m - z^k)}{z^m [F^*(\lambda - \lambda z)]^{-1} - 1}$$

where $p_k = p[\tilde{q} = k]$ $(k = 0, \ldots, m - 1)$.

(c) The $\{p_k\}$ can be determined from the condition that within the unit disk of the z-plane, the numerator must vanish when the denominator does. Hence show that if $F(x) = 1 - e^{-\mu x}$,

$$Q(z) = \frac{z_m - 1}{z_m - z}$$

where z_m is the zero of $z^m[1 + \lambda(1 - z)/\mu] - 1$ outside the unit disk.

5.14. Consider an M/G/1 system with bulk service. Whenever the server becomes free, he accepts *two* customers from the queue into service simultaneously, or, if only one is on queue, he accepts that one; in either case, the service time for the group (of size 1 or 2) is taken from $B(x)$. Let q_n be the number of customers remaining after the nth service instant. Let v_n be the number of arrivals during the nth service. Define $B^*(s)$, $Q(z)$, and $V(z)$ as transforms associated with the random variables $\tilde{x}$, $\tilde{q}$, and $\tilde{v}$ as usual. Let $\rho = \lambda\bar{x}/2$.

(a) Using the method of imbedded Markov chains, find

$$E(\tilde{q}) = \lim_{n\to\infty} E(q_n)$$

in terms of ρ, σ_b^2, and $P(\tilde{q} = 0) \triangleq p_0$.

(b) Find $Q(z)$ in terms of $B^*(\cdot)$, p_0, and $p_1 \triangleq P(\tilde{q} = 1)$.

(c) Express p_1 in terms of p_0.

5.15. Consider an M/G/1 queueing system with the following variation. The server refuses to serve any customers unless at least two customers are ready for service, at which time both are "taken into" service. These two customers are served individually and independently, one after the other. The instant at which the second of these two is finished is called a "critical" time and we shall use these critical times as the points in an imbedded Markov chain. Immediately following a critical time, if there are two more ready for service, they are both "taken into" service as above. If one or none are ready, then the server waits until a pair is ready, and so on. Let

q_n = number of customers left behind in the system immediately following the nth critical time

v_n = number of customers arriving during the combined service time of the nth *pair* of customers

(a) Derive a relationship between q_{n+1}, q_n, and v_{n+1}.

(b) Find

$$V(z) = \sum_{k=0}^{\infty} P[v_n = k]z^k$$

(c) Derive an expression for $Q(z) = \lim Q_n(z)$ as $n \to \infty$ in terms of $p_0 = P[\tilde{q} = 0]$, where

$$Q_n(z) = \sum_{k=0}^{\infty} P[q_n = k]z^k$$

(d) How would you solve for p_0?

(e) Describe (do *not* calculate) two methods for finding $\bar{q}$.

5.16. Consider an M/G/1 queueing system in which service is given as follows. Upon entry into service, a coin is tossed, which has probability p of giving Heads. If the result is Heads, then the service time for that customer is zero seconds. If Tails, his service time is drawn from the following exponential distribution:

$$pe^{-px} \qquad x \geq 0$$

(a) Find the average service time $\bar{x}$.

(b) Find the variance of service time σ_b^2.

(c) Find the expected waiting time W.

(d) Find $W^*(s)$.

(e) From (d), find the expected waiting time W.

(f) From (d), find $W(t) = P[\text{waiting time} \leq t]$.

5.17. Consider an M/G/1 queue. Let E be the event that T sec have elapsed since the arrival of the last customer. We begin at a random time and measure the time w until event E next occurs. This measurement may involve the observation of many customer arrivals before E occurs.

(a) Let $\hat{A}(t)$ be the interarrival-time distribution for those intervals during which E does *not* occur. Find $\hat{A}(t)$.

(b) Find $\hat{A}^*(s) = \int_0^\infty e^{-st}\, d\hat{A}(t)$.

(c) Find $W^*(s \mid n) = \int_0^\infty e^{-sw}\, dW(w \mid n)$, where $W(w \mid n) = P$[time to event $E \le w \mid n$ arrivals occur before E].

(d) Find $W^*(s) = \int_0^\infty e^{-sw}\, dW(w)$, where $W(w) = P$[time to event $E \le w$].

(e) Find the mean time to event E.

5.18. Consider an M/G/1 system in which time is divided into intervals of length q sec each. Assume that *arrivals* are Bernoulli, that is,

$$P[\text{1 arrival in any interval}] = \lambda q$$
$$P[\text{0 arrivals in any interval}] = 1 - \lambda q$$
$$P[> \text{1 arrival in any interval}] = 0$$

Assume that a customer's *service time* $\tilde{x}$ is some multiple of q sec such that

$$P[\text{service time} = nq \text{ sec}] = g_n \qquad n = 0, 1, 2, \ldots$$

(a) Find E[number of arrivals in an interval].

(b) Find the average arrival rate.

(c) Express $E[\tilde{x}] \triangleq \bar{x}$ and $E[\tilde{x}(\tilde{x} - q)] \triangleq \overline{x^2} - \bar{x}q$ in terms of the moments of the g_n distribution (i.e., let $\overline{g^k} \triangleq \sum_{n=0}^\infty n^k g_n$).

(d) Find $y_{mn} = P[m$ customers arrive in nq sec].

(e) Let $v_m = P[m$ customers arrive during the service of a customer] and let

$$V(z) = \sum_{m=0}^\infty v_m z^m \qquad \text{and} \qquad G(z) = \sum_{m=0}^\infty g_m z^m$$

Express $V(z)$ in terms of $G(z)$ and the system parameters λ and q.

(f) Find the mean number of arrivals during a customer service time from (*e*).

5.19. Suppose that in an M/G/1 queueing system the *cost* of making a customer wait t sec is $c(t)$ dollars, where $c(t) = \alpha e^{\beta t}$. Find the average cost of queueing for a customer. Also determine the conditions necessary to keep the average cost finite.

5.20. We wish to find the *interdeparture* time probability density function $d(t)$ for an M/G/1 queueing system.

(a) Find the Laplace transform $D^*(s)$ of this density conditioned first on a nonempty queue left behind, and second on an empty queue left behind by a departing customer. Combine these results to get the Laplace transform of the interdeparture time density and from this find the density itself.

(b) Give an explicit form for the probability distribution $D(t)$, or density $d(t) = dD(t)/dt$, of the interdeparture time when we have a constant service time, that is

$$B(x) = \begin{cases} 0 & x < T \\ 1 & x \geq T \end{cases}$$

5.21. Consider the following modified order of service for M/G/1. Instead of LCFS as in Figure 5.11, assume that after the interval x_1, the sub-busy period generated by C_2 occurs, which is followed by the sub-busy period generated by C_3, and so on, until the busy period terminates. Using the sequence of arrivals and service times shown in the upper contour of Figure 5.11*a*, redraw parts *a*, *b*, and *c* to correspond to the above order of service.

5.22. Consider an M/G/1 system in which a departing customer immediately joins the queue again with probability p, or departs forever with probability $q = 1 - p$. Service is FCFS, and the service time for a returning customer is independent of his previous service times. Let $B^*(s)$ be the transform for the service time pdf and let $B_T^*(s)$ be the transform for a customer's *total* service time pdf.

(a) Find $B_T^*(s)$ in terms of $B^*(s)$, p, and q.

(b) Let $\overline{x_T^n}$ be the nth moment of the total service time. Find $\overline{x_T^1}$ and $\overline{x_T^2}$ in terms of $\bar{x}$, $\overline{x^2}$, p, and q.

(c) Show that the following recurrence formula holds:

$$\overline{x_T^n} = \overline{x^n} + \frac{p}{q}\sum_{k=1}^{n}\binom{n}{k}\overline{x^k}\,\overline{x_T^{n-k}}$$

(d) Let

$$Q_T(z) = \sum_{k=0}^{\infty} p_{kT} z^k$$

where $p_{kT} = P[\text{number in system} = k]$. For $\lambda\bar{x} < q$ prove that

$$Q_T(z) = \left(1 - \frac{\lambda\bar{x}}{q}\right)\frac{q(1-z)B^*[\lambda(1-z)]}{(q+pz)B^*[\lambda(1-z)] - z}$$

(e) Find $\bar{N}$, the average number of customers in the system.

5.23. Consider a first-come–first-served M/G/1 queue with the following changes. The server serves the queue as long as someone is in the system. Whenever the system empties the server goes away on vacation for a certain length of time, which may be a random variable. At the end of his vacation the server returns and begins to serve customers again; if he returns to an empty system then he goes away on vacation again. Let $F(z) = \sum_{j=1}^{\infty} f_j z^j$ be the z-transform for the number of customers awaiting service when the server returns from vacation to find at least one customer waiting (that is, f_j is the probability that at the initiation of a busy period the server finds j customers awaiting service).

(a) Derive an expression which gives q_{n+1} in terms of q_n, v_{n+1}, and j (the number of customer arrivals during the server's vacation).

(b) Derive an expression for $Q(z)$ where $Q(z) = \lim E[z^{q_n}]$ as $n \to \infty$ in terms of p_0 (equal to the probability that a departing customer leaves 0 customers behind). (HINT: condition on j.)

(c) Show that $p_0 = (1 - \rho)/F^{(1)}(1)$ where $F^{(1)}(1) = \partial F(z)/\partial z\big|_{z=1}$ and $\rho = \lambda\bar{x}$.

(d) Assume now that the service vacation will end whenever a new customer enters the empty system. For this case find $F(z)$ and show that when we substitute it back into our answer for (b) then we arrive at the classical M/G/1 solution.

5.24. We recognize that an arriving customer who finds k others in the system is delayed by the remaining service time for the customer in service plus the sum of $(k - 1)$ complete service times.

(a) Using the notation and approach of Exercise 5.7, show that we may express the transform of the waiting time pdf as

$$W^*(s) = p_0 + \int_0^\infty \sum_{k=1}^{\infty} p_k(x_0)[B^*(s)]^{k-1} \times \int_0^\infty e^{-sy}\, r(y + x_0)\, e^{-\int_0^{y+x_0} r(u)\,du}\, dy \times e^{\int_0^{x_0} r(u)\,du}\, dx_0$$

(b) Show that the expression in (a) reduces to $W^*(s)$ as given in Eq. (5.106).

5.25. Let us relate $\overline{s^k}$, the kth moment of the time in system to $\overline{N^k}$, the kth moment of the number in system.

(a) Show that Eq. (5.98) leads directly to Little's result, namely

$$\bar{N} = \lambda\bar{s} \triangleq \lambda T$$

(b) From Eq. (5.98) establish the second-moment relationship

$$\overline{N^2} - \bar{N} = \lambda^2\, \overline{s^2}$$

(c) Prove that the general relationship is

$$\overline{N(N-1)(N-2)\cdots(N-k+1)} = \lambda^k\, \overline{s^k}$$

6

The Queue G/M/m

We have so far studied systems of the type M/M/1 and its variants (elementary queueing theory) and M/G/1 (intermediate queueing theory). The next natural system to study is G/M/1, in which we have an arbitrary interarrival time distribution $A(t)$ and an exponentially distributed service time. It turns out that the m-server system G/M/m is almost as easy to study as is the single-server system G/M/1, and so we proceed directly to the m-server case. This study falls within intermediate queueing theory along with M/G/1, and it too may be solved using the method of the imbedded Markov chain, as elegantly presented by Kendall [KEND 51].

6.1. TRANSITION PROBABILITIES FOR THE IMBEDDED MARKOV CHAIN (G/M/m)

The system under consideration contains m servers, who render service in order of arrival. Customers arrive singly with interarrival times identically and independently distributed according to $A(t)$ and with a mean time between arrivals equal to $1/\lambda$. Service times are distributed exponentially with mean $1/\mu$, the same distribution applying to each server independently. We consider steady-state results only (see discussion below).

As was the case in M/G/1, where the state variable became a continuous variable, so too in the system G/M/m we have a continuous-state variable in which we are required to keep track of *the elapsed time since the last arrival*, as well as the number in system. This is true since the probability of an arrival in any particular time interval depends upon the elapsed time (the "age") since the last arrival. It is possible to proceed with the analysis by considering the two-dimensional state description consisting of the age since the last arrival and the number in system; such a procedure is again referred to as the method of supplementary variables. A second approach, very much like that which we used for M/G/1, is the method of the imbedded Markov chain, which we pursue below. We have already seen a third approach, namely, the method of stages from Chapter 4.

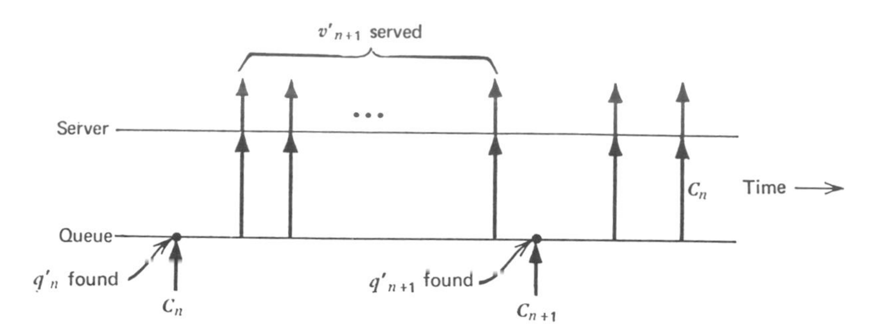

Figure 6.1 The imbedded Markov points.

If we are to use the imbedded Markov chain approach then it must be that the points we select as the regeneration points implicity inform us of the elapsed time since the last arrival in an analogous way as for the expended service time in the case M/G/1. The natural set of points to choose for this purpose is the set of *arrival* instants. It is certainly clear that at these epochs the elapsed time since the last arrival is zero. Let us therefore define

q_n' = number of customers found in the system immediately prior to the arrival of C_n

We use q_n' for this random variable to distinguish it from q_n, the number of customers left behind by the departure of C_n in the M/G/1 system. In Figure 6.1 we show a sequence of arrival times and identify them as critical points imbedded in the time axis. It is clear that the sequence $\{q_n'\}$ forms a discrete-state Markov chain. Defining

v'_{n+1} = the number of customers *served* between the *arrival* of C_n and C_{n+1},

we see immediately that the following fundamental relation must hold:

$$q'_{n+1} = q_n' + 1 - v'_{n+1} \qquad \blacksquare(6.1)$$

We must now calculate the transition probabilities associated with this Markov chain, and so we define

$$p_{ij} = P[q'_{n+1} = j \mid q_n' = i] \qquad (6.2)$$

It is clear that p_{ij} is merely the probability that $i + 1 - j$ customers are served during an interarrival time. It is further clear that

$$p_{ij} = 0 \qquad \text{for} \qquad j > i + 1 \qquad \blacksquare(6.3)$$

since there are at most $i + 1$ present between the arrival of C_n and C_{n+1}. The Markov state-transition-probability diagram has transitions such as shown in Figure 6.2; in this figure we show only the transitions *out* of state E_i.

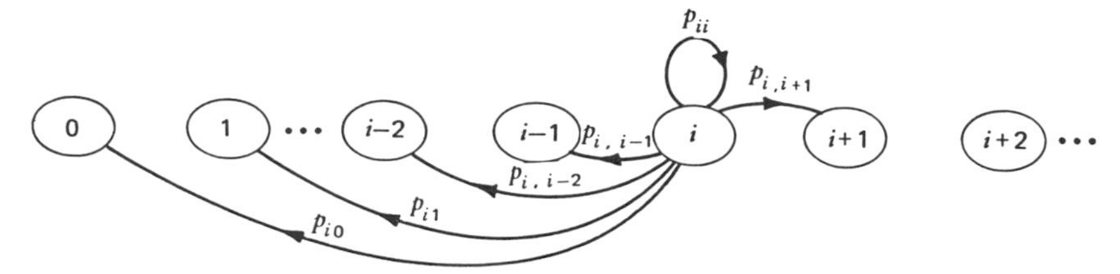

Figure 6.2 State-transition-probability diagram for the G/M/m imbedded Markov chain.

We are concerned with steady-state results only and so we must inquire as to the conditions under which this Markov chain will be ergodic. It may easily be shown that the condition for ergodicity is, as we would expect, $\lambda < m\mu$, where λ is the average arrival rate associated with our input distribution and μ is the parameter associated with our exponential service time (that is, $\bar{x} = 1/\mu$). As defined in Chapter 2 and as used in Section 3.5, we define the utilization factor for this system as

$$\rho \triangleq \frac{\lambda}{m\mu} \tag{6.4}$$

Once again this is the average rate at which work enters the system ($\lambda\bar{x} = \lambda/\mu$ sec of work per elapsed second) divided by the maximum rate at which the system can do work (m sec of work per elapsed second). Thus our condition for ergodicity is simply $\rho < 1$. In the ergodic case we are assured that an equilibrium probability distribution will exist describing the number of customers present at the arrival instants, thus we define

$$r_k = \lim_{n\to\infty} P[q_n' = k] \tag{6.5}$$

and it is this probability distribution we seek for the system G/M/m. As we know from Chapter 2, the direct method of solution for this equilibrium distribution requires that we solve the following system of linear equations:

$$\mathbf{r} = \mathbf{r}\mathbf{P} \tag{6.6}$$

where

$$\mathbf{r} = [r_0, r_1, r_2, \ldots] \tag{6.7}$$

and $\mathbf{P}$ is the matrix whose elements are the one-step transition probabilities p_{ij}.

Our first task then is to find these one-step transition probabilities. We must consider four regions in the i, j plane as shown in Figure 6.3, which gives the case $m = 6$. Regarding the region labeled 1, we already know from Eq. (6.3) that $p_{ij} = 0$ for $i + 1 < j$. Now for region 2 let us consider the range $j \le i + 1 \le m$, which is the case in which no customers are waiting and all present are engaged with their own server. During the interarrival period, we

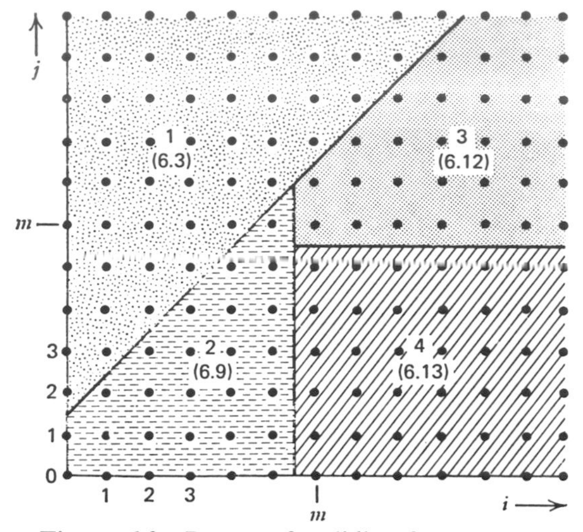

Figure 6.3 Range of validity for p_{ij} equations (equation numbers are also given in parentheses).

see that $i + 1 - j$ customers will complete their service. Since service times are exponentially distributed, the probability that any given customer will depart within t sec after the arrival of C_n is given by $1 - e^{-\mu t}$; similarly the probability that a given customer will not depart by this time is $e^{-\mu t}$. Therefore, in this region we have

$$P[i + 1 - j \text{ departures within } t \text{ sec after } C_n \text{ arrives} \mid q_n' = i]$$
$$= \binom{i+1}{i+1-j}[1 - e^{-\mu t}]^{i+1-j}[e^{-\mu t}]^j \quad (6.8)$$

where the binomial coefficient

$$\binom{i+1}{i+1-j} = \binom{i+1}{j}$$

merely counts the number of ways in which we can choose the $i + 1 - j$ customers to depart out of the $i + 1$ that are available in the system. With t_{n+1} as the interarrival time between C_n and C_{n+1}, Eq. (6.8) gives $P[q'_{n+1} = j \mid q_n' = i, t_{n+1} = t]$. Removing the condition on t_{n+1} we then have the one-step transition probability in this range, namely,

$$p_{ij} = \int_0^\infty \binom{i+1}{j}[1 - e^{-\mu t}]^{i+1-j} e^{-\mu t j}\, dA(t) \qquad j \le i + 1 \le m \quad (6.9)$$

Next consider the range $m \le j \le i + 1$, $i \ge m$ (region 3),* which corresponds to the simple case in which all m servers are busy throughout the

* The point $i = m - 1, j = m$ can properly lie either in region 2 or region 3.

interarrival interval. Under this assumption (that all m servers remain busy), since each service time is exponentially distributed (memoryless), then the number of customers served during this interval will be Poisson distributed (in fact it is a pure Poisson death process) with parameter $m\mu$; that is defining "t, all m busy" as the event that $t_{n+1} = t$ and all m servers remain busy during t_{n+1}, we have

$$P[k \text{ customers served} \mid t, \text{ all } m \text{ busy}] = \frac{(m\mu t)^k}{k!} e^{-m\mu t}$$

As pointed out earlier, if we are to go from state i to state j, then exactly $i + 1 - j$ customers must have been served during the interarrival time; taking account of this and removing the condition on t_{n+1} we have

$$p_{ij} = \int_{t=0}^{\infty} P[i + 1 - j \text{ served} \mid t, \text{ all } m \text{ busy}] \, dA(t)$$

or

$$p_{ij} = \int_{t=0}^{\infty} \frac{(m\mu t)^{i+1-j}}{(i + 1 - j)!} e^{-m\mu t} \, dA(t) \qquad m \le j \le i + 1 \tag{6.10}$$

Note that in Eq. (6.10) the indices i and j appear only as the *difference* $i + 1 - j$, and so it behooves us to define a new quantity with a single index

$$\beta_{i+1-j} \triangleq p_{ij} \qquad m \le j \le i + 1, m \le i \tag{6.11}$$

where β_n = the probability of serving n customers during an interarrival time given that all m servers remain busy during this interval; thus, with $n = i + 1 - j$, we have

$$\beta_n = p_{i,i+1-n} = \int_{t=0}^{\infty} \frac{(m\mu t)^n}{n!} e^{-m\mu t} \, dA(t) \qquad 0 \le n \le i + 1 \quad m, m \le i \tag{6.12}$$

The last case we must consider (region 4) is $j < m < i + 1$, which describes the situation where C_n arrives to find m customers in service and $i - m$ waiting in queue (which he joins); upon the arrival of C_{n+1} there are exactly j customers, all of whom are in service. If we assume that it requires y sec until the queue empties then one may calculate p_{ij} in a straightforward manner to yield (see Exercise 6.1)

$$p_{ij} = \int_0^{\infty} \binom{m}{j} e^{-j\mu t} \times \left[\int_0^t \frac{(m\mu y)^{i-m}}{(i - m)!} (e^{-\mu y} - e^{-\mu t})^{m-j} m\mu \, dy \right] dA(t) \qquad j < m < i + 1 \tag{6.13}$$

Thus Eqs. (6.3), (6.9), (6.12), and (6.13) give the complete description of the one-step transition probabilities for the G/M/m system.

Having established the form for our one-step transition probabilities we may place them in the transition matrix

$$\mathbf{P} = \begin{bmatrix} p_{00} & p_{01} & 0 & 0 & & & & & & & \\ p_{10} & p_{11} & p_{12} & 0 & & & \cdot & \cdot & \cdot & & \\ p_{20} & p_{21} & p_{22} & p_{23} & & & & & & & \\ \vdots & & & \vdots & & & & & & & \\ p_{m-2,0} & p_{m-2,1} & & \cdots & p_{m-2,m-1} & 0 & 0 & 0 & \cdots & & \\ p_{m-1,0} & p_{m-1,1} & & \cdots & p_{m-1,m-1} & \beta_0 & 0 & 0 & \cdots & & \\ p_{m,0} & p_{m,1} & & \cdots & p_{m,m-1} & \beta_1 & \beta_0 & 0 & \cdots & & \\ & \vdots & & & & & & & & & \\ p_{m+n,0} & p_{m+n,1} & & \cdots & p_{m+n,m-1} & \beta_{n+1} & \beta_n & \cdots & \beta_0 & 0 & \cdots \\ & \vdots & & & & & & & & \vdots & \end{bmatrix}$$

In this matrix all terms above the upper diagonal are zero, and the terms β_n are given through Eq. (6.12). The "boundary" terms denoted in this matrix by their generic symbol p_{ij} are given either by Eqs. (6.9) or (6.13) according to the range of subscripts i and j. Of most importance to us are the transition probabilities β_n.

6.2. CONDITIONAL DISTRIBUTION OF QUEUE SIZE

Now we are in a position to find the equilibrium probabilities r_k, which must satisfy the system of linear equations given in Eq. (6.6). At this point we perhaps could guess at the form for r_k that satisfies these equations, but rather than that we choose to motivate the results that we obtain by the following intuitive arguments. In order to do this we define

$$N_k(t) = \text{number of arrival instants in the interval } (0, t) \text{ in which the arriving customer finds the system in state } E_k, \text{ given 0 customers at } t = 0 \tag{6.14}$$

Note from Figure 6.2 that the system can move up by at most one state, but may move down by many states in any single transition. We consider this motion between states and define (for $m - 1 \leq k$)

$$\sigma_k = E[\text{number of times state } E_{k+1} \text{ is reached between two successive visits to state } E_k] \tag{6.15}$$

We have that the probability of reaching state E_{k+1} no times between returns to state E_k is equal to $1 - \beta_0$ (that is, given we are in state E_k the only way we can reach state E_{k+1} before our next visit to state E_k is for no customers to be served, which has probability β_0, and so the probability of not getting to E_{k+1} first is $1 - \beta_0$, the probability of serving at least one). Furthermore, let

$$\begin{aligned}\gamma &= P[\text{leave state } E_{k+1} \text{ and return to it some time later without passing through state } E_j, \text{ where } j \leq k] \\ &= P[\text{leave state } E_{k+1} \text{ and return to it later without passing through state } E_k]\end{aligned}$$

This last is true since a visit to state E_j for $j \leq k$ must result in a visit to state E_k before next returning to state E_{k+1} (we move up only *one* state at a time). We note that γ is independent of k so long as $k \geq m - 1$ (i.e., all m servers are busy). We have the simple calculation

$$P[n \text{ occurrences of state } E_{k+1} \text{ between two successive visits to state } E_k] = \gamma^{n-1}(1 - \gamma)\beta_0$$

This last equation is calculated as the probability (β_0) of reaching state E_{k+1} at all, times the probability (γ^{n-1}) of returning to E_{k+1} a total of $n - 1$ times without first touching state E_k, times the probability $(1 - \gamma)$ of then visiting state E_k without first returning to state E_{k+1}. From this we may calculate

$$\sigma_k = \sum_{n=1}^{\infty} n\gamma^{n-1}(1 - \gamma)\beta_0$$

as the average number of visits to E_{k+1} between successive visits to state E_k. Thus

$$\sigma_k = \frac{\beta_0}{1 - \gamma} \qquad \text{for } k \geq m - 1$$

Note that σ_k is *independent* of k and so we may drop the subscript, in which case we have

$$\sigma \overset{\Delta}{=} \sigma_k = \frac{\beta_0}{1 - \gamma} \qquad \text{for } k \geq m - 1 \tag{6.16}$$

From the definition in Eq. (6.15), σ must be the limit of the ratio of the number of times we find ourselves in state E_{k+1} to the number of times we find ourselves in state E_k; thus we may write

$$\sigma = \lim_{t \to \infty} \frac{N_{k+1}(t)}{N_k(t)} = \frac{\beta_0}{1 - \gamma} \qquad k \geq m - 1 \tag{6.17}$$

However, the limit is merely the ratio of the steady-state probability of finding the system in state E_{k+1} to the probability of finding it in state E_k. Consequently, we have established

$$r_{k+1} = \sigma r_k \qquad k \geq m - 1 \tag{6.18}$$

The solution to this last set of equations is clearly

$$r_k = K\sigma^k \qquad k \geq m - 1 \tag{6.19}$$

for some constant K. This is a basic result, which says that the distribution of number of customers found at the arrival instants is *geometric* for the case $k \geq m - 1$. It remains for us to find σ and K, as well as r_k for $k < m - 1$.

Our intuitive reasoning (which may easily be made rigorous by results from renewal theory) has led us to the basic equation (6.19). We could have "pulled this out of a hat" by guessing that the solution to Eq. (6.6) for the probability vector $\mathbf{r} \triangleq [r_0, r_1, r_2, \ldots]$ might perhaps be of the form

$$\mathbf{r} = [r_0, r_1, r_2, \ldots, r_{m-2}, K\sigma^{m-1}, K\sigma^m, K\sigma^{m+1}, \ldots,] \tag{6.20}$$

This flash of brilliance would, of course, have been correct (as our calculations have just shown); once we suspect this result we may easily verify it by considering the kth equation ($k \geq m$) in the set (6.6), which reads

$$\begin{aligned} r_k = K\sigma^k &= \sum_{i=0}^{\infty} r_i p_{ik} \\ &= \sum_{i=k-1}^{\infty} r_i p_{ik} \\ &= \sum_{i=k-1}^{\infty} K\sigma^i \beta_{i+1-k} \end{aligned}$$

Canceling the constant K as well as common factors of σ we have

$$\sigma = \sum_{i=k-1}^{\infty} \sigma^{i+1-k} \beta_{i+1-k}$$

Changing the index of summation we finally have

$$\sigma = \sum_{n=0}^{\infty} \sigma^n \beta_n$$

Of course we know β_n from Eq. (6.12), which permits the following calculation:

$$\sigma = \sum_{n=0}^{\infty} \sigma^n \int_{t=0}^{\infty} \frac{(m\mu t)^n}{n!} e^{-m\mu t}\, dA(t)$$

$$= \int_0^{\infty} e^{-(m\mu - m\mu\sigma)t}\, dA(t)$$

This equation must be satisfied if our assumed ("calculated") guess is to be correct. However, we recognize this last integral as the Laplace transform for the pdf of interarrival times evaluated at a special point; thus we have

$$\sigma = A^*(m\mu - m\mu\sigma) \qquad \blacksquare(6.21)$$

This functional equation for σ must be satisfied if our assumed solution is to be acceptable. It can be shown [TAKA 62] that so long as $\rho < 1$ then there is a unique real solution for σ in the range $0 < \sigma < 1$, and it is this solution which we seek; note that $\sigma = 1$ must always be a solution of the functional equation since $A^*(0) = 1$.

We now have the defining equation for σ and it remains for us to find the unknown constant K as well as r_k for $k = 0, 1, 2, \ldots, m - 2$. Before we settle these questions, however, let us establish some additional important results for the G/M/m system using Eq. (6.19), our basic result so far. This basic result establishes that the distribution for number in system is geometrically distributed in the range $k \geq m - 1$. Working from there let us now calculate the probability that an arriving customer must wait for service. Clearly

$$P[\text{arrival queues}] = \sum_{k=m}^{\infty} r_k$$

$$= \sum_{k=m}^{\infty} K\sigma^k$$

$$= \frac{K\sigma^m}{1 - \sigma} \qquad (6.22)$$

(This operation is permissible since $0 < \sigma < 1$ as discussed above.) The conditional probability of finding a queue length of size n, given that a customer must queue, is

$$P[\text{queue size} = n \mid \text{arrival queues}] = \frac{r_{m+n}}{P[\text{arrival queues}]}$$

and so

$$P[\text{queue size} = n \mid \text{arrival queues}] = \frac{K\sigma^{n+m}}{K\sigma^m/(1 - \sigma)}$$

$$= (1 - \sigma)\sigma^n \qquad n \geq 0 \qquad \blacksquare(6.23)$$

Thus we conclude that the conditional queue length distribution (given that a queue exists) is geometric for any G/M/m system.

6.3. CONDITIONAL DISTRIBUTION OF WAITING TIME

Let us now seek the distribution of queueing time, given that a customer must queue. From Eq. (6.23), a customer who queues will find $m + n$ in the system with probability $(1 - \sigma)\sigma^n$. Under such conditions our arriving customer must wait until $n + 1$ customers depart from the system before he is allowed into service, and this interval will constitute his waiting time. Thus we are asking for the distribution of an interval whose length is made up of the sum of $n + 1$ independently and exponentially distributed random variables (each with parameter $m\mu$). The resulting convolution is most easily expressed as a transform, which gives rise to the usual product of transforms. Thus defining $W^*(s)$ to be the Laplace transform of the queueing time as in Eq. (5.103) (i.e., as $E[e^{-s\tilde{w}}]$), and defining

$$W^*(s \mid n) = E[e^{-s\tilde{w}} \mid \text{arrival queues and queue size} = n] \qquad (6.24)$$

we have

$$W^*(s \mid n) = \left(\frac{m\mu}{s + m\mu}\right)^{n+1} \qquad (6.25)$$

But clearly

$$W^*(s \mid \text{arrival queues}) = \sum_{n=0}^{\infty} W^*(s \mid n) P[\text{queue size} = n \mid \text{arrival queues}]$$

and so from Eqs. (6.25) and (6.23) we have

$$W^*(s \mid \text{arrival queues}) = \sum_{n=0}^{\infty} (1 - \sigma)\sigma^n \left(\frac{m\mu}{s + m\mu}\right)^{n+1}$$

$$= (1 - \sigma) \frac{m\mu}{s + m\mu - m\mu\sigma}$$

Luckily, we recognize the inverse of this Laplace transform by inspection, thereby yielding the following conditional pdf for queueing time,

$$w(y \mid \text{arrival queues}) = (1 - \sigma) m\mu e^{-m\mu(1-\sigma)y} \qquad y \geq 0 \qquad \blacksquare (6.26)$$

Quite a surprise! The conditional pdf for queueing time is *exponentially distributed* for the system G/M/m!

Thus far we have two principal results: first, that the conditional queue size is geometrically distributed with parameter σ as given in Eq. (6.23); and second, that the conditional pdf for queueing time is exponentially distributed with parameter $m\mu(1 - \sigma)$ as given in Eq. (6.26). The parameter σ is found as the unique root in the range $0 < \sigma < 1$ of the functional equation (6.21). We are still searching for the distribution r_k and have carried that solution to the point of Eq. (6.20); we have as yet to evaluate the constant K as well as the first $m - 1$ terms in that distribution. Before we proceed with these last steps let us study an important special case.

6.4. THE QUEUE G/M/1

This is perhaps the most important system and forms the "dual" to the system M/G/1. Since $m = 1$ then Eq. (6.19) gives us the solution for r_k for all values of k, that is,

$$r_k = K\sigma^k \qquad k = 0, 1, 2, \ldots$$

K is now easily evaluated since these probabilities must sum to unity. From this we obtain immediately

$$r_k = (1 - \sigma)\sigma^k \qquad k = 0, 1, 2, \ldots \qquad \blacksquare \; (6.27)$$

where, of course, σ is the unique root of

$$\sigma = A^*(\mu - \mu\sigma) \qquad \blacksquare \; (6.28)$$

in the range $0 < \sigma < 1$. *Thus the system G/M/1 gives rise to a geometric distribution for number of customers found in the system by an arrival; this applies as an unconditional statement regardless of the form for the interarrival distribution.* We have already seen an example of this in Eq. (4.42) for the system E_r/M/1. We comment that the state probabilities, $p_k = P[k \text{ in system}]$, *differ* from Eq. (6.27) in that $p_0 = 1 - \rho$ whereas $r_0 = (1 - \sigma)$ and $p_k = \rho(1 - \sigma)\sigma^{k-1} = \rho r_{k-1}$ for $k = 1, 2, \ldots$ {see Eq. (3.24), p. 209 of [COHE 69]}; in the M/G/1 queue we found $p_k = r_k$.

A customer will be forced to wait for service with probability $1 - r_0 = \sigma$, and so we may use Eq. (6.26) to obtain the unconditional distribution of waiting time as follows (where we define A to be the event "arrival queues" and A^c, the complementary event):

$$\begin{aligned} W(y) &= P[\text{queueing time} \le y] \\ &= 1 - P[\text{queueing time} > y \mid A]P[A] \\ &\qquad - P[\text{queueing time} > y \mid A^c]P[A^c] \end{aligned} \qquad (6.29)$$

Clearly, the last term in this equation is zero; the remaining conditional probability in this last expression may be obtained by integrating Eq. (6.26) from y to infinity for $m = 1$; this computation gives $e^{-\mu(1-\sigma)y}$ and since σ is the probability of queueing we have immediately from Eq. (6.29) that

$$W(y) = 1 - \sigma e^{-\mu(1-\sigma)y} \qquad y \geq 0 \tag{6.30}$$

We have the remarkable conclusion that the unconditional waiting-time distribution is exponential (with a jump of size $1 - \sigma$ at the origin) for the system G/M/1. If we compare this result to (5.123) and Figure 5.9, which gives the waiting-time distribution for M/M/1, we see that the results agree with ρ replacing σ. That is, the queueing-time distribution for G/M/1 is of the *same form* as for M/M/1!

By straightforward calculation, we also have that the mean wait in G/M/1 is

$$W = \frac{\sigma}{\mu(1 - \sigma)} \tag{6.31}$$

Example

Let us now illustrate this method for the example M/M/1. Since $A(t) = 1 - e^{-\lambda t} (t \geq 0)$ we have immediately

$$A^*(s) = \frac{\lambda}{s + \lambda} \tag{6.32}$$

Using Eq. (6.28) we find that σ must satisfy

$$\sigma = \frac{\lambda}{\mu - \mu\sigma + \lambda}$$

or

$$\mu\sigma^2 - (\mu + \lambda)\sigma + \lambda = 0$$

which yields

$$(\sigma - 1)(\mu\sigma - \lambda) = 0$$

Of these two solutions for σ, the case $\sigma = 1$ is unacceptable due to stability conditions $(0 < \sigma < 1)$ and therefore the only acceptable solution is

$$\sigma = \frac{\lambda}{\mu} = \rho \qquad \text{M/M/1} \tag{6.33}$$

which yields from Eq. (6.27)

$$r_k = (1 - \rho)\rho^k \qquad k \geq 0 \tag{6.34}$$

This, of course, is our usual solution for M/M/1. Further, using $\sigma = \rho$ as the value for σ in our waiting time distribution [Eq. (6.30)] we come up immediately with the known solution given in Eq. (5.123).

Example

As a second (slightly more interesting) example let us consider a G/M/1 system, with an interarrival time distribution such that

$$A^*(s) = \frac{2\mu^2}{(s + \mu)(s + 2\mu)} \tag{6.35}$$

Note that this corresponds to an E_2/M/1 system in which the two arrival stages have different death rates; we choose these rates to be linear multiples of the service rate μ. As always our first step is to evaluate σ from Eq. (6.28) and so we have

$$\sigma = \frac{2\mu^2}{(\mu - \mu\sigma + \mu)(\mu - \mu\sigma + 2\mu)}$$

This leads directly to the cubic equation

$$\sigma^3 - 5\sigma^2 + 6\sigma - 2 = 0$$

We know for sure that $\sigma = 1$ is always a root of Eq. (6.28), and this permits the straightforward factoring

$$(\sigma - 1)(\sigma - 2 - \sqrt{2})(\sigma - 2 + \sqrt{2}) = 0$$

Of these three roots it is clear that only $\sigma = 2 - \sqrt{2}$ is acceptable (since $0 < \sigma < 1$ is required). Therefore Eq. (6.27) immediately gives the distribution for number in system (seen by arrivals)

$$r_k = (\sqrt{2} - 1)(2 - \sqrt{2})^k \qquad k = 0, 1, 2, \ldots \tag{6.36}$$

Similarly we find

$$W(y) = 1 - (2 - \sqrt{2})e^{-\mu(\sqrt{2}-1)y} \qquad y \geq 0 \tag{6.37}$$

for the waiting-time distribution.

Let us now return to the more general system G/M/m.

6.5. THE QUEUE G/M/m

At the end of Section 6.3 we pointed out that the only remaining unknowns for the general G/M/m solution were: K, an unknown constant, and the $m - 1$ "boundary" probabilities $r_0, r_1, \ldots, r_{m-2}$. That is, our solution appears in the form of Eq. (6.20); we may factor out the term $K\sigma^{m-1}$ to obtain

$$\mathbf{r} = K\sigma^{m-1}[R_0, R_1, \ldots, R_{m-2}, 1, \sigma, \sigma^2, \sigma^3, \ldots] \tag{6.38}$$

where

$$R_k = \frac{r_k \sigma^{1-m}}{K} \qquad k = 0, 1, \ldots, m-2 \tag{6.39}$$

Furthermore, for convenience we define

$$J = K\sigma^{m-1} \tag{6.40}$$

We have as yet not used the first $m - 1$ equations represented by the matrix equation (6.6). We now require them for the evaluation of our unknown terms (of which there are $m - 1$). In terms of our one-step transition probabilities p_{ij} we then have

$$R_k = \sum_{i=k-1}^{\infty} R_i p_{ik} \qquad k = 0, 1, \ldots, m-2$$

where we may extend the definition for R_k in Eq. (6.39) beyond $k = m - 2$ by use of Eq. (6.19), that is, $R_i = \sigma^{i-m+1}$ for $i \geq m - 1$. The tail of the sum above may be evaluated to give

$$R_k = \sum_{i=k-1}^{m-2} R_i p_{ik} + \sum_{i=m-1}^{\infty} \sigma^{i+1-m} p_{ik}$$

Solving for R_{k-1}, the lowest-order term present, we have

$$R_{k-1} = \frac{R_k - \sum_{i=k}^{m-2} R_i p_{ik} - \sum_{i=m-1}^{\infty} \sigma^{i+1-m} p_{ik}}{p_{k-1,k}} \tag{6.41}$$

for $k = 1, 2, \ldots, m - 1$. The set of equations (6.41) is a triangular set in the unknowns R_k; in particular we may start with the fact that $R_{m-1} = 1$ [see Eq. (6.38)] and then solve recursively over the range $k = m - 1, m - 2, \ldots, 1, 0$ in order. Finally we may use the conservation of probability to evaluate the constant J (this being equivalent to evaluating K) as

$$J \sum_{k=0}^{m-2} R_k + J \sum_{k=m-1}^{\infty} \sigma^{k-m+1} = 1$$

or

$$J = \frac{1}{\dfrac{1}{1-\sigma} + \sum_{k=0}^{m-2} R_k} \tag{6.42}$$

This then provides a complete prescription for evaluating the distribution of the number of customers in the system. We point out that Takács [TAKA 62] gives an explicit (albeit complex) expression for these boundary probabilities.

Let us now determine the distribution of waiting time in this system [we already have seen the conditional distribution in Eq. (6.26)]. First we have the probability that an arriving customer need not queue, given by

$$W(0) = \sum_{k=0}^{m-1} r_k = J \sum_{k=0}^{m-1} R_k \tag{6.43}$$

On the other hand, if a customer arrives to find $k \geq m$ others in the system he must wait until exactly $k - m + 1$ customers depart before he may enter service. Since there are m servers working continuously during his wait then the interdeparture times must be exponentially distributed with parameter $m\mu$, and so his waiting time must be of the form of a $(k - m + 1)$-stage Erlangian distribution as given in Eq. (2.147). Thus for this case $(k \geq m)$ we may write

$$P[\tilde{w} \leq y \mid \text{customer finds } k \text{ in system}] = \int_0^y \frac{m\mu(m\mu x)^{k-m}}{(k-m)!} e^{-m\mu x} \, dx$$

If we now remove the condition on k we may write the unconditional distribution as

$$\begin{aligned} W(y) &= W(0) + J \sum_{k=m}^{\infty} \int_0^y \frac{(m\mu)(m\mu x)^{k-m} \sigma^{k-m+1}}{(k-m)!} e^{-m\mu x} \, dx \\ &= W(0) + J\sigma \int_0^y m\mu e^{-m\mu x(1-\sigma)} \, dx \end{aligned} \tag{6.44}$$

We may now use the expression for J in Eq. (6.42) and for $W(0)$ in Eq. (6.43) and carry out the integration in Eq. (6.44) to obtain

$$W(y) = 1 - \frac{\sigma}{1 + (1 - \sigma) \sum_{k=0}^{m-2} R_k} e^{-m\mu(1-\sigma)y} \qquad y \geq 0 \tag{6.45}$$

This is the final solution for our waiting-time distribution and shows that *in the general case G/M/m we still have the exponential distribution* (*with an accumulation point at the origin*) *for waiting time!*

We may calculate the average waiting time either from Eq. (6.45) or as follows. As we saw, a customer who arrives to find $k \geq m$ others in the system must wait until $k - m + 1$ services are complete, each of which takes on the average $1/m\mu$ sec. We now sum over all those cases where our customer must wait to obtain

$$E[\tilde{w}] \triangleq W = \sum_{k=m}^{\infty} \frac{1}{m\mu}(k - m + 1)r_k$$

But in this range we know that $r_k = K\sigma^k$ and so

$$W = \frac{K}{m\mu} \sum_{k=m}^{\infty} (k - m + 1)\sigma^k$$

and this is easily calculated to yield

$$W = \frac{K\sigma^m}{m\mu(1 - \sigma)^2} = \frac{J\sigma}{m\mu(1 - \sigma)^2} \qquad \blacksquare$$

6.6. THE QUEUE G/M/2

Let us see how far we can get with the system G/M/2. From Eq. (6.19) we have immediately

$$r_k = K\sigma^k \qquad k = 1, 2, \ldots$$

Conserving probability we find

$$\sum_{k=0}^{\infty} r_k = 1 = r_0 + \sum_{k=1}^{\infty} K\sigma^k$$

This yields the following relationship between K and r_0:

$$K = \frac{(1 - r_0)(1 - \sigma)}{\sigma} \tag{6.46}$$

Our task now is to find another relation between K and r_0. This we may do from Eq. (6.41), which states

$$R_0 = \frac{R_1 - \sum_{i=1}^{\infty} \sigma^{i-1} p_{i1}}{p_{01}} \tag{6.47}$$

But $R_1 = 1$. The denominator is given by Eq. (6.9), namely,

$$p_{01} = \int_0^{\infty} \binom{1}{1} [1 - e^{-\mu t}]^0 e^{-\mu t}\, dA(t)$$

This we recognize as

$$p_{01} = A^*(\mu) \tag{6.48}$$

Regarding the one-step transition probabilities in the numerator sum of Eq. (6.47) we find they break into two regions: the term p_{11} must be calculated from Eq. (6.9) and the terms p_{i1} for $i = 2, 3, 4, \ldots$ must be calculated from

Eq. (6.13). Proceeding we have

$$p_{11} = \int_0^\infty \binom{2}{1} [1 - e^{-\mu t}] e^{-\mu t}\, dA(t)$$

Again we recognize this as the transform

$$p_{11} = 2A^*(\mu) - 2A^*(2\mu) \tag{6.49}$$

Also for $i = 2, 3, 4, \ldots$, we have

$$p_{i1} = \int_0^\infty \binom{2}{1} e^{-\mu t} \left[\int_0^t \frac{(2\mu y)^{i-2}}{(i-2)!} (e^{-\mu y} - e^{-\mu t}) 2\mu\, dy \right] dA(t) \tag{6.50}$$

Substituting these last equations into Eq. (6.47) we then have

$$R_0 = \frac{1}{A^*(\mu)} \left[1 - 2A^*(\mu) + 2A^*(2\mu) - \sum_{i=2}^{\infty} \sigma^{i-1} p_{i1} \right] \tag{6.51}$$

The summation in this equation may be carried out within the integral signs of Eq. (6.50) to give

$$\sum_{i=2}^{\infty} \sigma^{i-1} p_{i1} = 2A^*(2\mu) + \frac{2A^*(2\mu - 2\mu\sigma) - 4\sigma A^*(\mu)}{2\sigma - 1} \tag{6.52}$$

But from Eq. (6.21) we recognize that $\sigma = A^*(2\mu - 2\mu\sigma)$ and so we have

$$\sum_{i=2}^{\infty} \sigma^{i-1} p_{i1} = 2A^*(2\mu) + \frac{2\sigma}{2\sigma - 1} [1 - 2A^*(\mu)]$$

Substituting back into Eq. (6.51) we find

$$R_0 = \frac{2A^*(\mu) - 1}{(2\sigma - 1)A^*(\mu)}$$

However from Eq. (6.39) we know that

$$R_0 = \frac{r_0}{K\sigma}$$

and so we may express r_0 as

$$r_0 = \frac{K\sigma[1 - 2A^*(\mu)]}{(1 - 2\sigma)A^*(\mu)} \tag{6.53}$$

Thus Eqs. (6.46) and (6.53) give us two equations in our two unknowns K and r_0, which when solved simultaneously lead to

$$r_0 = \frac{(1 - \sigma)[1 - 2A^*(\mu)]}{1 - \sigma - A^*(\mu)}$$

$$K = A^*(\mu) \frac{(1 - \sigma)(1 - 2\sigma)}{\sigma[1 - \sigma - A^*(\mu)]}$$

Thus, in conclusion, the distribution of customers in the G/M/2 system is given by

$$r_k = \begin{cases} \dfrac{(1-\sigma)[1-2A^*(\mu)]}{1-\sigma-A^*(\mu)} & k = 0 \\ A^*(\mu)\dfrac{(1-\sigma)(1-2\sigma)}{1-\sigma-A^*(\mu)}\sigma^{k-1} & k = 1, 2, \ldots \end{cases} \tag{6.54}$$

where σ is the unique root within the unit circle as determined by the equation $\sigma = A^*(2\mu - 2\mu\sigma)$. Similarly the distribution of waiting time in this system is given from Eq. (6.45) as

$$W(y) = 1 - \frac{\sigma(1-2\sigma)A^*(\mu)}{1-\sigma-A^*(\mu)} e^{-2\mu(1-\sigma)y} \qquad y \geq 0 \tag{6.55}$$

Example

As an example, let us consider the system M/M/2. From Eqs. (6.21) and (6.32) we have the expression for σ:

$$\sigma = \frac{\lambda}{2\mu - 2\mu\sigma + \lambda}$$

which gives us the quadratic

$$2\mu\sigma^2 - (\lambda + 2\mu)\sigma + \lambda = 0$$

Recalling the definition $\rho = \lambda/m\mu$, which in this case gives $\rho = \lambda/2\mu$, we find that the two roots of this quadratic are $\sigma = 1$, $\sigma = \rho$; clearly only the second of these two is permissible. Similarly we have

$$A^*(\mu) = \frac{\lambda}{\mu + \lambda} = \frac{2\rho}{1 + 2\rho}$$

We then have directly from Eq. (6.54) that the solution for the distribution of number in system is

$$r_k = \begin{cases} \dfrac{1-\rho}{1+\rho} & k = 0 \\ 2\dfrac{1-\rho}{1+\rho}\rho^k & k = 1, 2, \ldots \end{cases} \tag{6.56}$$

and for the waiting-time distribution we have

$$W(y) = 1 - \frac{2\rho^2}{1+\rho} e^{-2\mu(1-\rho)y} \qquad y \geq 0 \tag{6.57}$$

Comparing Eq. (6.56) with our results from Chapter 3 [Eqs. (3.37) and (3.39)] we find that they agree for $m = 2$.

This completes our study of the G/M/m queue. Some further results of interest may be found in [DESM 73]. In the next chapter, we view transforms as probabilities and gain considerable reduction in the analytic effort required to solve equilibrium and transient queueing problems.

REFERENCES

COHE 69 Cohen, J. W., *The Single Server Queue*, Wiley (New York) 1969.

DESM 73 De Smit, J. H. A., "On the Many Server Queue with Exponential Service Times," *Advances in Applied Probability*, **5**, 170–182 (1973).

KEND 51 Kendall, D. G., "Some Problems in the Theory of Queues," *Journal of the Royal Statistical Society, Ser. B.*, **13**, 151–185 (1951).

TAKA 62 Takács, L., *Introduction to the Theory of Queues*, Oxford University Press (New York) 1962.

EXERCISES

6.1. Prove Eq. (6.13). [HINT: condition on an interarrival time of duration t and then further condition on the time ($\leq t$) it will take to empty the queue.]

6.2. Consider $E_2/M/1$ (with infinite queueing room).
- **(a)** Solve for r_k in terms of σ.
- **(b)** Evaluate σ explicitly.

6.3. Consider M/M/m.
- **(a)** How do p_k and r_k compare?
- **(b)** Compare Eqs. (6.22) and (3.40).

6.4. Prove Eq. (6.31).

6.5. Show that Eq. (6.52) follows from Eq. (6.50).

6.6. Consider an $H_2/M/1$ system in which $\lambda_1 = 2$, $\lambda_2 = 1$, $\mu = 2$, and $\alpha_1 = 5/8$.
- **(a)** Find σ.
- **(b)** Find r_k.
- **(c)** Find $w(y)$.
- **(d)** Find W.

6.7. Consider a D/M/1 system with $\mu = 2$ and with the same ρ as in the previous exercise.
- **(a)** Find σ (correct to two decimal places).

(b) Find r_k.
(c) Find $w(y)$.
(d) Find W.

6.8. Consider a G/M/1 queueing system with room for at most two customers (one in service plus one waiting). Find r_k $(k = 0, 1, 2)$ in terms of μ and $A^*(s)$.

6.9. Consider a G/M/1 system in which the cost of making a customer wait y sec is

$$c(y) = ae^{by}$$

(a) Find the average cost of queueing for a customer.
(b) Under what conditions will the average cost be finite?

7

The Method of Collective Marks

When one studies stochastic processes such as in queueing theory, one finds that the work divides into two parts. The first part typically requires a careful *probabilistic argument* in order to arrive at expressions involving the random variables of interest.* The second part is then one of analysis in which the *formal manipulation* of symbols takes place either in the original domain or in some transformed domain. Whereas the probabilistic arguments typically must be made with great care, they nevertheless leave one with a comfortable feeling that the "physics" of the situation are constantly within one's understanding and grasp. On the other hand, whereas the analytic manipulations that one carries out in the second part tend to be rather straightforward (albeit difficult) formal operations, one is unfortunately left with the uneasy feeling that these manipulations relate back to the original problem in no clearly understandable fashion. This "nonphysical" aspect to problem solving typically is taken on when one moves into the domain of transforms, (either Laplace or z-transforms).

In this chapter we demonstrate that one may deal with transforms and still maintain a handle on the probabilistic arguments taking place as these transforms are manipulated. There are two separate operations involved: the "marking" of customers; and the observation of "catastrophe" processes. Together these methods are referred to as the method of *collective marks*. Both operations need not necessarily be used simultaneously, and we study them separately below. This material is drawn principally from [RUNN 65]; these ideas were introduced by van Dantzig [VAN 48] in order to expose the probabilistic interpretation for transforms.

7.1. THE MARKING OF CUSTOMERS

Assume that, at the entrance to a queueing system, there is a gremlin who marks (i.e., tags) arriving customers with the following probabilities:

$$P[\text{customer is marked}] = 1 - z \tag{7.1}$$

$$P[\text{customer is not marked}] = z \tag{7.2}$$

* As, for example, the arguments leading up to Eqs. (5.31) and (6.1).

where $0 \leq z \leq 1$. We assume that the gremlin marks customers with these probabilities independent of *all* other aspects of the queueing process. As we shall see below, this marking process allows us to create generating functions in a very natural way.

It is most instructive if we illustrate the use of this marking process by examples:

Example 1: Poisson Arrivals

We first consider a Poisson arrival process with a mean arrival rate of λ customers per second. Assume that customers are marked as above. Let us consider the probability

$$q(z, t) \triangleq P[\text{no marked customers arrive in } (0, t)] \tag{7.3}$$

It is clear that k customers will arrive in the interval $(0, t)$ with the probability $(\lambda t)^k e^{-\lambda t}/k!$. Moreover, with probability z^k, none of these k customers will be marked; this last is true since marking takes place independently among customers. Now summing over all values of k we have immediately that

$$\begin{aligned} q(z, t) &= \sum_{k=0}^{\infty} \frac{(\lambda t)^k e^{-\lambda t}}{k!} z^k \\ &= e^{\lambda t(z-1)} \end{aligned} \tag{7.4}$$

Going back to Eq. (2.134) we see that Eq. (7.4) is merely the generating function for a Poisson arrival process. We thus conclude that the generating function for this arrival process may also be interpreted as the probabilistic quantity expressed in Eq. (7.3). This will not be the first time we may give a probabilistic interpretation for a generating function!

Example 2: M/M/∞

We consider the birth–death queueing system with an infinite number of servers. We also assume at time $t = 0$ that there are i customers present. The parameters of our system as usual are λ and μ [i.e., $A(t) = 1 - e^{-\lambda t}$ and $B(x) = 1 - e^{-\mu x}$].

We are interested in the quantity

$$P_k(t) = P[k \text{ customers in the system at time } t] \tag{7.5}$$

and we define its generating function as we did in Eq. (2.153) to be

$$P(z, t) = \sum_{k=0}^{\infty} P_k(t) z^k \tag{7.6}$$

Once again we mark customers according to Eqs. (7.1) and (7.2). In analogy with Example 1, we recognize that Eq. (7.6) may be interpreted as the probability that the system contains no marked customers at time t (where the term z^k again represents the probability that none of the k customers present is marked). Here then is our crucial observation: We may calculate $P(z, t)$ *directly* by finding the probability that there are no marked customers in the system at time t, rather than calculating $P_k(t)$ and then finding its z-transform!

We proceed as follows: We need merely find the probability that none of the customers still present in the system at time t is marked and this we do by accounting for all customers present at time 0 as well as all customers who arrive in the interval $(0, t)$. For any customer present at time 0 we may calculate the probability that he is still present at time t *and* is marked as $(1 - z)[1 - B(t)]$ where the first factor gives the probability that our customer was marked in the first place and the second factor gives the probability that his service time is greater than t. Clearly, then, this quantity subtracted from unity is the probability that a customer originally present is not a marked customer present at time t; and so we have

$$P[\text{customer present initially is } \textit{not} \text{ a marked customer present at time } t] = 1 - (1 - z)e^{-\mu t}$$

Now for the new customers who enter in the interval $(0, t)$, we have as before $P[k \text{ arrivals in } (0, t)] = (\lambda t)^k e^{-\lambda t}/k!$. Given that k have arrived in this interval then their arrival instants are uniformly distributed over this interval [see Eq. (2.136)]. Let us consider one such arriving customer and assume that he arrives at a time $\tau < t$. Such a customer will not be a marked customer present at time t with probability

$$P[\text{new arrival is not a marked customer present at time } t \text{ given he arrived at } \tau \le t] = 1 - (1 - z)[1 - B(t - \tau)] \quad (7.7)$$

However, we have that

$$P[\text{arrival time} \le \tau] = \frac{\tau}{t} \qquad \text{for } 0 \le \tau \le t$$

and so

$$P[\text{new arrival still in system at } t] = \int_{\tau=0}^{t} e^{-\mu(t-\tau)} \frac{d\tau}{t} = \frac{1 - e^{-\mu t}}{\mu t} \quad (7.8)$$

Unconditioning the arrival time from Eq. (7.7) as shown in Eq. (7.8) we have

$$P[\text{new arrival is not a marked customer present at } t] = 1 - (1 - z)\frac{1 - e^{-\mu t}}{\mu t}$$

Thus we may calculate the probability that there are no marked customers at time t as follows:

$$P(z, t) = \sum_{k=0}^{\infty} P[k \text{ arrive in } (0, t)]$$
$$\times \{P[\text{new arrival is not a marked customer present at } t]\}^k$$
$$\times \{P[\text{initial customer is not a marked customer present at } t]\}^i$$

Using our established relationships we arrive at

$$P(z, t) = \sum_{k=0}^{\infty} \frac{(\lambda t)^k}{k!} e^{-\lambda t} \left[1 - (1 - z) \frac{1 - e^{-\mu t}}{\mu t}\right]^k [1 - (1 - z)e^{-\mu t}]^i$$

which then gives the known result

$$P(z, t) = [1 - (1 - z)e^{-\mu t}]^i e^{-(\lambda/\mu)(1-z)[1-e^{-\mu t}]} \tag{7.9}$$

It should be clear to the student that the usual method for obtaining this result would have been extremely complex.

Example 3: M/G/1

In this example we consider the FCFS M/G/1 system. Recall that the random variables $w_n, t_{n+1}, t_{n+2}, \ldots, x_n, x_{n+1}, \ldots$ are all independent of each other. As usual, we define $B^*(s)$ and $W_n^*(s)$ as the Laplace transforms for the service-time pdf $b(x)$ and the waiting-time pdf $w_n(y)$ for C_n, respectively. We define the event

$$\{\text{no } M \text{ in } w_n\} \triangleq \left\{\begin{matrix}\text{no customers who arrive during the waiting} \\ \text{time of } C_n \text{ are marked}\end{matrix}\right\} \tag{7.10}$$

We wish to find the probability of this event, that is, $P[\text{no } M \text{ in } w_n]$. Conditioning on the number of arriving customers and on the waiting time w_n, and then removing these conditions, we have

$$P[\text{no } M \text{ in } w_n] = \sum_{k=0}^{\infty} \int_0^{\infty} \frac{(\lambda y)^k}{k!} e^{-\lambda y} z^k \, dW_n(y)$$
$$= \int_0^{\infty} e^{-\lambda y(1-z)} \, dW_n(y)$$

We recognize the integral as $W_n^*(\lambda - \lambda z)$ and so

$$P[\text{no } M \text{ in } w_n] = W_n^*(\lambda - \lambda z) \tag{7.11}$$

Thus once again we have a very simple probabilistic interpretation for the (Laplace) transform of an important distribution. By identical arguments we may arrive at

$$P[\text{no } M \text{ in } x_n] = B^*(\lambda - \lambda z) \tag{7.12}$$

This last gives us another interpretation for an old expression we have seen in Chapter 5.

Now comes a startling insight! It is clear that the arrival of customers during the waiting time of C_n and the arrival of customers during the service time of C_n must be independent events since these are nonoverlapping intervals and our arrival process is memoryless. Thus the events of no marked customers arriving in each of these two disjoint intervals of time must be independent, and so the probability that no marked customers arrive in the union of these two disjoint intervals must be the product of the probabilities that none such arrive in each of the intervals separately. Thus we may write

$$P[\text{no } M \text{ in } w_n + x_n] = P[\text{no } M \text{ in } w_n]P[\text{no } M \text{ in } x_n] \tag{7.13}$$

$$= W_n^*(\lambda - \lambda z)B^*(\lambda - \lambda z) \tag{7.14}$$

This last result is pleasing in two ways. First, because it says that the probability of two independent joint events is equal to the product of the probabilities of the individual events [Eq. (7.13)]. Second, because it says that the transform of the pdf of the sum of two independent random variables is equal to the product of the transforms of the pdf of the individual random variables [Eq. (7.14)]. Thus two familiar results (regarding disjoint events and regarding sums of independent random variables) have led to a meaningful new insight, namely, that multiplication of transforms implies not only the sum of two independent random variables, but also implies the product of the probabilities of two independent events! Not often are we privileged to see such fundamental principles related.

Let us now continue with the argument. At the moment we have Eq. (7.14) as one means for expressing the probability that no marked customers arrive during the interval $w_n + x_n$. We now proceed to calculate this probability by a second argument. Of course we have

$$P[\text{no } M \text{ in } w_n + x_n] = P[\text{no } M \text{ in } w_n + x_n \textit{ and } C_{n+1} \text{ marked}] + P[\text{no } M \text{ in } w_n + x_n \textit{ and } C_{n+1} \text{ not marked}] \tag{7.15}$$

Furthermore, we have

$$P[\text{no } M \text{ in } w_n + x_n \text{ and } C_{n+1} \text{ marked}] = 0 \qquad \text{if} \quad w_{n+1} > 0$$

since if C_{n+1} must wait, then he must arrive in the interval $w_n + x_n$ and it is impossible for him to be marked and still to have the event $\{\text{no } M \text{ in } w_n + x_n\}$. Thus the first term on the right-hand side of Eq. (7.15) must be $P[w_{n+1} = 0](1 - z)$ where this second factor is merely the probability that C_{n+1} is marked. Now consider the second term on the right-hand side of Eq. (7.15); as shown in Figure 7.1 it is clear that no customers arrive between C_n and C_{n+1}, and therefore the customers of interest (namely, those arriving after

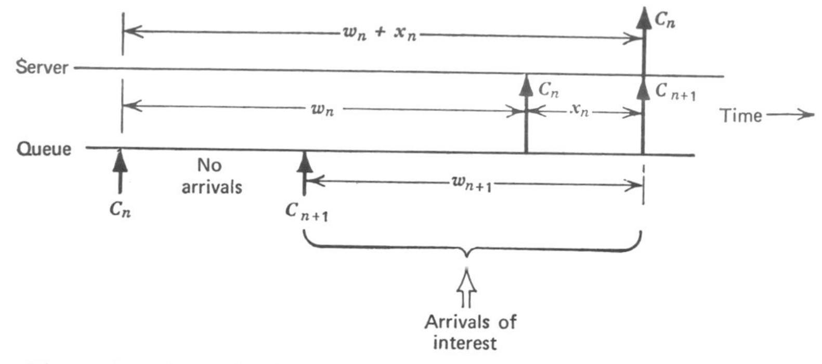

Figure 7.1 Arrivals of interest during $w_n + x_n$.

C_{n+1} does, but yet in the interval $w_n + x_n$) must arrive in the interval w_{n+1} since this interval will end when $w_n + x_n$ ends. Thus this second term must be

$$P[\text{no } M \text{ in } w_n + x_n \text{ and } C_{n+1} \text{ not marked}] = P[\text{no } M \text{ in } w_{n+1}] \times P[C_{n+1} \text{ not marked}] = P[\text{no } M \text{ in } w_{n+1}]\, z$$

From these observations and the result of Eq. (7.11) we may write Eq. (7.15) as

$$P[\text{no } M \text{ in } w_n + x_n] = (1 - z)P[C_{n+1} \text{ arrives after } w_n + x_n] + zW_{n+1}^*(\lambda - \lambda z)$$

Now if we think of a second *separate* marking process in which *all* the customers are marked (with an additional tag) with probability one, and ask that no such marked customers arrive during the interval $w_n + x_n$, then we are asking that no customers at all arrive during this interval (which is the same as asking that C_{n+1} arrive after $w_n + x_n$); we may calculate this using Eq. (7.14) with $z = 0$ (since this guarantees that all customers be marked) and obtain $W_n^*(\lambda)B^*(\lambda)$ for this probability. Thus we arrive at

$$P[\text{no } M \text{ in } w_n + x_n] = (1 - z)W_n^*(\lambda)B^*(\lambda) + zW_{n+1}^*(\lambda - \lambda z) \quad (7.16)$$

We now have two expressions for $P[\text{no } M \text{ in } w_n + x_n]$, which may be equated to obtain

$$W_n^*(\lambda - \lambda z)B^*(\lambda - \lambda z) = (1 - z)W_n^*(\lambda)B^*(\lambda) + zW_{n+1}^*(\lambda - \lambda z) \quad (7.17)$$

The interesting part is over. The use of the method of collective marks has brought us to Eq. (7.17), which is not easily obtained by other methods, but which in fact checks with the result due to other methods. Rather than dwell on the techniques required to carry this equation further we refer the reader to Runnenburg [RUNN 65] for additional details of the time-dependent solution.

Now, for $\rho < 1$, we have an ergodic process with $W^*(s) = \lim W_n^*(s)$ as $n \to \infty$. Equation (7.17) then reduces to

$$W^*(\lambda - \lambda z)B^*(\lambda - \lambda z) = (1 - z)W^*(\lambda)B^*(\lambda) + zW^*(\lambda - \lambda z)$$

If we make the change variable $s = \lambda - \lambda z$ and solve for $W^*(s)$, we obtain

$$W^*(s) = \frac{sW^*(\lambda)B^*(\lambda)}{s - \lambda + \lambda B^*(s)}$$

Since $W^*(0) = 1$, we evaluate $W^*(\lambda)B^*(\lambda) = (1 - \rho)$ and arrive at

$$W^*(s) = \frac{s(1 - \rho)}{s - \lambda + \lambda B^*(s)} \tag{7.18}$$

which, of course, is the P–K transform equation for waiting time.

We have demonstrated three examples where the marking of customers has allowed us to argue purely with probabilistic reasoning to derive expressions relating transforms. What we have here traded has been straightforward but tedious analysis for deep but physical probabilistic reasoning. We now consider the catastrophe process.

7.2. THE CATASTROPHE PROCESS

Let us pursue the method of collective marks a bit further by observing "catastrophe" processes. Measuring from time 0 let us consider that some event occurs at time t ($t \geq 0$), where the pdf associated with the time of occurrence of this event is given by $f(t)$. Furthermore, let there be an independent "catastrophe" process taking place simultaneously which generates catastrophes† at a rate γ according to a Poisson process.

We wish to calculate the probability that the event at time t takes place before the first catastrophe (measuring from time 0). Conditioning on t and integrating over all t, we get

$$\begin{aligned} P[\text{event occurs before catastrophe}] &= \int_0^\infty e^{-\gamma t} f(t)\, dt \\ &= F^*(\gamma) \end{aligned} \tag{7.19}$$

where, as usual, $f(t) \Leftrightarrow F^*(s)$ are Laplace transform pairs. Thus we have a probabilistic interpretation for the Laplace transform (evaluated at the point γ) of the pdf for the time of occurrence of the event, namely, it is the probability that an event with this pdf occurs before a Poisson catastrophe at rate γ occurs.

† A catastrophe is merely an impressive name given to these generated times to distinguish them from the "event" of interest at time t.

As a second illustration using catastrophe processes, consider a sequence of events (that is, a point process) on the interval $(0, \infty)$. Measuring from time 0 we would like to calculate the pdf of the time until the nth event, which we denote† by $f_{(n)}(t)$, and with distribution $F_{(n)}(t)$, where the time between events is given as before with density $f(t)$. That is,

$$F_{(n)}(t) = P[n\text{th event has occurred by time } t]$$

We are interested in deriving an expression for the renewal function $H(t)$, which we recall from Section 5.2 is equal to the expected number of events (renewals) in an interval of length t. We proceed by defining

$$P_n(t) = P[\text{exactly } n \text{ events occur in } (0, t)] \tag{7.20}$$

The renewal function may therefore be calculated as

$$\begin{aligned} H(t) &= E[\text{number of events in } (0, t)] \\ &= \sum_{n=0}^{\infty} nP_n(t) \end{aligned}$$

But from its definition we see that $P_n(t) = F_{(n)}(t) - F_{(n+1)}(t)$ and so we have

$$\begin{aligned} H(t) &= \sum_{n=0}^{\infty} n[F_{(n)}(t) - F_{(n+1)}(t)] \\ &= \sum_{n=1}^{\infty} F_{(n)}(t) \end{aligned} \tag{7.21}$$

If we now permit a Poisson catastrophe process (at rate γ) to develop we may ask for the expectation of the following random variable:

$$N_c \triangleq \text{number of events occurring before the first catastrophe} \tag{7.22}$$

With probability $\gamma e^{-\gamma t}\, dt$ the first catastrophe will occur in the interval $(t, t + dt)$ and then $H(t)$ will give the expected number of events occurring before this first catastrophe, that is,

$$H(t) = E[N_c \mid \text{first catastrophe occurs in } (t, t + dt)]$$

Summing over all possibilities we may then write

$$E[N_c] = \int_0^{\infty} H(t)\gamma e^{-\gamma t}\, dt \tag{7.23}$$

In Section 5.2 we had defined $H^*(s)$ to be the Laplace transform of the renewal density $h(t)$ defined as $h(t) \triangleq dH(t)/dt$, that is,

$$H^*(s) \triangleq \int_0^{\infty} h(t)e^{-st}\, dt \tag{7.24}$$

† We use the subscript (n) to remind the reader of the definition in Eq. (5.110) denoting the n-fold convolution. We see that $f_{(n)}(t)$ is indeed the n-fold convolution of the lifetime density $f(t)$.

If we integrate this last equation by parts, we see that the right-hand side of Eq. (7.24) is merely $\int_0^\infty sH(t)e^{-st}\,dt$ and so from Eq. (7.23) we have (making the substitution $s = \gamma$)

$$E[N_c] = H^*(\gamma) \tag{7.25}$$

Let us now calculate $E[N_c]$ by an alternate means. From Eq. (7.19) we see that the catastrophe will occur before the first event with probability $1 - F^*(\gamma)$ and in this case $N_c = 0$. On the other hand, with probability $F^*(\gamma)$ we will get at least one event occurring before the catastrophe. Let N_c' be the random variable $N_c - 1$ conditioned on at least one event; then we have $N_c = 1 + N_c'$. Because of the memoryless property of the Poisson process as well as the fact that the event occurrences generate an imbedded Markov process we see that N_c' must have the same distribution as N_c itself. Forming expectations on N_c we may therefore write

$$E[N_c] = 0[1 - F^*(\gamma)] + \{1 + E[N_c]\}F^*(\gamma)$$

This gives immediately

$$E[N_c] = \frac{F^*(\gamma)}{1 - F^*(\gamma)} \tag{7.26}$$

We now have two expressions for $E[N_c]$ and so by equating them (and making the change of variable $s = \gamma$) we have the final result

$$H^*(s) = \frac{F^*(s)}{1 - F^*(s)} \tag{7.27}$$

This last we recognize as the transform expression for the integral equation of renewal theory [see Eq. (5.21)]; its integral formulation is given in Eq. (5.22).

It is fair to say that the method of collective marks is a rather elegant way to get some useful and important results in the theory of stochastic processes. On the other hand, this method has as yet yielded no results that were not previously known through the application of other methods. Thus at present its principal use lies in providing an alternative way for viewing the fundamental relationships, thereby enhancing one's insight into the probabilistic structure of these processes.

Thus ends our treatment of intermediate queueing theory. In the next part, we venture into the kingdom of the G/G/1 queue.

REFERENCES

RUNN 65 Runnenburg, J. Th., "On the Use of the Method of Collective Marks in Queueing Theory," *Proc. Symposium on Congestion Theory*, eds. W. L. Smith and W. E. Wilkinson, University of North Carolina Press (1965).

VAN 48 van Dantzig, D., "Sur la méthode des fonctions génératrices," *Colloques internationaux du CNRS*, **13**, 29–45 (1948).

EXERCISES

7.1. Consider the M/G/1 system shown in the figure below with average arrival rate λ and service-time distribution $= B(x)$. Customers are served first-come–first-served from queue A until they either leave or receive a sec of service, at which time they join an entrance box as shown in the figure. Customers continue to collect in the entrance box forming

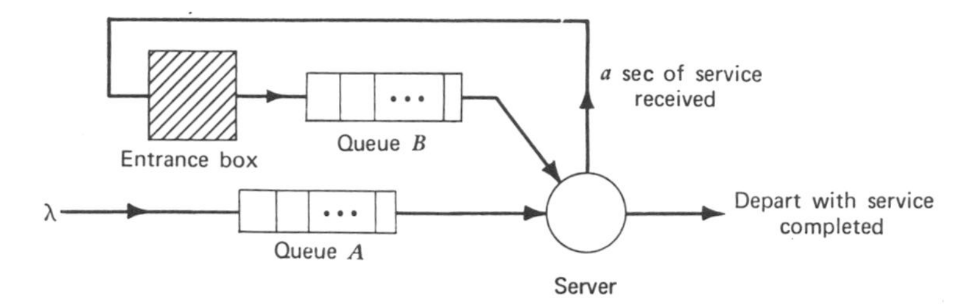

a group until queue A empties and the server becomes free. At this point, the entrance box "dumps" all it has collected as a *bulk arrival* to queue B. Queue B will receive service until a new arrival (to be referred to as a "starter") joins queue A at which time the server switches from queue B to serve queue A and the customer who is preempted returns to the head of queue B. The entrance box then begins to fill and the process repeats. Let

$$g_n = P[\text{entrance box delivers bulk of size } n \text{ to queue } B]$$

$$G(z) = \sum_{n=0}^{\infty} g_n z^n$$

(a) Give a probabilistic interpretation for $G(z)$ using the method of collective marks.

(b) Given that the "starter" reaches the entrance box, and using the method of collective marks find [in terms of λ, a, $B(\cdot)$, *and* $G(z)$]

$$P_k = P[k \text{ customers arrive to queue } A \text{ during the "starter's" service time } \textit{and} \text{ no marked customers arrive to the entrance box from the } k \text{ sub-busy periods created in queue } A \text{ by each of these customers}]$$

(c) Given that the "starter" does *not* reach the entrance box, find P_k as defined above.

(d) From (b) and (c), give an expression (involving an integral) for $G(z)$ in terms of λ, a, $B(\cdot)$, and itself.

(e) From (d) find the average bulk size $\bar{n} = \sum_{n=0}^{\infty} n g_n$.

7.2. Consider the M/G/∞ system. We wish to find $P(z, t)$ as defined in Eq. (7.6). Assume the system contains $i = 0$ customers at $t = 0$. Let $p(t)$ be the probability that a customer who arrived in the interval $(0, t)$ is still present at t. Proceed as in Example 2 of Section 7.1.

(a) Express $p(t)$ in terms of $B(x)$.

(b) Find $P(z, t)$ in terms of λ, t, z, and $p(t)$.

(c) From (b) find $P_k(t)$ defined in Eq. (7.5).

(d) From (c), find $\lim P_k(t) = p_k$ as $t \to \infty$.

7.3. Consider an M/G/1 queue, which is idle at time 0. Let $p = P$[no catastrophe occurs during the time the server is busy with those customers who arrived during $(0, t)$] and let $q = P$[no catastrophe occurs during $(0, t + U(t))$] where $U(t)$ is the unfinished work at time t. Catastrophes occur at a rate γ.

(a) Find p.

(b) Find q.

(c) Interpret $p - q$ as a probability and find an independent expression for it. We may then use (a) and (b) to relate the distribution of unfinished work to $B^*(s)$.

7.4. Consider the G/M/m system. The root σ, which is defined in Eq. (6.21) plays a central role in the solution. Examine Eq. (6.21) from the viewpoint of collective marks and give a probabilistic interpretation for σ.

PART IV

ADVANCED MATERIAL

We find ourselves in difficult terrain as we enter the foothills of G/G/1. Not even the average waiting time is known for this queue! In Chapter 8, we nevertheless develop a "spectral" method for handling these systems which often leads to useful results. The difficult part of this method reduces to locating the roots of a function, as we have so often seen before. The spectral method suffers from the disadvantage of not providing one with the general behavior pattern of the system; each new queue must be studied by itself. However, we do discuss Kingman's algebra for queues, which so nicely exposes the common framework for all of the various methods so far used to attack the G/G/1 queue. Finally, we introduce the concept of a dual queue, and express some of our principal results in terms of idle times and dual queues.

8

The Queue G/G/1

We have so far made effective use of the Markovian property in the queueing systems M/M/1, M/G/1, and G/M/m. We must now leave behind many (but not all) of the simplifications that derive from the Markovian property and find new methods for studying the more difficult system G/G/1.

In this chapter we solve the G/G/1 system equations by *spectral* methods, making use of transform and complex-variable techniques. There are, however, numerous other approaches: In Section 5.11 we introduced the ladder indices and pointed out the way in which they were related to important events in queueing systems; these ideas can be extended and applied to the general system G/G/1. *Fluctuations* of sums of random variables (i.e., the ladder indices) have been studied by Andersen [ANDE 53*a*, ANDE 53*b*, ANDE 54] and also by Spitzer [SPIT 56, SPIT 60], who simplified and expanded Andersen's work. This led, among other things, to *Spitzer's identity*, of great importance in that approach to queueing theory. Much earlier (in the 1930's) Pollaczek considered a formalism for solving these systems and his approach (summarized in 1957 [POLL 57]) is now referred to as *Pollaczek's method*. More recently, Kingman [KING 66] has developed an *algebra for queues*, which places all these methods in a common framework and exposes the underlying similarity among them; he also identifies where the problem gets difficult and why, but unfortunately he shows that this method does not extend to the multiple server system. Keilson [KEIL 65] applies the method of *Green's function*. Beneš [BENE 63] studied G/G/1 through the *unfinished work* and its "relatives."

Let us now establish the basic equations for this system.

8.1. LINDLEY'S INTEGRAL EQUATION

The system under consideration is one in which the interarrival times between customers are independent and are given by an arbitrary distribution $A(t)$. The service times are also independently drawn from an arbitrary distribution given by $B(x)$. We assume there is one server available and that service is offered in a first-come–first-served order. The basic relationship among the pertinent random variables is derived in this section and leads to Lindley's integral equation, whose solution is given in the following section.

We consider a sequence of arriving customers indexed by the subscript n and remind the reader of our earlier notation:

C_n = the nth customer arriving to the system
$t_n = \tau_n - \tau_{n-1}$ = interarrival time between C_{n-1} and C_n
x_n = service time for C_n
w_n = waiting time (in queue) for C_n

We assume that the random variables $\{t_n\}$ and $\{x_n\}$ are independent and are given, respectively, by the distribution functions $A(t)$ and $B(x)$ independent of the subscript n. As always, we look for a Markov process to simplify our analysis. Recall for M/G/1, that the unfinished work $U(t)$ is a Markov process for all t. For G/G/1, it should be clear that although $U(t)$ is no longer Markovian, imbedded within $U(t)$ is a crucial Markov process defined at the *customer-arrival times*. At these regeneration points, all of the past history that is pertinent to future behavior is completely summarized in the current value of $U(t)$. That is, for FCFS systems, the value of the unfinished work just prior to the arrival of C_n is exactly equal to his waiting time (w_n) *and this Markov process is the object of our study*. In Figures 8.1 and 8.2 we use the time-diagram notation for queues (as defined in Figure 2.2) to illustrate the history of C_n in two cases: Figure 8.1 displays the case where C_{n+1} arrives to the system before C_n departs from the service facility; and Figure 8.2 shows the case in which C_{n+1} arrives to an empty system. For the conditions of Figure 8.1 it is clear that

$$t_{n+1} + w_{n+1} = w_n + x_n$$

That is,

$$w_{n+1} = w_n + x_n - t_{n+1} \qquad \text{if} \qquad w_n + x_n - t_{n+1} \geq 0 \tag{8.1}$$

The condition expressed in Eq. (8.1) assures that C_{n+1} arrives to find a busy

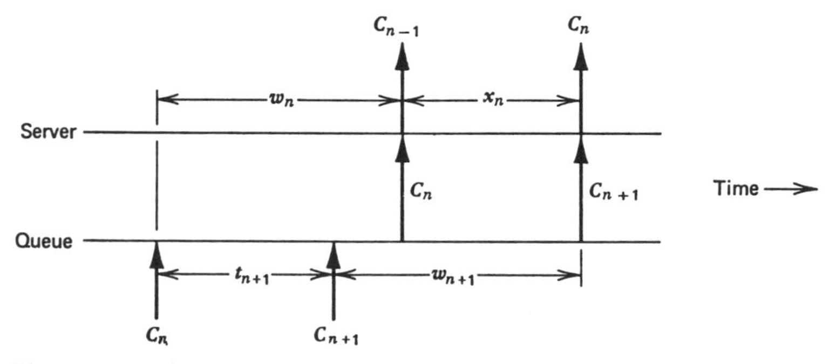

Figure 8.1 The case where C_{n+1} arrives to find a busy system.

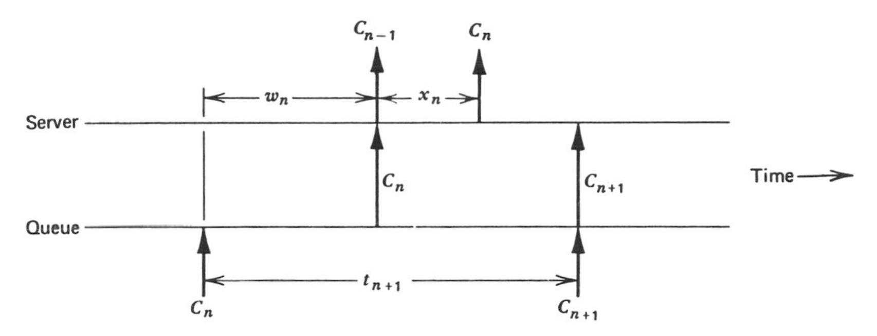

Figure 8.2 The case where C_{n+1} arrives to find an idle system.

system. From Figure 8.2 we see immediately that

$$w_{n+1} = 0 \qquad \text{if} \qquad w_n + x_n - t_{n+1} \leq 0 \tag{8.2}$$

where the condition in Eq. (8.2) assures that C_{n+1} arrives to find an idle system. For convenience we now define a new (key) random variable u_n as

$$u_n \triangleq x_n - t_{n+1} \tag{8.3}$$

This random variable is merely the difference between the service time for C_n and the interarrival time between C_{n+1} and C_n (for a stable system we will require that the expectation of u_n be negative). We may thus combine Eqs. (8.1)–(8.3) to obtain the following fundamental and yet elementary relationship, first established by Lindley [LIND 52];

$$w_{n+1} = \begin{cases} w_n + u_n & \text{if} \quad w_n + u_n \geq 0 \\ 0 & \text{if} \quad w_n + u_n \leq 0 \end{cases} \tag{8.4}$$

The term $w_n + u_n$ is merely the sum of the unfinished work (w_n) found by C_n plus the service time (x_n), which he now adds to the unfinished work, less the time duration (t_{n+1}) until the arrival of the next customer C_{n+1}; if this quantity is nonnegative then it represents the amount of unfinished work found by C_{n+1} and therefore represents his waiting time w_{n+1}. However, if this quantity goes negative it indicates that an interval of time has elapsed since the arrival of C_n, which exceeds the amount of unfinished work present in the system just after the arrival of C_n, thereby indicating that the system has gone idle by the time C_{n+1} arrives.

We may write Eq. (8.4) as

$$w_{n+1} = \max\,[0, w_n + u_n] \tag{8.5}$$

We introduce the notation $(x)^+ \triangleq \max\,[0, x]$; we then have

$$w_{n+1} = (w_n + u_n)^+ \qquad \blacksquare \tag{8.6}$$

Since the random variables $\{t_n\}$ and $\{x_n\}$ are independent among themselves and each other, then one observes that the sequence of random variables $\{w_0, w_1, w_2, \ldots\}$ forms a Markov process with stationary transition probabilities. This can be seen immediately from Eq. (8.4) since the new value w_{n+1} depends upon the previous sequence of random variables w_i ($i = 0, 1, \ldots, n$) only through the most recent value w_n plus a random variable u_n, which is independent of the random variables w_i for all $i \leq n$.

Let us solve Eq. (8.5) recursively beginning with w_0 as an initial condition. We have (defining C_0 to be our initial arrival)

$$\begin{aligned}
w_1 &= (w_0 + u_0)^+ \\
w_2 &= (w_1 + u_1)^+ = \max[0, w_1 + u_1] \\
&\qquad = \max[0, u_1 + \max(0, w_0 + u_0)] \\
&\qquad = \max[0, u_1, u_1 + u_0 + w_0] \\
w_3 &= (w_2 + u_2)^+ = \max[0, w_2 + u_2] \\
&\qquad = \max[0, u_2 + \max(0, u_1, u_1 + u_0 + w_0)] \\
&\qquad = \max[0, u_2, u_2 + u_1, u_2 + u_1 + u_0 + w_0] \\
&\;\;\vdots \\
w_n &= (w_{n-1} + u_{n-1})^+ = \max[0, w_{n-1} + u_{n-1}] \\
&\qquad = \max[0, u_{n-1}, u_{n-1} + u_{n-2}, \ldots, u_{n-1} + \cdots + u_1, \\
&\qquad\qquad u_{n-1} + \cdots + u_1 + u_0 + w_0]
\end{aligned} \tag{8.7}$$

However, since the sequence of random variables $\{u_i\}$ is a sequence of independent and identically distributed random variables, then they are "interchangeable" and we may consider a new random variable w_n' with the same distribution as w_n, where

$$\begin{aligned}
w_n' \triangleq \max[0, u_0, u_0 + u_1, u_0 + u_1 + u_2, \ldots, u_0 + u_1 + \cdots + u_{n-2}, \\
u_0 + u_1 + \cdots + u_{n-2} + u_{n-1} + w_0]
\end{aligned} \tag{8.8}$$

Equation (8.8) is obtained from Eq. (8.7) by relabeling the random variables u_i. It is now convenient to define the quantities U_n as

$$\begin{aligned}
U_n &= \sum_{i=0}^{n-1} u_i \\
U_0 &= 0
\end{aligned} \tag{8.9}$$

We thus have from Eq. (8.8)

$$w_n' = \max[U_0, U_1, U_2, \ldots, U_{n-1}, U_n + w_0] \tag{8.10}$$

From this last form we see for $w_0 = 0$ that w_n' can only increase with n. Therefore the limiting random variable lim w_n' as $n \to \infty$ must converge to the (possibly infinite) random variable $\tilde{w}$

$$\tilde{w} = \sup_{n \geq 0} U_n \tag{8.11}$$

Our imbedded Markov chain is ergodic if, with probability one, $\tilde{w}$ is finite, and if so, then the distribution of w_n' and of w_n both converge to the distribution of $\tilde{w}$; in this case, the distribution of $\tilde{w}$ is the waiting-time distribution. Lindley [LIND 52] has shown that for $0 < E[|u_n|] < \infty$ then the system is stable if and only if $E[u_n] < 0$. Therefore, we will henceforth assume

$$E[u_n] < 0 \tag{8.12}$$

Equation (8.12) is our usual condition for stability as may be seen from the following:

$$\begin{aligned} E[u_n] &= E[x_n - t_{n+1}] \\ &= E[x_n] - E[t_{n+1}] \\ &= \bar{x} - \bar{t} \\ &= \bar{t}(\rho - 1) \end{aligned} \tag{8.13}$$

where as usual we assume that the expected service time is $\bar{x}$ and the expected interarrival time is $\bar{t}$ (and we have $\rho = \bar{x}/\bar{t}$). From Eqs. (8.12) and (8.13) we see we have required that $\rho < 1$, as is our usual condition for stability. Let us denote (as usual) the stationary distribution for w_n (and also therefore for w_n') by

$$\lim_{n \to \infty} P[w_n \leq y] = \lim_{n \to \infty} P[w_n' \leq y] = W(y) \tag{8.14}$$

which must exist for $\rho < 1$ [LIND 52]. Thus $W(y)$ will be our assumed stationary distribution for time spent in queue; we will not dwell upon the proof of its existence but rather upon the method for its calculation. As we know for such Markov processes, this limiting distribution is independent of the initial state w_0.

Before proceeding to the formal derivation of results let us investigate the way in which Eq. (8.7) in fact produces the waiting time. This we do by example; consider Figure 8.3, which represents the unfinished work $U(t)$. For the sequence of arrivals and departures given in this figure, we present the table below showing the interarrival times t_{n+1}, service times x_n, the random variables u_n, and the waiting time w_n as measured from the diagram; in the last row of this table we give the waiting times w_n as calculated from Eq. (8.7) as follows.

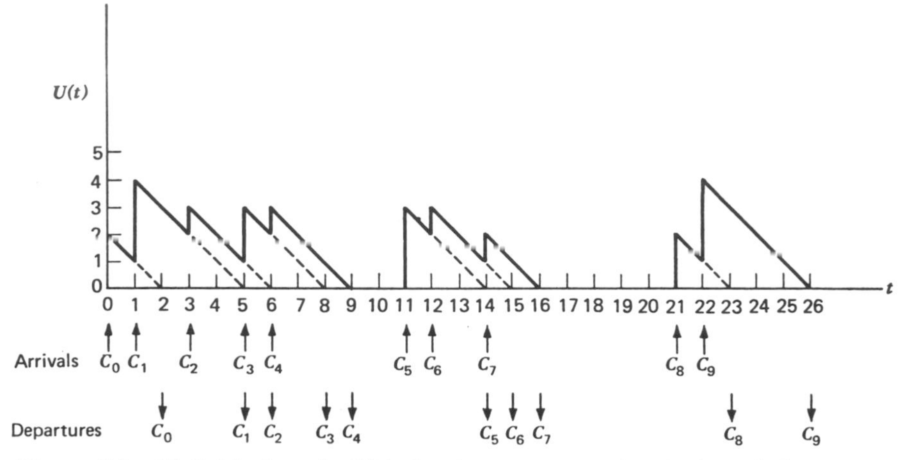

Figure 8.3 Unfinished work $U(t)$ showing sequence of arrivals and departures.

Table of values from Figure 8.3.

n	0	1	2	3	4	5	6	7	8	9	$\cdots$	
t_{n+1}	1	2	2	1	5	1	2	7	1	$\cdots$		
x_n	2	3	1	2	1	3	1	1	2	3	$\cdots$	
u_n	1	1	−1	1	−4	2	−1	−6	1	$\cdots$		
w_n	0	1	2	1	2	0	2	1	0	1	$\cdots$	measured from Fig. 8.3
w_n	0	1	2	1	2	0	2	1	0	1	$\cdots$	calculated from Eq. 8.7

$$w_0 = 0$$
$$w_1 = \max(0, w_0 + u_0) = \max(0, 1) = 1$$
$$w_2 = \max(0, u_1, u_1 + u_0 + w_0) = \max(0, 1, 2) = 2$$
$$w_3 = \max(0, u_2, u_2 + u_1, u_2 + u_1 + u_0 + w_0) = \max(0, -1, 0, 1) = 1$$
$$w_4 = \max(0, u_3, u_3 + u_2, u_3 + u_2 + u_1, u_3 + u_2 + u_1 + u_0 + w_0)$$
$$= \max(0, 1, 0, 1, 2) = 2$$
$$w_5 = \max(0, u_4, u_4 + u_3, u_4 + u_3 + u_2, u_4 + u_3 + u_2 + u_1, u_4 + u_3 + u_2 + u_1 + u_0 + w_0)$$
$$= \max(0, -4, -3, -4, -3, -2) = 0$$
$$w_6 = \max(0, 2, -2, -1, -2, -1, 0) = 2$$
$$w_7 = \max(0, -1, 1, -3, -2, -3, -2, -1) = 1$$
$$w_8 = \max(0, -6, -7, -5, -9, -8, -9, -8, -7) = 0$$
$$w_9 = \max(0, 1, -5, -6, -4, -8, -7, -8, -7, -6) = 1$$

These calculations are quite revealing. For example, whenever we find an m for which $w_m = 0$, then the m rightmost calculations in Eq. (8.7) need be made no more in calculating w_n for all $n > m$; this is due to the fact that a busy period has ended and the service times and interarrival times from that busy period cannot affect the calculations in future busy periods. Thus we see the isolating effect of idle periods which ensue between busy periods. Furthermore, when $w_m = 0$, then the rightmost term $(U_m + w_0)$ gives the (negative of the) total accumulated idle time of the system during the interval $(0, \tau_m)$.

Let us now proceed with the theory for calculating $W(y)$. We define $C_n(u)$ as the PDF for the random variable u_n, that is,

$$C_n(u) \triangleq P[u_n = x_n - t_{n+1} \leq u] \tag{8.15}$$

and we note that u_n is *not* restricted to a half line. We now derive the expression for $C_n(u)$ in terms of $A(t)$ and $B(x)$:

$$\begin{aligned} C_n(u) &= P[x_n - t_{n+1} \leq u] \\ &= \int_{t=0}^{\infty} P[x_n \leq u + t \mid t_{n+1} = t] \, dA(t) \end{aligned}$$

However, the service time for C_n is independent of t_{n+1} and therefore

$$C_n(u) = \int_{t=0}^{\infty} B(u + t) \, dA(t) \tag{8.16}$$

Thus, as we expected, $C_n(u)$ is independent of n and we therefore write

$$C_n(u) \triangleq C(u) = \int_{t=0}^{\infty} B(u + t) \, dA(t) \tag{8.17}$$

Also, let $\tilde{u}$ denote the random variable

$$\tilde{u} = \tilde{x} - \tilde{t}$$

Note that the integral given in Eq. (8.17) is very much like a convolution form for $a(t)$ and $B(x)$; it is not quite a straight convolution since the distribution $C(u)$ represents the *difference* between x_n and t_{n+1} rather than the *sum.* Using our convolution notation ($\circledast$), and defining $c_n(u) \triangleq dC_n(u)/du$ we have

$$c_n(u) = c(u) = a(-u) \circledast b(u) \qquad \blacksquare \tag{8.18}$$

It is (again) convenient to define the waiting-time distribution for customer C_n as

$$W_n(y) = P[w_n \leq y] \tag{8.19}$$

For $y \geq 0$ we have from Eq. (8.4)

$$W_{n+1}(y) = P[w_n + u_n \leq y]$$
$$= \int_{0^-}^{\infty} P[u_n \leq y - w \mid w_n = w] \, dW_n(w)$$

And now once again, since u_n is independent of w_n we have

$$W_{n+1}(y) = \int_{0^-}^{\infty} C_n(y - w) \, dW_n(w) \qquad \text{for } y \geq 0 \tag{8.20}$$

However, as postulated in Eq. (8.14) this distribution has a limit $W(y)$ and therefore we have the following integral equation, which defines the limiting distribution of waiting time for customers in the system G/G/1:

$$W(y) = \int_{0^-}^{\infty} C(y - w) \, dW(w) \qquad \text{for } y \geq 0$$

Further, it is clear that

$$W(y) = 0 \qquad \text{for } y < 0$$

Combining these last two we have *Lindley's integral equation* [LIND 52], which is seen to be an integral equation of the Wiener–Hopf type [SPIT 57].

$$W(y) = \begin{cases} \int_{0^-}^{\infty} C(y - w) \, dW(w) & y \geq 0 \\ 0 & y < 0 \end{cases} \tag{8.21}$$

Equation (8.21) may be rewritten in at least two other useful forms, which we now proceed to derive. Integrating by parts, we have (for $y \geq 0$)

$$W(y) = C(y - w)W(w)\Big|_{w=0^-}^{\infty} - \int_{0^-}^{\infty} W(w) \, dC(y - w)$$
$$= \lim_{w \to \infty} C(y - w)W(w) - C(y)W(0^-) - \int_{0^-}^{\infty} W(w) \, dC(y - w)$$

We see that $\lim C(y - w) = 0$ as $w \to \infty$ since the limit of $C(u)$ as $u \to -\infty$ is the probability that an interarrival time approaches infinity, which clearly must go to zero if the interarrival time is to have finite moments. Similarly, we have $W(0^-) = 0$ and so our form for Lindley's integral equation may be rewritten as

$$W(y) = \begin{cases} -\int_{0^-}^{\infty} W(w) \, dC(y - w) & y \geq 0 \\ 0 & y < 0 \end{cases} \tag{8.22}$$

Let us now show a third form for this equation. By the simple variable change $u = y - w$ for the argument of our distributions we finally arrive at

$$W(y) = \begin{cases} \int_{-\infty}^{y} W(y - u)\, dC(u) & y \geq 0 \\ 0 & y < 0 \end{cases} \quad \blacksquare (8.23)$$

Equations (8.21), (8.22), and (8.23) all describe the basic integral equation which governs the behavior of G/G/1. These integral equations, as mentioned above, are Weiner–Hopf-type integral equations and are not unfamiliar in the theory of stochastic processes.

One observes from these forms that Lindley's integral equation is almost, but not quite, a convolution integral. The important distinction between a convolution integral and that given in Lindley's equation is that the latter integral form holds only when the variable is nonnegative; the distribution function is identically zero for values of negative argument. Unfortunately, since the integral holds only for the half-line we must borrow techniques from the theory of complex variables and from contour integration in order to solve our system. We find a similar difficulty in the design of optimal linear filters in the mathematical theory of communication; there too, a Weiner–Hopf integral equation describes the optimal solution, except that for linear filters, the unknown appears as one factor in the integrand rather than as in our case in queueing theory, where the unknown appears on both sides of the integral equation. Nevertheless, the solution techniques are amazingly similar and the reader acquainted with the theory of optimal realizable linear filters will find the following arguments familiar.

In the next section, we give a fairly general solution to Lindley's integral equation by the use of spectral (transform) methods. In Exercise 8.6 we examine a solution approach by means of an example that does not require transforms; the example chosen is the system $D/E_r/1$ considered by Lindley. In that (direct) approach it is required to assume the solution form. We now consider the spectral solution to Lindley's equation in which such assumed solution forms will not be necessary.

8.2. SPECTRAL SOLUTION TO LINDLEY'S INTEGRAL EQUATION

In this section we describe a method for solving Lindley's integral equation by means of spectrum factorization [SMIT 53]. Our point of departure is the form for this equation given by (8.23). As mentioned earlier it would be rather straightforward to solve this equation if the right-hand side were a true convolution (it is, in fact, a convolution for the nonnegative half-line on the variable y but not so otherwise). In order to get around this difficulty we use the following ingenious device whereby we define a "complementary" waiting time, which completes the convolution, and which takes on the value of the integral for negative y only, that is,

$$W_-(y) \triangleq \begin{cases} 0 & y \geq 0 \\ \int_{-\infty}^{y} W(y-u)\, dC(u) & y < 0 \end{cases} \tag{8.24}$$

Note that the left-hand side of Eq. (8.23) might consistently be written as $W_+(y)$ in the same way in which we defined the left-hand side of Eq. (8.24). We now observe that if we add Eqs. (8.23) and (8.24) then the right-hand side takes on the integral expression for all values of the argument, that is,

$$W(y) + W_-(y) = \int_{-\infty}^{y} W(y-u)c(u)\, du \qquad \text{for all real } y \tag{8.25}$$

where we have denoted the pdf for $\tilde{u}$ by $c(u)$ [$\triangleq dC(u)/du$].

To proceed, we assume that the pdf of the interarrival time is* $0(e^{-Dt})$ as $t \to \infty$ (where D is any real number greater than zero), that is,

$$\lim_{t\to\infty} \frac{a(t)}{e^{-Dt}} < \infty \tag{8.26}$$

The condition (8.26) really insists that the pdf associated with the interarrival time drops off at least as fast as an exponential for very large interarrival times. From this condition it may be seen from Eq. (8.17) that the behavior of $C(u)$ as $u \to -\infty$ is governed by the behavior of the interarrival time; this is true since as u takes on large negative values the argument for the service-time distribution can be made positive only for large values of t, which also appears as the argument for the interarrival time density. Thus we can show

$$\lim_{u\to-\infty} \frac{C(u)}{e^{Du}} < \infty$$

That is, $C(u)$ is $0(e^{Du})$ as $u \to -\infty$. If we now use this fact in Eq. (8.24) it is easy to establish that $W_-(y)$ is also $0(e^{Dy})$ as $y \to -\infty$.

* The notation $0(g(x))$ as $x \to x_0$ refers to any function that (as $x \to x_0$) decays to zero at least as rapidly as $g(x)$ [where $g(x) > 0$], that is,

$$\lim_{x\to x_0} \left| \frac{0(g(x))}{g(x)} \right| = K < \infty$$

Let us now define some (bilateral) transforms for various of our functions. For the Laplace transform of $W_-(y)$ we define

$$\Phi_-(s) \triangleq \int_{-\infty}^{\infty} W_-(y)e^{-sy}\,dy \tag{8.27}$$

Due to the condition we have established regarding the asymptotic property of $W_-(y)$, it is clear that $\Phi_-(s)$ is analytic in the region Re $(s) < D$. Similarly, for the distribution of our waiting time $W(y)$ we define

$$\Phi_+(s) \triangleq \int_{-\infty}^{\infty} W(y)e^{-sy}\,dy \tag{8.28}$$

Note that $\Phi_+(s)$ is the Laplace transform of the PDF for waiting time, whereas in previous chapters we have defined $W^*(s)$ as the Laplace transform of the pdf for waiting time; thus by entry 11 of Table I.3, we have

$$s\Phi_+(s) = W^*(s) \tag{8.29}$$

Since there are regions for Eqs. (8.23) and (8.24) in which the functions drop to zero, we may therefore rewrite these transforms as

$$\Phi_-(s) = \int_{-\infty}^{0^-} W_-(y)e^{-sy}\,dy \tag{8.30}$$

$$\Phi_+(s) = \int_{0^-}^{\infty} W(y)e^{-sy}\,dy \tag{8.31}$$

Since $W(y)$ is a true distribution function (and therefore it remains bounded as $y \to \infty$) then $\Phi_+(s)$ is analytic for Re $(s) > 0$. As usual, we define the transform for the pdf of the interarrival time and for the pdf of the service time as $A^*(s)$ and $B^*(s)$, respectively. Note for the condition (8.26) that $A^*(-s)$ is analytic in the region Re $(s) < D$ just as was $\Phi_-(s)$.

From Appendix I we recall that the Laplace transform for the convolution of two functions is the product of the transforms of each. Equation (8.18) is almost the convolution of the service-time density with the interarrival-time density; the only difficulty is the negative argument for the interarrival-time density. Nevertheless, the above-mentioned fact regarding products of transforms goes through merely with the negative argument (this is Exercise 8.1). Thus for the Laplace transform of $c(u)$ we find

$$C^*(s) = A^*(-s)B^*(s) \tag{8.32}$$

Let us now return to Eq. (8.25), which expresses the fundamental relationship among the variables of our problem and the waiting-time distribution $W(y)$. Clearly, the time spent in queue must be a nonnegative random variable, and so we recognize the right-hand side of Eq. (8.25) as a convolution between the waiting time PDF and the pdf for the random variable $\tilde{u}$. The Laplace transform of this convolution must therefore give the product of the Laplace transform $\Phi_+(s)$ (for the waiting-time distribution) and $C^*(s)$ (for the density on $\tilde{u}$). The transform of the left-hand side we recognize from Eqs. (8.30) and (8.31) as being $\Phi_+(s) + \Phi_-(s)$, thus

$$\Phi_+(s) + \Phi_-(s) = \Phi_+(s)C^*(s)$$

From Eq. (8.32) we therefore obtain

$$\Phi_+(s) + \Phi_-(s) = \Phi_+(s)A^*(-s)B^*(s)$$

which gives us

$$\Phi_-(s) = \Phi_+(s)[A^*(-s)B^*(s) - 1] \tag{8.33}$$

We have already established that both $\Phi_-(s)$ and $A^*(-s)$ are analytic in the region $\text{Re}(s) < D$. Furthermore, since $\Phi_+(s)$ and $B^*(s)$ are transforms of bounded functions of nonnegative variables then both functions must be analytic in the region $\text{Re}(s) > 0$.

We now come to the *spectrum factorization*. The purpose of this factorization is to find a suitable representation for the term

$$A^*(-s)B^*(s) - 1 \tag{8.34}$$

in the form of two factors. Let us pause for a moment and recall the method of stages whereby Erlang conceived the ingenious idea of approximating a distribution by means of a collection of series and parallel exponential stages. The Laplace transform for the pdf's obtainable in this fashion was generally given in Eq. (4.62) or Eq. (4.64); we immediately recognize these to be rational functions of s (that is, a ratio of a polynomial in s divided by a polynomial in s). We may similarly conceive of approximating the Laplace transforms $A^*(-s)$ and $B^*(s)$ each in such forms; if we so approximate, then the term given by Eq. (8.34) will also be a rational function of s. We thus choose to consider those queueing systems for which $A^*(s)$ and $B^*(s)$ may be suitably approximated with (or which are given initially as) such rational functions of s, in which case we then propose to form the following spectrum factorization

$$A^*(-s)B^*(s) - 1 = \frac{\Psi_+(s)}{\Psi_-(s)} \tag{8.35}$$

Clearly $\Psi_+(s)/\Psi_-(s)$ will be some rational function of s, and we are now desirous of finding a particular factored form for this expression. We specifically wish to find a factorization such that:

- For $\text{Re}(s) > 0$, $\Psi_+(s)$ is an analytic function of s with no zeroes in this half-plane.
- For $\text{Re}(s) < D$, $\Psi_-(s)$ is an analytic function of s with no zeroes in this half-plane.

(8.36)

Furthermore, we wish to find these functions with the *additional* properties:

$$\begin{aligned} &\bullet \quad \text{For } \operatorname{Re}(s) > 0, \ \lim_{|s|\to\infty} \frac{\Psi_+(s)}{s} = 1. \\ &\bullet \quad \text{For } \operatorname{Re}(s) < D, \ \lim_{|s|\to\infty} \frac{\Psi_-(s)}{s} = -1. \end{aligned} \tag{8.37}$$

The conditions in (8.37) are convenient and must have opposite polarity in the limit since we observe that as s runs off to infinity along the imaginary axis, both $A^*(-s)$ and $B^*(s)$ must decay to 0 [if they are to have finite moments and if $A(t)$ and $B(x)$ do not contain a sequence of discontinuities, which we will not permit] leaving the left-hand side of Eq. (8.35) equal to -1, which we have suitably matched by the ratio of limits given by Conditions (8.37). We shall find that this spectrum factorization, which requires us to find $\Psi_+(s)$ and $\Psi_-(s)$ with the appropriate properties, contains the difficult part of this method of solution. Nevertheless, assuming that we have found such a factorization it is then clear that we may write Eq. (8.33) as

$$\Phi_-(s) = \Phi_+(s)\frac{\Psi_+(s)}{\Psi_-(s)}$$

or

$$\Phi_-(s)\Psi_-(s) = \Phi_+(s)\Psi_+(s) \tag{8.38}$$

where the common region of analyticity for both sides of Eq. (8.38) is within the strip

$$0 < \operatorname{Re}(s) < D \tag{8.39}$$

That this last is true may be seen as follows. We have already assumed that $\Psi_+(s)$ is analytic for $\operatorname{Re}(s) > 0$ and it is further true that $\Phi_+(s)$ is analytic in this same region since it is the Laplace transform of a function that is identically zero for negative arguments; the product of these two must therefore be analytic for $\operatorname{Re}(s) > 0$. Similarly, $\Psi_-(s)$ has been given to be analytic for $\operatorname{Re}(s) < D$ and we have that $\Phi_-(s)$ is analytic here as explained earlier following Eq. (8.27); thus the product of these two will be analytic in $\operatorname{Re}(s) < D$. Thus the common region is as stated in Eq. (8.39). Now, Eq. (8.38) establishes that these two functions are equal in the common strip and so they must represent functions which, when continued in the region $\operatorname{Re}(s) < 0$, are analytic and when continued in the region $\operatorname{Re}(s) > D$, are also analytic; therefore their analytic continuation contains no singularities in the entire finite s-plane. Since we have established the behavior of the function $\Phi_+(s)\Psi_+(s) = \Phi_-(s)\Psi_-(s)$ to be analytic and bounded in the finite s-plane, and since we assume Condition (8.37), we may then apply Liouville's theorem*

* Liouville's theorem states, "If $f(z)$ is analytic and bounded for all finite values of z, then $f(z)$ is a constant."

[TITC 52], which immediately establishes that this function must be a constant (say, K). We thus have

$$\Phi_-(s)\Psi_-(s) = \Phi_+(s)\Psi_+(s) = K \tag{8.40}$$

This immediately yields

$$\Phi_+(s) = \frac{K}{\Psi_+(s)} \tag{8.41}$$

The reader should recall that what we are seeking in this development is an expression for the distribution of queueing time whose Laplace transform is exactly the function $\Phi_+(s)$, which is now given through Eq. (8.41). It remains for us to demonstrate a method for evaluating the constant K.

Since $s\Phi_+(s) = W^*(s)$, we have

$$s\Phi_+(s) = W^*(s) \triangleq \int_{0^-}^{\infty} e^{-sy}\, dW(y)$$

Let us now consider the limit of this equation as $s \to 0$; working with the right-hand side we have

$$\lim_{s\to 0} \int_{0^-}^{\infty} e^{-sy}\, dW(y) = \int_{0^-}^{\infty} dW(y) = 1$$

We have thus established

$$\lim_{s\to 0} s\,\Phi_+(s) = 1 \tag{8.42}$$

This is nothing more than the final value theorem (entry 18, Table I.3) and comes about since $W(\infty) = 1$. From Eq. (8.41) and this last result we then have

$$\lim_{s\to 0} s\Phi_+(s) = \lim_{s\to 0} \frac{sK}{\Psi_+(s)} = 1$$

and so we may write

$$K = \lim_{s\to 0} \frac{\Psi_+(s)}{s} \tag{8.43}$$

Equation (8.43) provides a means of calculating the constant K in our solution for $\Phi_+(s)$ as given in Eq. (8.41). If we make a Taylor expansion of the function $\Psi_+(s)$ around $s = 0$ [viz., $\Psi_+(s) = \Psi_+(0) + s\Psi_+^{(1)}(0) + (s^2/2!)\Psi_+^{(2)}(0) + \cdots$] and note from Eqs. (8.35) and (8.36) that $\Psi_+(0) = 0$, we then recognize that this limit may also be written as

$$K = \lim_{s\to 0} \frac{d\Psi_+(s)}{ds} \tag{8.44}$$

and this provides us with an alternate way for calculating the constant K. We may further explore this constant K by examining the behavior of $\Phi_+(s)\Psi_+(s)$ anywhere in the region $\text{Re}(s) > 0$ [i.e., see Eq. (8.40)]; we choose to examine this behavior in the limit as $s \to \infty$ where we know from Eq. (8.37) that $\Psi_+(s)$ behaves as s does; that is,

$$\begin{aligned} K &= \lim_{s\to\infty} \Phi_+(s)\Psi_+(s) \\ &= \lim_{s\to\infty} \Phi_+(s)s \\ &= \lim_{s\to\infty} s\int_{0^-}^{\infty} e^{-sy}W(y)\,dy \end{aligned}$$

Making the change of variable $sy = x$ we have

$$K = \lim_{s\to\infty} \int_{0^-}^{\infty} e^{-x}W\left(\frac{x}{s}\right)dx$$

As $s \to \infty$ we may pull the constant term $W(0^+)$ outside the integral and then obtain the value of the remaining integral, which is unity. We thus obtain

$$K = W(0^+) \tag{8.45}$$

This establishes that the constant K is merely the probability that an arriving customer need not queue†.

In conclusion then, assuming that we can find the appropriate spectrum factorization in Eq. (8.35) we may immediately solve for the Laplace transform of the waiting-time distribution through Eq. (8.41), where the constant K is given in either of the three forms Eq. (8.43), (8.44), or (8.45). Of course it then remains to invert the transform but the problems involved in that calculation have been faced before in numerous of our other solution forms.

It is possible to carry out the solution of this problem by concentrating on $\Psi_-(s)$ rather than $\Psi_+(s)$, and in some cases this simplifies the calculations. In such cases we may proceed from Eq. (8.35) to obtain

$$\Psi_+(s) = \Psi_-(s)[A^*(-s)B^*(s) - 1] \tag{8.46}$$

From Eq. (8.41) we then have

$$\Phi_+(s) = \frac{K}{[A^*(-s)B^*(s) - 1]\Psi_-(s)} \tag{8.47}$$

† Note that $W(0^+)$ is *not* necessarily equal to $1 - \rho$, which is the fraction of time the server is idle. (These two *are* equal for the system M/G/1.)

In order to evaluate the constant K in this case we differentiate Eq. (8.46) at $s = 0$, that is,

$$\Psi_+^{(1)}(0) = [A^*(0)B^*(0) - 1]\Psi_-^{(1)}(0) + \Psi_-(0)[A^*(0)B^{*(1)}(0) - A^{*(1)}(0)B^*(0)] \tag{8.48}$$

From Eq. (8.44) we recognize the left-hand side of Eq. (8.48) as the constant K and we may now evaluate the right-hand side to obtain

$$K = 0 + \Psi_-(0)[-\bar{x} + \bar{t}]$$

giving

$$K = \Psi_-(0)(1 - \rho)\bar{t} \tag{8.49}$$

Thus, if we wish to use $\Psi_-(s)$ in our solution form, we obtain the transform of the waiting-time distribution from Eq. (8.47), where the unknown constant K is evaluated in terms of $\Psi_-(s)$ through Eq. (8.49).

Summarizing then, once we have carried out the spectrum factorization as indicated in Eq. (8.35), we may proceed in one of two directions in solving for $\Phi_+(s)$, the transform of the waiting-time distribution. The first method gives us

$$\Phi_+(s) = \frac{1}{\Psi_+(s)} \lim_{s\to 0} \frac{\Psi_+(s)}{s} = \frac{W(0^+)}{\Psi_+(s)} \qquad \blacksquare(8.50)$$

and the second provides us with

$$\Phi_+(s) = \frac{\Psi_-(0)(1-\rho)\bar{t}}{[A^*(-s)B^*(s) - 1]\Psi_-(s)} \qquad \blacksquare(8.51)$$

We now proceed to demonstrate the use of these results in some examples.

Example 1: M/M/1

Our old friend M/M/1 is extremely straightforward and should serve to clarify the meaning of spectrum factorization. Since both the interarrival time and the service time are exponentially distributed random variables, we immediately have $A^*(s) = \lambda/(s + \lambda)$ and $B^*(s) = \mu/(s + \mu)$, where $\bar{x} = 1/\mu$ and $\bar{t} = 1/\lambda$. In order to solve for $\Phi_+(s)$ (the transform of the waiting time distribution), we must first form the expression given in Eq. (8.34), that is,

$$\begin{aligned} A^*(-s)B^*(s) - 1 &= \left(\frac{\lambda}{\lambda - s}\right)\left(\frac{\mu}{s + \mu}\right) - 1 \\ &= \frac{s^2 + s(\mu - \lambda)}{(\lambda - s)(s + \mu)} \end{aligned}$$

Thus, from Eq. (8.35), we obtain

$$\frac{\Psi_+(s)}{\Psi_-(s)} = A^*(-s)B^*(s) - 1$$

$$= \frac{s(s + \mu - \lambda)}{(s + \mu)(\lambda - s)} \tag{8.52}$$

In Figure 8.4 we show the location of the zeroes (denoted by a circle) and poles (denoted by a cross) in the complex s-plane for the function given in Eq. (8.52). Note that in this particular example the roots of the numerator (zeroes of the expression) and the roots of the denominator (poles of the expression) are especially simple to find; in general, one of the most difficult parts of this method of spectrum factorization is to solve for the roots. In order to factorize we require that conditions (8.36) and (8.37) maintain. Inspecting the pole–zero plot in Figure 8.4 and remembering that $\Psi_+(s)$ must be analytic and zero-free for Re $(s) > 0$, we may collect together the two zeroes (at $s = 0$ and $s = -\mu + \lambda$) and one pole (at $s = -\mu$) and still satisfy this required condition. Similarly, $\Psi_-(s)$ must be analytic and free from zeroes for the Re $(s) < D$ for some $D > 0$; we can obtain such a condition if we allow this function to contain the remaining pole (at $s = \lambda$) and choose $D = \lambda$. This we show in Figure 8.5.

Thus we have

$$\Psi_+(s) = \frac{s(s + \mu - \lambda)}{s + \mu} \tag{8.53}$$

$$\Psi_-(s) = \lambda - s \tag{8.54}$$

Note that Condition (8.37) is satisfied for the limit as $s \to \infty$.

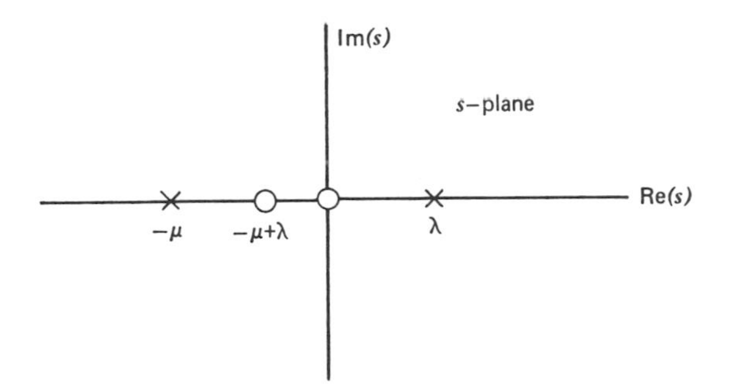

Figure 8.4 Zeroes (O) and poles (×) of $\Psi_+(s)/\Psi_-(s)$ for M/M/1.

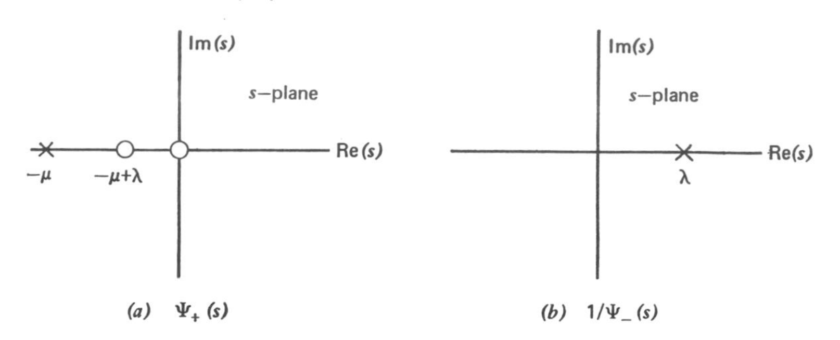

Figure 8.5 Factorization into $\Psi_+(s)$ and $1/\Psi_-(s)$ for M/M/1.

We are now faced with finding K. From Eq. (8.43) we have

$$\begin{aligned} K &= \lim_{s\to 0} \frac{\Psi_+(s)}{s} \\ &= \lim_{s\to 0} \frac{s+\mu-\lambda}{s+\mu} \\ &= 1-\rho \end{aligned} \tag{8.55}$$

Our expression for the Laplace transform of the waiting time PDF for M/M/1 is therefore from Eq. (8.41),

$$\Phi_+(s) = \frac{(1-\rho)(s+\mu)}{s(s+\mu-\lambda)} \tag{8.56}$$

At this point, typically, we attempt to invert the transform to get the waiting-time distribution. However, for this M/M/1 example, we have already carried out this inversion for $W^*(s) = s\Phi_+(s)$ in going from Eq. (5.120) to Eq. (5.123). The solution we obtain is the familiar form,

$$W(y) = 1 - \rho e^{-\mu(1-\rho)y} \qquad y \geq 0 \tag{8.57}$$

Example 2: G/M/1†

In this case $B^*(s) = \mu/(s+\mu)$ but now $A^*(s)$ is completely arbitrary, giving us

$$A^*(-s)B^*(s) - 1 = \frac{A^*(-s)\mu}{s+\mu} - 1$$

† This example forces us to locate roots using Rouche's theorem in a way often necessary for specific G/G/1 problems when the spectrum factorization method is used. Of course, we have already studied this system in Section 6.4 and will compare the results for both methods.

and so we have

$$\frac{\Psi_+(s)}{\Psi_-(s)} = \frac{\mu A^*(-s) - s - \mu}{s + \mu} \tag{8.58}$$

In order to factorize we must find the roots of the numerator in this equation. We need not concern ourselves with the poles due to $A^*(-s)$ since they must lie in the region Re $(s) > 0$ [i.e., $A(t) = 0$ for $t < 0$] and we are attempting to find $\Psi_+(s)$, which cannot include any such poles. Thus we only study the zeroes of the function

$$s + \mu - \mu A^*(-s) = 0 \tag{8.59}$$

Clearly, one root of this equation occurs at $s = 0$. In order to find the remaining roots, we make use of Rouche's theorem (given in Appendix I but which we repeat here):

Rouche's Theorem *If $f(s)$ and $g(s)$ are analytic functions of s inside and on a closed contour C, and also if $|g(s)| < |f(s)|$ on C, then $f(s)$ and $f(s) + g(s)$ have the same number of zeroes inside C.*

In solving for the roots of Eq. (8.59) we make the identification

$$f(s) = s + \mu$$
$$g(s) = -\mu A^*(-s)$$

We have by definition

$$A^*(-s) = \int_{0^-}^{\infty} e^{st}\, dA(t)$$

We now choose C to be the contour that runs up the imaginary axis and then forms an infinite-radius semicircle moving counterclockwise and surrounding the left half of the s-plane, as shown in Figure 8.6. We consider this contour since we are concerned about all the poles and zeroes in Re $(s) < 0$ so that we may properly include them in $\Psi_+(s)$ [recall that $\Psi_-(s)$ may contain none such]; Rouche's theorem will give us information concerning the number of zeroes in Re $(s) < 0$, which we must consider. As usual, we assume that the real and imaginary parts of the complex variable s are given by σ and ω, respectively, that is, for $j = \sqrt{-1}$

$$s = \sigma + j\omega$$

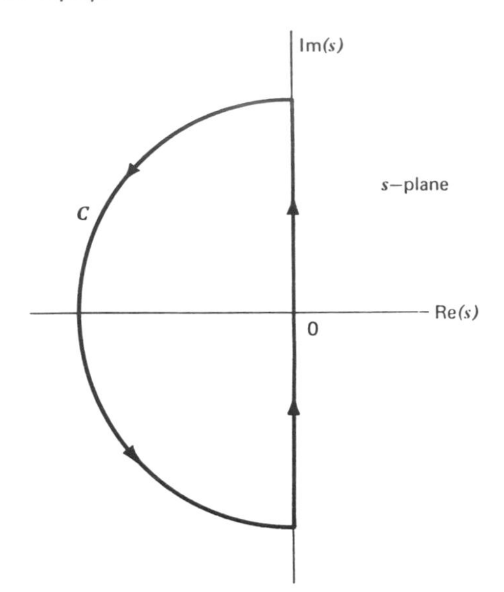

Figure 8.6 The contour C for G/M/1.

Now for the Re $(s) = \sigma \leq 0$ we have $e^{\sigma t} \leq 1$ $(t \geq 0)$ and so

$$
\begin{aligned}
|g(s)| &= \left|\mu \int_{0^-}^{\infty} e^{st}\, dA(t)\right| \\
&= \left|\mu \int_{0^-}^{\infty} e^{\sigma t} e^{j\omega t}\, dA(t)\right| \\
&\leq \left|\mu \int_{0^-}^{\infty} e^{j\omega t}\, dA(t)\right| \\
&\leq \left|\mu \int_{0^-}^{\infty} dA(t)\right| \\
&= \mu
\end{aligned}
\tag{8.60}
$$

Similarly we have

$$|f(s)| = |s + \mu| \tag{8.61}$$

Now, examining the contour C as shown in Figure 8.6, we observe that for all points *on* the contour, except at $s = 0$, we have from Eqs. (8.60) and (8.61) that

$$|f(s)| = |s + \mu| > \mu \geq |g(s)| \tag{8.62}$$

This follows since $s + \mu$ (for s on C) is a vector whose length is the distance from the point $-\mu$ to the point on C where s is located. We are almost in a

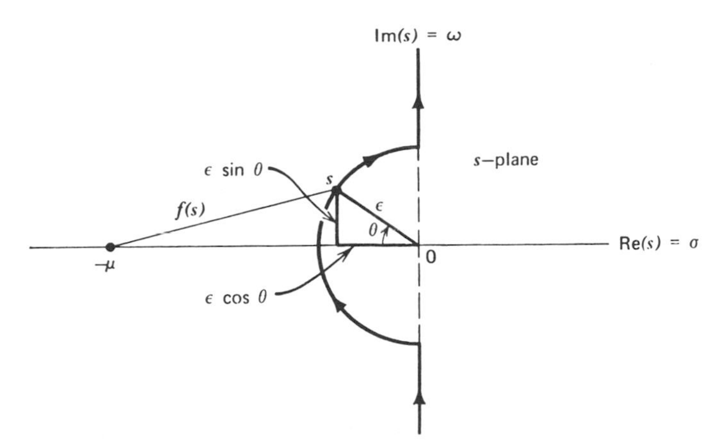

Figure 8.7 The excursion around the origin.

position to apply Rouche's theorem; the only remaining consideration is to show that $|f(s)| > |g(s)|$ in the vicinity $s = 0$. For this purpose we allow the contour C to make a small semicircular excursion to the left of the origin as shown in Figure 8.7. We note at $s = 0$ that $|g(0)| = |f(0)| = \mu$, which does not satisfy the conditions for Rouche's theorem. The small semicircular excursion of radius $\epsilon(\epsilon > 0)$ that we take to the left of the origin overcomes this difficulty as follows. Considering an arbitrary point s on this semicircle (see the figure), which lies at an angle θ with the σ-axis, we may write $s = \sigma + j\omega = -\epsilon \cos\theta + j\epsilon \sin\theta$ and so we have

$$|f(s)|^2 = |s + \mu|^2 = |-\epsilon \cos\theta + j\,\epsilon \sin\theta + \mu|^2$$

Forming the product of $(s + \mu)$ and its complex conjugate, we get

$$\begin{aligned} |f(s)|^2 &= (\mu - \epsilon \cos\theta)^2 + o(\epsilon) \\ &= \mu^2 - 2\mu\epsilon \cos\theta + o(\epsilon) \end{aligned} \tag{8.63}$$

Note that the smallest value for $|f(s)|$ occurs for $\theta = 0$. Evaluating $g(s)$ on this same semicircular excursion we have

$$|g(s)|^2 = \mu^2 \left| \int_{0-}^{\infty} e^{(-\epsilon \cos\theta + j\epsilon \sin\theta)t}\, dA(t) \right|^2$$

From the power-series expansion of the exponential inside the integral we have

$$|g(s)|^2 = \mu^2 \left| \int_{0-}^{\infty} [1 + (-\,\epsilon \cos\theta + j\epsilon \sin\theta)t + \cdots]\, dA(t) \right|^2$$

We recognize the integrals in this series as proportional to the moments of the interarrival time, and so

$$|g(s)|^2 = \mu^2 \, |1 - \epsilon\bar{t}\cos\theta + j\epsilon\bar{t}\sin\theta + o(\epsilon)|^2$$

Forming $|g(s)|^2$ by multiplying $g(s)$ by its complex conjugate, we have

$$\begin{aligned} |g(s)|^2 &= \mu^2(1 - 2\epsilon\bar{t}\cos\theta + o(\epsilon)) \\ &= \mu^2 - \frac{2\mu\epsilon}{\rho}\cos\theta + o(\epsilon) \end{aligned} \tag{8.64}$$

where, as usual, $\rho = \bar{x}/\bar{t} = 1/\mu\bar{t}$. Now since θ lies in the range $-\pi/2 \le \theta \le \pi/2$, which gives $\cos\theta \ge 0$, we have as $\epsilon \to 0$ that on the shrinking semicircle surrounding the origin

$$\mu^2 - 2\mu\epsilon\cos\theta > \mu^2 - \frac{2\mu\epsilon}{\rho}\cos\theta \tag{8.65}$$

This last is true since $\rho < 1$ for our stable system. The left-hand side of Inequality (8.65) is merely the expression given in Eq. (8.63) for $|f(s)|^2$ correct up to the first order in ϵ, and the right-hand side is merely the expression in Eq. (8.64) for $|g(s)|^2$, again correct up to the first order in ϵ. Thus we have shown that in the vicinity $s = 0$, $|f(s)| > |g(s)|$. This fact now having been established for all points on the contour C, we may apply Rouche's theorem and state that $f(s)$ and $f(s) + g(s)$ have the same number of zeroes *inside* the contour C. Since $f(s)$ has only one zero (at $s = -\mu$) it is clear that the expression given in Eq. (8.59) $[f(s) + g(s)]$ has only one zero for Re $(s) < 0$; let this zero occur at the point $s = -s_1$. As discussed above, the point $s = 0$ is also a root of Eq. (8.59).

We may therefore write Eq. (8.58) as

$$\frac{\Psi_+(s)}{\Psi_-(s)} = \left[\frac{\mu A^*(-s) - s - \mu}{s(s + s_1)}\right]\left[\frac{s(s + s_1)}{s + \mu}\right] \tag{8.66}$$

where the first bracketed term contains no poles and no zeroes in Re $(s) \le 0$ (we have divided out the only two zeroes at $s = 0$ and $s = -s_1$ in this half-plane). We now wish to extend the region Re $(s) \le 0$ into the region Re $(s) < D$ and we choose D (> 0) such that no new zeroes or poles of Eq. (8.59) are introduced as we extend to this new region. The first bracket qualifies for $[\Psi_-(s)]^{-1}$, and we see immediately that the second bracket qualifies for $\Psi_+(s)$ since none of its zeroes $(s = 0, s = -s_1)$ or poles $(s = -\mu)$ are in

Re $(s) > 0$. We may then factorize Eq. (8.66) in the following form:

$$\Psi_+(s) = \frac{s(s + s_1)}{s + \mu} \tag{8.67}$$

$$\Psi_-(s) = \frac{-s(s + s_1)}{s + \mu - \mu A^*(-s)} \tag{8.68}$$

We have now assured that the functions given in these last two equations satisfy Conditions (8.36) and (8.37). We evaluate the unknown constant K as follows:

$$K = \lim_{s\to 0} \frac{\Psi_+(s)}{s} = \lim_{s\to 0} \frac{s + s_1}{s + \mu}$$

$$= \frac{s_1}{\mu} = W(0^+) \tag{8.69}$$

Thus we have from Eq. (8.41)

$$\Phi_+(s) = \frac{s_1(\mu + s)}{\mu s(s + s_1)}$$

The partial-fraction expansion for this last function gives us

$$\Phi_+(s) = \frac{1}{s} - \frac{1 - s_1/\mu}{s + s_1} \tag{8.70}$$

Inverting by inspection we obtain the final solution for G/M/1:

$$W(y) = 1 - \left(1 - \frac{s_1}{\mu}\right)e^{-s_1 y} \qquad y \geq 0 \tag{8.71}$$

The reader is urged to compare this last result with that given in Eq. (6.30), also for the system G/M/1; the comparison is clear and in both cases there is a single constant that must be solved for. In the solution given here that constant is solved as the root of Eq. (8.59) with Re $(s) < 0$; in the equation given in Chapter 6, one must solve Eq. (6.28), which is equivalent to Eq. (8.59).

Example 3:

The example for G/M/1 can be carried no further in the general case. We find it instructive therefore to consider a more specific G/M/1 example and finish the calculations; the example we choose is the one we used in Chapter 6, for which $A^*(s)$ is given in Eq. (6.35) and corresponds to an E_2/M/1 system, where the two arrival stages have different death rates. For that example we

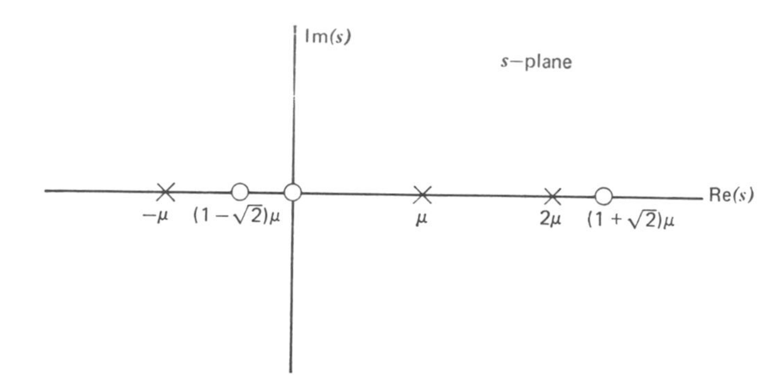

Figure 8.8 Pole-zero pattern for $E_2/M/1$ example.

note that the poles of $A^*(-s)$ occur at the points $s = \mu$, $s = 2\mu$, which as promised lie in the region Re $(s) > 0$. As our first step in factorizing we form

$$\frac{\Psi_+(s)}{\Psi_-(s)} = A^*(-s)B^*(s) - 1$$

$$= \left[\frac{2\mu^2}{(\mu - s)(2\mu - s)}\right]\left(\frac{\mu}{s + \mu}\right) - 1$$

$$= \frac{-s(s - \mu + \mu\sqrt{2})(s - \mu - \mu\sqrt{2})}{(s + \mu)(\mu - s)(2\mu - s)} \tag{8.72}$$

The spectrum factorization is considerably simplified if we plot these poles and zeroes in the complex plane as shown in Figure 8.8. It is clear that the two poles and one zero in the right half-plane must be associated with $\Psi_-(s)$. Furthermore, since the strip $0 < \text{Re}(s) < \mu$ contains no zeroes and no poles we choose $D = \mu$ and identify the remaining two zeroes and the single pole in the region Re $(s) < D$ as being associated with $\Psi_+(s)$. Note well that the zero located at $s = (1 - \sqrt{2})\mu$ is in fact the single root of the expression $\mu A^*(-s) - s - \mu$ located in the left half-plane, as discussed above, and therefore $s_1 = -(1 - \sqrt{2})\mu$. Of course, we need go no further to solve our problem since the solution is now given through Eq. (8.71); however, let us continue identifying various forms in our solution to clarify the remaining steps. With this factorization we may rewrite Eq. (8.72) as

$$\frac{\Psi_+(s)}{\Psi_-(s)} = \left[\frac{-(s - \mu - \mu\sqrt{2})}{(\mu - s)(2\mu - s)}\right]\left[\frac{s(s - \mu + \mu\sqrt{2})}{s + \mu}\right]$$

In this form we recognize the first bracket as $1/\Psi_-(s)$ and the second bracket as $\Psi_+(s)$. Thus we have

$$\Psi_+(s) = \frac{s(s - \mu + \mu\sqrt{2})}{s + \mu} \tag{8.73}$$

We may evaluate the constant K from Eq. (8.69) to find

$$K = \frac{s_1}{\mu} = -1 + \sqrt{2} \tag{8.74}$$

and this of course corresponds to $W(0^+)$, which is the probability that a new arrival must wait for service. Finally then we substitute these values into Eq. (8.71) to find

$$W(y) = 1 - (2 - \sqrt{2})e^{-\mu(\sqrt{2}-1)y} \qquad y \geq 0 \tag{8.75}$$

which as expected corresponds exactly to Eq. (6.37).

This method of spectrum factorization has been used successfully by Rice [RICE 62], who considers the busy period for the G/G/1 system. Among the interesting results available, there is one corresponding to the limiting distribution of long waiting times in the heavy-traffic case (which we develop in Section 2.1 of Volume II); Rice gives a similar approximation for the duration of a busy period in the heavy traffic case.

8.3. KINGMAN'S ALGEBRA FOR QUEUES

Let us once again state the fundamental relationships underlying the G/G/1 queue. For $u_n = x_n - t_{n+1}$ we have the basic relationship

$$w_{n+1} = (w_n + u_n)^+ \tag{8.76}$$

and we have also seen that

$$w_n = \max[0, u_{n-1}, u_{n-1} + u_{n-2}, \ldots, u_{n-1} + \cdots + u_1, u_{n-1} + \cdots + u_0 + w_0]$$

We observed earlier that $\{w_n\}$ is a Markov process with stationary transition probabilities; its total stochastic structure is given by $P[w_{m+n} \leq y \mid w_m = x]$, which may be calculated as an n-fold integral over the n-dimensional joint distribution of the n random variables $w_{m+1}, \ldots, w_{m+n}$ over that region of the space which results in $w_{m+n} \leq y$. This calculation is much too complicated and so we look for alternative means to solve this problem. Pollaczek [POLL 57] used a spectral approach and complex integrals to carry out the solution. Lindley [LIND 52] observed that w_n has the same distribution as w_n', defined earlier as

$$w_n' = \max[U_0, U_1, \ldots, U_{n-1}, U_n + w_0]$$

If we have the case $E[u_n] < 0$, which corresponds to $\rho = \bar{x}/\bar{t} < 1$, then a stable solution exists for the limiting random variable $\tilde{w}$ such that

$$\tilde{w} = \sup_{n \geq 0} U_n \tag{8.77}$$

independent of w_0. The method of spectrum factorization given in the previous section is Smith's [SMIT 53] approach to the solution of Lindley's Wiener Hopf integral equation. Another approach due to Spitzer using combinatorial methods leads to Spitzer's identity [SPIT 57]. Many proofs for this identity exist and Wendel [WEND 58] carried it out by exposing the underlying algebraic structure of the problem. Keilsen [KEIL 65] demonstrated the application of Green's functions to the solution of G/G/1. Beneš [BENE 63] also considered the G/G/1 system by investigating the unfinished work and its variants. These many approaches, each of which is rather complicated, forces one to inquire whether or not there is a larger underlying structure, which places these solution methods in a common framework. In 1966 Kingman [KING 66] addressed this problem and introduced his algebra for queues to expose the common structure; we study this algebra briefly in this section.

From Eq. (8.76) we clearly could solve for the pdf of w_{n+1} iteratively starting with $n = 0$ and with a given pdf for w_0; recall that the pdf for u_n [i.e., $c(u)$] is independent of n. Our iterative procedure would proceed as follows. Suppose we had already calculated the pdf for w_n, which we denote by $w_n(y) \triangleq dW_n(y)/dy$, where $W_n(y) = P[w_n \leq y]$. To find $w_{n+1}(y)$ we follow the prescription given in Eq. (8.76) and begin by forming the pdf for the sum $w_n + u_n$, which, due to the independence of these two random variables, is clearly the convolution $w_n(y) \circledast c(y)$. This convolution will result in a density function that has nonnegative values for negative as well as positive values of its argument. However Eq. (8.76) requires that our next step in the calculation of $w_{n+1}(y)$ is to calculate the pdf associated with $(w_n + u_n)^+$; this requires that we take the total probability associated with all negative arguments for this density just found [i.e., for $w_n(y) \circledast c(y)$] and collect it together as an impulse of probability located at the origin for $w_{n+1}(y)$. The value of this impulse will just be the integral of our former density on the negative half line. We say in this case that "we sweep the probability in the negative half line up to the origin." The values found from the convolution on the positive half line are correct for w_{n+1} in that region. The algebra that describes this operation is that which Kingman introduces for studying the system G/G/1. Our iterative procedure continues by next forming the convolution of $w_{n+1}(y)$ with $c(y)$, sweeping the probability in the negative half line up to the origin to form $w_{n+2}(y)$ and then proceeds to form $w_{n+3}(y)$ in a like fashion, and so on.

The elements of this algebra consist of all finite signed measures on the real line (for example, a pdf on the real line). For any two such measures, say h_1 and h_2, the sum $h_1 + h_2$ and also all scalar multiples of either belong to this algebra. The product operation $h_1 \circledast h_2$ is defined as the convolution of h_1 with h_2. It can be shown that this algebra is a real commutative algebra. There also exists an identity element denoted by e such that $e \circledast h = h$ for any h in the algebra, and it is clear that e will merely be a unit impulse located at the origin. We are interested in operators that map real functions into other real functions and that are measurable. Specifically we are interested in the operator that takes a value x and maps it into the value $(x)^+$, where as usual we have $(x)^+ \triangleq \max\,[0, x]$. Let us denote this operator by π, which is not to be confused with the matrix of the transition probabilities used in Chapter 2; thus, if we let A denote some event which is measurable, and let $h(A) = P\{\omega\colon X(\omega)\epsilon A\}$ denote the measure of this event, then π is defined through

$$\pi[h(A)] = P\{\omega\colon X^+(\omega)\epsilon A\}$$

We note the linearity of this operator, that is, $\pi(ah) = a\pi(h)$ and $\pi(h_1 + h_2) = \pi(h_1) + \pi(h_2)$. Thus we have a commutative algebra (with identity) along with the linear operator π that maps this algebra into itself.

Since $[(x)^+]^+ = (x)^+$ we see that an important property of this operator π is that

$$\pi^2 = \pi$$

A linear operator satisfying such a condition is referred to as a *projection*. Furthermore a projection whose range and null space are both subalgebras of the underlying algebra is called a Wendel projection; it can be shown that π has this property, and it is this that makes the solution for G/G/1 possible.

Now let us return to considerations of the queue G/G/1. Recall that the random variable u_n has pdf $c(u)$ and that the waiting time for the nth customer w_n has pdf $w_n(y)$. Again since u_n and w_n are independent then $w_n + u_n$ has pdf $c(y) \circledast w_n(y)$. Furthermore, since $w_{n+1} = (w_n + u_n)^+$ we have therefore

$$w_{n+1}(y) = \pi(c(y) \circledast w_n(y)) \qquad n = 0, 1, \ldots \tag{8.78}$$

and this equation gives the pdf for waiting times by induction. Now if $\rho < 1$ the limiting pdf $w(y)$ exists and is independent of w_0. *That is, $\tilde{w}$ must have the same pdf as* $(\tilde{w} + \tilde{u})^+$ (a remark due to Lindley [LIND 52]). This gives us the basic equation defining the stationary pdf for waiting time in G/G/1:

$$w(y) = \pi(c(y) \circledast w(y)) \tag{8.79}$$

The solution of this equation is of main interest in solving G/G/1. The remaining portion of this section gives a succinct summary of some elegant results involving this algebra; only the courageous are encouraged to continue.

The particular formalism used for constructing this algebra and carrying out the solution of Eq. (8.79) is what distinguishes the various methods we have mentioned above. In order to see the relationship among the various approaches we now introduce *Spitzer's identity*. In order to state this identity, which involves the recurrence relation given in Eq. (8.78), we must introduce the following z-transform:

$$X(z, y) = \sum_{n=0}^{\infty} w_n(y) z^n \tag{8.80}$$

Addition and scalar multiplication may be defined in the obvious way for this power series and "multiplication" will be defined as corresponding to convolution as is the usual case for transforms. Spitzer's identity is then given as

$$X(z, y) = e^{-\pi(\gamma)}\pi(w_0(y)e^{\pi(\gamma)-\gamma}) \tag{8.81}$$

where

$$\gamma \triangleq \log\,[e - zc(y)] \tag{8.82}$$

Thus $w_n(y)$ may be found by expanding $X(z, y)$ as a power series in z and picking out the coefficient of z^n. It is not difficult to show that

$$X(z, y) = w_0(y) + z\pi(c(y) \circledast X(z, y)) \tag{8.83}$$

We may also form a generating function on the sequence $E[e^{-sw_n}] \triangleq W_n^*(s)$, which permits us to find the transform of the limiting waiting time; that is,

$$\lim_{n\to\infty} W_n^*(s) = W^*(s) \triangleq E[e^{-s\tilde{w}}]$$

so long as $\rho < 1$. This leads us to the following equation, which is also referred to as Spitzer's identity and is directly applicable to our queueing problem:

$$W^*(s) = \exp\left(-\sum_{n=1}^{\infty} \frac{1}{n} E[1 - e^{-s(U_n)^+}]\right) \tag{8.84}$$

We never claimed it would be simple!†

If we deal with $W_n^*(s)$ it is possible to define another real commutative algebra (in which the product is defined as multiplication rather than convolution as one might expect). The algebraic solution to our basic equation (8.79) may be carried out in either of these two algebras; in the transformed

† From this identity we easily find that

$$E[\tilde{w}] \triangleq W = \sum_{n=1}^{\infty} \frac{1}{n} E[(U_n)^+]$$

case one deals with the power series

$$X^*(z, s) \triangleq \sum_{n=0}^{\infty} W_n^*(s) z^n \tag{8.85}$$

rather than with the series given in Eq. (8.80).

Pollaczek considers this latter case and for G/G/1 obtains the following equation which serves to define the system behavior:

$$X^*(z, s) = W_0^*(s) + \frac{zs}{2\pi j} \int_{jc-\infty}^{jc+\infty} \frac{C^*(s') X^*(z, s')\, ds'}{s'(s - s')} \tag{8.86}$$

and he then shows after considerable complexity that this solution must be of the form

$$X^*(z, s) = e^{-\hat{\pi}(\hat{\gamma}(s))} \hat{\pi}(W_0^*(s) e^{\hat{\pi}(\hat{\gamma}(s)) - \hat{\gamma}(s)}) \tag{8.87}$$

where

$$\hat{\gamma}(s) \triangleq \log(1 - zC^*(s))$$

and

$$\hat{\pi}(\hat{\gamma}(s)) \triangleq \frac{s}{2\pi j} \int_{jc-\infty}^{jc+\infty} \frac{\hat{\gamma}(s')\, ds'}{s'(s - s')}$$

When $C^*(s)$ is simple enough then these expressions can be evaluated by contour integrals.

On the other hand, the method we have described in the previous section using spectrum factorization may be phrased in terms of this algebra as follows. If we replace $s\Phi_+(s)$ by $W^*(s)$ and $s\Phi_-(s)$ by $W_-^*(s)$ then our basic equation reads

$$W^*(s) + W_-^*(s) = C^*(s) W^*(s)$$

Corresponding to Eq. (8.83) the transformed version becomes

$$\hat{\pi}(X^*(z, s) - W_0^*(s) - zC^*(s) X^*(z, s)) = 0$$

and the spectrum factorization takes the form

$$1 - zC^*(s) = e^{\hat{\pi}(\hat{\gamma}(s))} e^{\gamma(s) - \hat{\pi}(\hat{\gamma}(s))} \tag{8.88}$$

This spectrum factorization, of course, is the critical step.

This unification as an algebra for queues is elegant but as yet has provided little in the way of extending the theory. In particular, Kingman points out that this approach does not easily extend to the system G/G/m since whereas the range of this algebra is a subalgebra, its null space is not; therefore, we do not have a Wendel projection. Perhaps the most enlightening aspect of this discussion is the significant equation (8.79), which gives the basic condition that must be satisfied by the pdf of waiting time. We take advantage of its recurrence form, Eq. (8.78), in Chapter 2, Volume II.

8.4. THE IDLE TIME AND DUALITY

Here we obtain an expression for $W^*(s)$ in terms of the transform of the idle-time pdf and interpret this result in terms of duality in queues.

Let us return to the basic equation given in (8.5), that is,

$$w_{n+1} = \max [0, w_n + u_n]$$

We now define a new random variable which is the "other half" of the waiting time, namely,

$$y_n = -\min [0, w_n + u_n] \tag{8.89}$$

This random variable in some sense corresponds to the random variable whose distribution is $W_-(y)$, which we studied earlier. Note from these last two equations that when $y_n > 0$ then $w_{n+1} = 0$ in which case y_n is merely the length of the *idle* period, which is terminated with the arrival of C_{n+1}. Moreover, since either w_{n+1} or y_n must be 0, we have that

$$w_{n+1}y_n = 0 \tag{8.90}$$

We adopt the convention that in order for an idle period to exist, it must have nonzero length, and so if y_n and w_{n+1} are both 0, then we say that the busy period continues (an annoying triviality).

From the definitions we observe the following to be true in all cases:

$$w_{n+1} - y_n = w_n + u_n \tag{8.91}$$

From this last equation we may obtain a number of important results and we proceed here as we did in Chapter 5, where we derived the expected queue size for the system M/G/1 using the imbedded Markov chain approach. In particular, let us take the expectation of both sides of Eq. (8.91) to give

$$E[w_{n+1}] - E[y_n] = E[w_n] + E[u_n]$$

We assume $E[u_n] < 0$, which (except for D/D/1 where ≤ 0 will do) is the necessary and sufficient condition for there to be a stationary (and unique) waiting-time distribution independent of n; this is the same as requiring $\rho = \bar{x}/\bar{t} < 1$. In this case we have*

$$\lim_{n\to\infty} E[w_{n+1}] = \lim_{n\to\infty} E[w_n]$$

* One must be cautious in claiming that

$$\lim_{n\to\infty} E[w_n^k] = \lim_{n\to\infty} E[w_{n+1}^k]$$

since these are distinct random variables. We permit that step here, but refer the interested reader to Wolff [WOLF 70] for a careful treatment.

and so our earlier equation gives

$$E[\tilde{y}] = -E[\tilde{u}] \tag{8.92}$$

where $y_n \to \tilde{y}$ and $u_n \to \tilde{u}$. (We note that the idle periods are independent and identically distributed, but the duration of an idle period does depend upon the duration of the previous busy period.) Now from Eq. (8.13) we have $E[\tilde{u}] = \bar{t}(\rho - 1)$ and so

$$E[\tilde{y}] = \bar{t}(1 - \rho) \tag{8.93}$$

Let us now square Eq. (8.91) and then take expected values as follows:

$$w_{n+1}^2 - 2w_{n+1}y_n + y_n^2 = w_n^2 + 2w_nu_n + u_n^2$$

Using Eq. (8.90) and recognizing that the moments of the limiting distribution on w_n must be independent of the subscript we have

$$E[(\tilde{y})^2] = 2E[\tilde{w}\tilde{u}] + E[(\tilde{u})^2]$$

We now revert to the simpler notation for moments, $\overline{w^k} \triangleq E[(\tilde{w})^k]$, etc. Since w_n and u_n are independent random variables we have $E[\tilde{w}\tilde{u}] = \bar{w}\bar{u}$; using this and Eq. (8.92) we find

$$\bar{w} \triangleq W = -\frac{\overline{u^2}}{2\bar{u}} - \frac{\overline{y^2}}{2\bar{y}} \tag{8.94}$$

Recalling that the mean residual life of a random variable X is given by $\overline{X^2}/2\bar{X}$, we observe that W is merely the mean residual life of $-\tilde{u}$ less the mean residual life of $\tilde{y}$! We must now evaluate the second moment of $\tilde{u}$. Since $\tilde{u} = \tilde{x} - \tilde{t}$, then $\overline{u^2} = \overline{(\tilde{x} - \tilde{t})^2}$, which gives

$$\overline{u^2} = \sigma_a^2 + \sigma_b^2 + (\bar{t})^2(1 - \rho)^2 \tag{8.95}$$

where σ_a^2 and σ_b^2 are the variance of the interarrival-time and service-time densities, respectively. Using this expression and our previous result for $\bar{u}$ we may thus convert Eq. (8.94) to

$$W = \frac{\sigma_a^2 + \sigma_b^2 + (\bar{t})^2(1 - \rho)^2}{2\bar{t}(1 - \rho)} - \frac{\overline{y^2}}{2\bar{y}} \tag{8.96}$$

We must now calculate the first two moments of $\tilde{y}$ [we already know that $\bar{y} = \bar{t}(1 - \rho)$ but wish to express it differently to eliminate a constant]. This we do by conditioning these moments with respect to the occurrence of an idle period. That is, let us define

$$\begin{aligned} a_0 &= P[\tilde{y} > 0] \\ &= P[\text{arrival finds the system idle}] \end{aligned} \tag{8.97}$$

It is clear that we have a stable system when $a_0 > 0$. Furthermore, since we have defined an idle period to occur only when the system remains idle for a nonzero interval of time, we have that

$$P[\tilde{y} \leq y \mid \tilde{y} > 0] = P[\text{idle period} \leq y] \tag{8.98}$$

and this last is just the idle-period distribution earlier denoted by $F(y)$. We denote by I the random variable representing the idle period. Now we may calculate the following:

$$\begin{aligned}\bar{y} &= E[\tilde{y} \mid \tilde{y} = 0]P[\tilde{y} = 0] + E[\tilde{y} \mid \tilde{y} > 0]P[\tilde{y} > 0] \\ &= 0 + a_0 E[\tilde{y} \mid \tilde{y} > 0]\end{aligned}$$

The expectation in this last equation is merely the expected value of I and so we have

$$\bar{y} = a_0 \bar{I} \tag{8.99}$$

Similarly, we find

$$\overline{y^k} = a_0 \overline{I^k} \tag{8.100}$$

Thus, in particular, $\overline{y^2}/2\bar{y} = \overline{I^2}/2\bar{I}$ (a_0 cancels!) and so we may rewrite the expression for W in Eq. (8.96) as

$$W = \frac{\sigma_a^{\,2} + \sigma_b^{\,2} + (\bar{t})^2(1 - \rho)^2}{2\bar{t}(1 - \rho)} - \frac{\overline{I^2}}{2\bar{I}} \quad \blacksquare \tag{8.101}$$

Unfortunately this is as far as we can go in establishing W for G/G/1. The calculation now involves the determination of the first two moments of the idle period. In general, for G/G/1 we cannot easily solve for these moments since the idle period depends upon the particular way in which the previous busy period terminated. However in Chapter 2, Volume II, we place bounds on the second term in this equation, thereby bounding the mean wait W.

As we did for M/G/1 in Chapter 5 we now return to our basic equation (8.91) relating the important random variables and attempt to find the transform of the waiting time density $W^*(s) \triangleq E[e^{-s\tilde{w}}]$ for G/G/1. As one might expect this will involve the idle-time distribution as well. Forming the transform on both sides of Eq. (8.91) we have

$$E[e^{-s(w_{n+1}-y_n)}] = E[e^{-s(w_n+u_n)}]$$

However since w_n and u_n are independent, we find

$$E[e^{-s(w_{n+1}-y_n)}] = E[e^{-sw_n}]E[e^{-su_n}] \tag{8.102}$$

In order to evaluate the left-hand side of this transform expression we take advantage of the fact that only one or the other of the random variables w_{n+1} and y_n may be nonzero. Accordingly, we have

$$E[e^{-s(w_{n+1}-y_n)}] = E[e^{-s(-y_n)} \mid y_n > 0]P[y_n > 0] + E[e^{-sw_{n+1}} \mid y_n = 0]P[y_n = 0] \qquad (8.103)$$

To determine the right-hand side of this last equation we may use the following similar expansion:

$$E[e^{-sw_{n+1}}] = E[e^{-sw_{n+1}} \mid y_n = 0]P[y_n = 0] + E[e^{-sw_{n+1}} \mid y_n > 0]P[y_n > 0] \qquad (8.104)$$

However, since $w_{n+1}y_n = 0$, we have $E[e^{-sw_{n+1}} \mid y_n > 0] = 1$. Making use of the definition for a_0 in Eq. (8.97) and allowing the limit as $n \to \infty$ we obtain the following transform expression from Eq. (8.104):

$$E[e^{-s\tilde{w}} \mid \tilde{y} = 0]P[\tilde{y} = 0] = W^*(s) - a_0$$

We may then write the limiting form of Eq. (8.103) as

$$E[e^{-s(\tilde{w}-\tilde{y})}] = I^*(-s)a_0 + W^*(s) - a_0 \qquad (8.105)$$

where $I^*(s)$ is the Laplace transform of the idle-time pdf [see Eq. (8.98) for the definition of this distribution]. Thus, from this last and from Eq. (8.102), we obtain immediately

$$W^*(s)C^*(s) = a_0 I^*(-s) + W^*(s) - a_0$$

where as in the past $C^*(s)$ is the Laplace transform for the density describing the random variable $\tilde{u}$. This last equation finally gives us [MARS 68]

$$W^*(s) = \frac{a_0[1 - I^*(-s)]}{1 - C^*(s)} \qquad ■ (8.106)$$

which represents the generalization of the Pollaczek–Khinchin transform equation given in Chapter 5 and which now applies to the system G/G/1. Clearly this equation holds at least along the imaginary axis of the complex s-plane, since in that case it becomes the characteristic function of the various distributions which are known to exist.

Let us now consider some examples.

Example 1: M/M/1

For this system we know that the idle-period distribution is the same as the interarrival-time distribution, namely,

$$F(y) = P[I \le y] = 1 - e^{-\lambda y} \qquad y \ge 0 \qquad (8.107)$$

And so we have the first two moments $\bar{I} = 1/\lambda$, $\overline{I^2} = 2/\lambda^2$; we also have $\sigma_a^2 = 1/\lambda^2$ and $\sigma_b^2 = 1/\mu^2$. Using these values in Eq. (8.101) we find

$$W = \frac{\lambda^2(1/\lambda^2 + 1/\mu^2) + (1-\rho)^2}{2\lambda(1-\rho)} - \frac{1}{\lambda}$$

and so

$$W = \frac{\rho/\mu}{1-\rho} \tag{8.108}$$

which of course checks with our earlier results for M/M/1.

We know that $I^*(s) = \lambda/(s+\lambda)$ and $C^*(s) = \lambda\mu/(\lambda - s)(s+\mu)$. Moreover, since the probability that a Poisson arrival finds the system empty is the same as the long-run proportion of time the system is empty, we have that $a_0 = 1 - \rho$ and so Eq. (8.106) yields

$$\begin{aligned} W^*(s) &= \frac{(1-\rho)[1-\lambda/(\lambda-s)]}{1-\lambda\mu/(\lambda-s)(s+\mu)} \\ &= \frac{-(1-\rho)s(s+\mu)}{(\lambda-s)(s+\mu)-\lambda\mu} \\ &= \frac{(1-\rho)(s+\mu)}{s+\mu-\lambda} \end{aligned} \tag{8.109}$$

which is the same as Eq. (5.120).

Example 2: *M/G/1*

In this case the idle-time distribution is as in M/M/1; however, we must leave the variance for the service-time distribution as an unknown. We obtain

$$\begin{aligned} W &= \frac{\lambda^2[(1/\lambda^2) + \sigma_b^2] + (1-\rho)^2}{2\lambda(1-\rho)} - \frac{1}{\lambda} \\ &= \rho\frac{(1+C_b^2)}{2\mu(1-\rho)} \end{aligned} \tag{8.110}$$

which is the P–K formula. Also, $C^*(s) = B^*(s)\lambda/(\lambda - s)$ and again $a_0 = (1-\rho)$. Equation (8.106) then gives

$$\begin{aligned} W^*(s) &= \frac{(1-\rho)[1-\lambda/(\lambda-s)]}{1-[\lambda/(\lambda-s)]B^*(s)} \\ &= \frac{s(1-\rho)}{s-\lambda+\lambda B^*(s)} \end{aligned} \tag{8.111}$$

which is the P–K transform equation for waiting time!

Example 3: $D/D/1$

In this case we have that the length of the idle period is a constant and is given by $\tilde{I} = \bar{t} - \bar{x} = \bar{t}(1 - \rho)$; therefore $\bar{I} = \bar{t}(1 - \rho)$, and $\overline{I^2} = (\bar{I})^2$. Moreover, $\sigma_a^2 = \sigma_b^2 = 0$. Therefore Eq. (8.101) gives

$$W = \frac{0 + (\bar{t})^2(1 - \rho)^2}{2\bar{t}(1 - \rho)} - \tfrac{1}{2}\bar{t}(1 - \rho)$$

and so

$$W = 0 \tag{8.112}$$

This last is of course correct since the equilibrium waiting time in the (stable) system D/D/1 is always zero.

Since $\bar{x}$, $\bar{t}$, and I are all constants, we have $B^*(s) = e^{-s\bar{x}}$, $A^*(s) = e^{-s\bar{t}}$ and $I^*(s) = e^{-s\bar{I}} = e^{-s\bar{t}(1-\rho)}$. Also, with probability one an arrival finds the system empty; thus $a_0 = 1$. Then Eq. (8.106) gives

$$\begin{aligned} W^*(s) &= \frac{1[1 - e^{s\bar{t}(1-\rho)}]}{1 - e^{s\bar{t}}e^{-s\bar{x}}} \\ &= 1 \end{aligned} \tag{8.113}$$

and so $w(y) = u_0(y)$, an impulse at the origin which of course checks with the result that no waiting occurs.

Considerations of the idle-time distribution naturally lead us to the study of *duality* in queues. This material is related to the ladder indices we had defined in Section 5.11. The random walk we are interested in is the sequence of values taken on by U_n [as given in Eq. (8.9)]. Let us denote by U_{n_k} the value taken on by U_n at the kth *ascending* ladder index (instants when the function first drops below its latest *maximum*). Since $\bar{u} < 0$ it is clear that $\lim U_n = -\infty$ as $n \to \infty$. Therefore, there will exist a (finite) integer K such that K is the largest ascending ladder index for U_n. Now from Eq. (8.11) repeated below

$$\tilde{w} = \sup_{n \geq 0} U_n$$

It is clear that

$$\tilde{w} = U_{n_K}$$

Now let us define the random variable $\hat{I}_k$ (which as we shall see is related to an idle time) as

$$\hat{I}_k = U_{n_k} - U_{n_{k-1}} \tag{8.114}$$

for $k \le K$. That is $\hat{I}_k$ is merely the amount by which the new ascending ladder height exceeds the previous ascending ladder height. Since all of the random variables u_n are independent then the random variables $\hat{I}_k$ conditioned on K are independent and identically distributed. If we now let $1 - \sigma = P[U_n \le U_{n_k}$ for all $n > n_k]$ then we may easily calculate the distribution for K as

$$P[K = k] = (1 - \sigma)\sigma^k \tag{8.115}$$

In exercise 8.16, we show that $(1 - \sigma) = P[\tilde{w} = 0]$. Also it is clear that

$$\begin{aligned}\hat{I}_1 + \hat{I}_2 + \cdots + \hat{I}_K &= U_{n_1} - U_{n_0} + U_{n_2} - U_{n_1} + \cdots + U_{n_K} - U_{n_{K-1}} \\ &= U_{n_K}\end{aligned}$$

where $n_0 \triangleq 0$ and $U_0 \triangleq 0$. Thus we see that $\tilde{w}$ has the same distribution as $\hat{I}_1 + \cdots + \hat{I}_K$ and so we may write

$$\begin{aligned}E[e^{-s\tilde{w}}] &= E[E[e^{-s(\hat{I}_1+\cdots+\hat{I}_K)} \mid K]] \\ &= E[(\hat{I}^*(s))^K]\end{aligned} \tag{8.116}$$

where $\hat{I}^*(s)$ is the Laplace transform for the pdf of each of the $\hat{I}_k$ (each of which we now denote simply by $\hat{I}$). We may now evaluate the expectation in Eq. (8.116) by using the distribution for K in Eq. (8.115) finally to yield

$$W^*(s) = \frac{1 - \sigma}{1 - \sigma\hat{I}^*(s)} \tag{8.117}$$

Here then, is yet another expression for $W^*(s)$ in the G/G/1 system.

We now wish to interpret the random variable $\hat{I}$ by considering a "dual" queue (whose variables we will distinguish by the use of the symbol ^). The dual queue for the G/G/1 system considered above is the queue in which the service times x_n in the original system become the interarrival times $\hat{t}_{n+1}$ in the dual queue and also the interarrival times t_{n+1} from the original queue become the service times $\hat{x}_n$ in the dual queue.† It is clear then that the random variable $\hat{u}_n$ for the dual queue will merely be $\hat{u}_n = \hat{x}_n - \hat{t}_{n+1} = t_{n+1} - x_n = -u_n$ and defining $\hat{U}_n = \hat{u}_0 + \cdots + \hat{u}_{n-1}$ for the dual queue we have

$$\hat{U}_n = -U_n \tag{8.118}$$

† Clearly, if the original queue is stable, the dual must be unstable, and conversely (except that both may be unstable if $\rho = 1$).

as the relationship among the dual and the original queues. It is then clear from our discussion in Section 5.11 that the ascending and descending ladder indices are interchanged for the original and the dual queue (the same is true of the ladder heights). Therefore the first ascending ladder index n_1 in the original queue will correspond to the first descending ladder index in the dual queue; however, we recall that descending ladder indices correspond to the arrival of a customer who terminates an idle period. We denote this customer by C_{n_1}. Clearly the length of the idle period that he terminates in the dual queue is the difference between the accumulated interarrival times and the accumulated service times for all customers up to his arrival (these services must have taken place in the first busy period), that is, for the dual queue,

$$
\begin{aligned}
\left\{\begin{matrix}\text{Length of first idle period} \\ \text{following first busy period}\end{matrix}\right\} &= \sum_{n=0}^{n_1-1} \hat{t}_{n+1} - \sum_{n=0}^{n_1-1} \hat{x}_n \\
&= \sum_{n=0}^{n_1-1} x_n - \sum_{n=0}^{n_1-1} t_{n+1} \\
&= u_0 + u_1 + \cdots + u_{n_1-1} \\
&= U_{n_1} \\
&= \hat{I}_1
\end{aligned} \qquad (8.119)
$$

where we have used Eq. (8.114) at the last step. Thus we see that *the random variable $\hat{I}$ is merely the idle period in the dual queue* and so our Eq. (8.117) relates the transform of the waiting time in the original queue to the transform of the idle time pdf in the dual queue [contrast this with Eq. (8.106), which relates this waiting-time transform to the transform of the idle time in its own queue].

This duality observation permits some rather powerful conclusions to be drawn in simple fashion (and these are discussed at length in [FELL 66], especially Sections VI.9 and XII.5). Let us discuss two of these.

Example 4: *G/M/1*

If we have a stable G/M/1 queue (with $\bar{t} = 1/\lambda$ and $\bar{x} = 1/\mu$) then the dual is an unstable queue of the type M/G/1 (with $\bar{\hat{t}} = 1/\mu$ and $\bar{\hat{x}} = 1/\lambda$ and so $\hat{I}$ (the distribution of idle time in the dual queue) will be of exponential form; therefore $\hat{I}^*(s) = \mu/(s + \mu)$, which gives from Eq. (8.117) the following result for the original G/M/1 queue:

$$W^*(s) = \frac{(1 - \sigma)(s + \mu)}{s + \mu - \sigma\mu}$$

Inverting this and forming the PDF for waiting time we have

$$W(y) = 1 - \sigma e^{-\mu(1-\sigma)y} \qquad y \geq 0 \tag{8.120}$$

which corresponds exactly to Eq. (6.30).

Example 5: *M/G/1*

As a second example let the original queue be of the form M/G/1 and therefore the dual is of the form G/M/1. Since $\sigma = P[\tilde{w} > 0]$ it must be that $\sigma = \rho$ for M/G/1. Now in the dual system, since a busy period ends at a random point in time (and since the service time in this dual queue is memoryless), an idle period will have a duration equal to the residual life of an interarrival time; therefore from Eq. (5.11) we see that

$$\hat{I}^*(s) = \frac{1 - B^*(s)}{s\bar{x}} \tag{8.121}$$

and when these calculations are applied to Eq. (8.117) we have

$$W^*(s) = \frac{1 - \rho}{1 - \rho\{[1 - B^*(s)]/s\bar{x}\}} \tag{8.122}$$

which is the P–K transform equation for waiting time rewritten as in Eq. (5.106).

This concludes our study of G/G/1. Sad to say, we have been unable to give analytic expressions for the waiting-time distribution explicitly in terms of known quantities. In fact, we have not even succeeded for the mean wait W! Nevertheless, we have given a method for handling the rational case by spectrum factorization, which is quite effective. In Chapter 2, Volume II, we return to G/G/1 and succeed in extracting many of its important properties through the use of bounds, inequalities, and approximations.

REFERENCES

ANDE 53a Andersen, S. E., "On Sums of Symmetrically Dependent Random Variables," *Skan. Aktuar.*, **36,** 123–138 (1953).

ANDE 53b Andersen, S. E., "On the Fluctuations of Sums of Random Variables I," *Math. Scand.*, **1,** 263–285 (1953).

ANDE 54 Andersen, S. E., "On the Fluctuations of Sums of Random Variables II," *Math. Scand.*, **2,** 195–223 (1954).

BENE 63 Beneš, V. E., *General Stochastic Processes in the Theory of Queues*, Addison-Wesley (Reading, Mass.), 1963.

FELL 66 Feller, W., *Probability Theory and its Applications*, Vol. II, Wiley (New York), 1966.

KEIL 65 Keilson, J., "The Role of Green's Functions in Congestion Theory," *Proc. Symp. on Congestion Theory* (edited by W. L. Smith and W. E. Wilkinson) Univ. of North Carolina Press (Chapel Hill), 43–71 (1965).

KING 66 Kingman, J. F. C., "On the Algebra of Queues," *Journal of Applied Probability*, **3,** 285–326 (1966).

LIND 52 Lindley, D. V., "The Theory of Queues with a Single Server," *Proc. Cambridge Philosophical Society*, **48,** 277–289 (1952).

MARS 68 Marshall, K. T., "Some Relationships between the Distributions of Waiting Time, Idle Time, and Interoutput Time in the GI/G/1 Queue," *SIAM Journal Applied Math.*, **16,** 324–327 (1968).

POLL 57 Pollaczek, F., *Problèmes Stochastiques Posés par le Phénomène de Formation dune' Queue d'Attente à un Guichet et par de Phénomènes Apparentés*, Gauthiers Villars (Paris), 1957.

RICE 62 Rice, S. O., "Single Server Systems," *Bell System Technical Journal*, **41,** Part I: "Relations Between Some Averages," 269–278 and Part II: "Busy Periods," 279–310 (1962).

SMIT 53 Smith, W. L., "On the Distribution of Queueing Times," *Proc. Cambridge Philosophical Society*, **49,** 449–461 (1953).

SPIT 56 Spitzer, F. "A Combinatorial Lemma and its Application to Probability Theory," *Transactions of the American Mathematical Society*, **82,** 323–339 (1956).

SPIT 57 Spitzer, F., "The Wiener–Hopf Equation whose Kernel is a Probability Density," *Duke Mathematics Journal*, **24,** 327–344 (1957).

SPIT 60 Spitzer, F. "A Tauberian Theorem and its Probability Interpretation," *Transactions of the American Mathematical Society*, **94,**150–160 (1960).

SYSK 62 Syski, R., *Introduction to Congestion Theory in Telephone Systems*, Oliver and Boyd (London), 1962.

TITC 52 Titchmarsh, E. C., *Theory of Functions*, Oxford Univ. Press (London), 1952.

WEND 58 Wendel, F. G., "Spitzer's Formula; a Short Proof," *Proc. American Mathematical Society*, **9,** 905–908 (1958).

WOLF 70 Wolff, R. W., "Bounds and Inequalities in Queueing," unpublished notes, Department of Industrial Engineering and Operations Research, University of California (Berkeley), 1970.

EXERCISES

8.1. From Eq. (8.18) show that $C^*(s) = A^*(-s)B^*(s)$.

8.2. Find $C(u)$ for M/M/1.

8.3. Consider the system M/D/1 with a fixed service time of $\bar{x}$ sec.
(a) Find

$$C(u) = P[u_n \leq u]$$

and sketch its shape.
(b) Find $E[u_n]$.

8.4. For the sequence of random variables given below, generate the figure corresponding to Figure 8.3 and complete the table.

n	0	1	2	3	4	5	6	7	8	9	
t_{n+1}	2	1	1	5	7	2	2	1	1	6	···
x_n	3	4	2	3	3	4	2	1	1	3	···
u_n											
w_n measured											
w_n calculated											

8.5. Consider the case where $\rho = 1 - \epsilon$ for $0 < \epsilon \ll 1$. Let us expand $W(y - u)$ in Eq. (8.23) as

$$W(y - u) = W(y) - uW^{(1)}(y) + \frac{u^2}{2}W^{(2)}(y) + R(u, y)$$

where $W^{(n)}(y)$ is the nth derivative of $W(y)$ and $R(u, y)$ is such that $\int_{-\infty}^{y} R(u, y)\, dC(u)$ is negligible due to the slow variation of $W(y)$ when $\rho = 1 - \epsilon$. Let $\overline{u^k}$ denote the kth moment of $\tilde{u}$.
(a) Under these conditions convert Lindley's integral equation to a second-order linear differential equation involving $\overline{u^2}$ and $\bar{u}$.
(b) With the boundary condition $W(0) = 0$, solve the equation found in (a) and express the mean wait W in terms of the first two moments of $\tilde{t}$ and $\tilde{x}$.

8.6. Consider the D/E$_r$/1 queueing system, with a constant interarrival time (of $\bar{t}$ sec) and a service-time pdf given as in Eq. (4.16).
(a) Find $C(u)$.

(b) Show that Lindley's integral equation yields $W(y - \bar{t}) = 0$ for $y < \bar{t}$ and

$$W(y - \bar{t}) = \int_0^y W(y - w)\, dB(w) \qquad \text{for } y \geq \bar{t}$$

(c) Assume the following solution for $W(y)$:

$$W(y) = 1 + \sum_{i=1}^{r} a_i e^{\alpha_i y} \qquad y \geq 0$$

where a_i and α_i may both be complex, but where Re $(\alpha_i) < 0$ for $i = 1, 2, \ldots, r$. Using this assumed solution, show that the following equations must hold:

$$e^{-\alpha_i \bar{t}} = \left(\frac{r\mu}{r\mu + \alpha_i}\right)^r \qquad i = 1, 2, \ldots, r$$

$$\sum_{i=0}^{r} \frac{a_i}{(r\mu + \alpha_i)^{j+1}} = 0 \qquad j = 0, 1, \ldots, r - 1$$

where $a_0 = 1$ and $\alpha_0 = 0$. Note that $\{\alpha_i\}$ may be found from the first set of (transcendental) equations, and then the second set gives $\{a_i\}$. It can be shown that the α_i are distinct. See [SYSK 62].

8.7. Consider the following queueing systems in which *no queue* is permitted. Customers who arrive to find the system busy must leave without service.

(a) M/M/1: Solve for $p_k = P[k \text{ in system}]$.

(b) $M/H_2/1$: As in Figure 4.10 with $\alpha_1 = \alpha$, $\alpha_2 = 1 - \alpha$, $\mu_1 = 2\mu\alpha$ and $\mu_2 = 2\mu(1 - \alpha)$.

(i) Find the mean service time $\bar{x}$.

(ii) Solve for p_0 (an empty system), p_α (a customer in the $2\mu\alpha$ box) and $p_{1-\alpha}$ (a customer in the $2\mu(1 - \alpha)$ box).

(c) $H_2/M/1$: Where $A(t)$ is hyperexponential as in (b), but with parameters $\mu_1 = 2\lambda\alpha$ and $\mu_2 = 2\lambda(1 - \alpha)$ instead. Draw the state-transition diagram (with labels on branches) for the following four states: E_{ij} is state with "arriving" customer in arrival stage i and j customers in service $i = 1, 2$ and $j = 0, 1$.

(d) M/E_r/1: Solve for $P_j = P[j$ stages of service left to go$]$.

(e) M/D/1: With all service times equal to $\bar{x}$
(i) Find the probability of an empty system.
(ii) Find the fraction of lost customers.

(f) E_2/M/1: Define the four states as E_{ij} where i is the number of "arrival" stages left to go and j is the number of customers in service. Draw the labeled state-transition diagram.

8.8. Consider a single-server queueing system in which the interarrival time is chosen with probability α from an exponential distribution of mean $1/\lambda$ and with probability $1 - \alpha$ from an exponential distribution with mean $1/\mu$. Service is exponential with mean $1/\mu$.
(a) Find $A^*(s)$ and $B^*(s)$.
(b) Find the expression for $\Psi_+(s)/\Psi_-(s)$ and show the pole–zero plot in the s-plane.
(c) Find $\Psi_+(s)$ and $\Psi_-(s)$.
(d) Find $\Phi_+(s)$ and $W(y)$.

8.9. Consider a G/G/1 system in which

$$A^*(s) = \frac{2}{(s+1)(s+2)}$$

$$B^*(s) = \frac{1}{s+1}$$

(a) Find the expression for $\Psi_+(s)/\Psi_-(s)$ and show the pole–zero plot in the s-plane.
(b) Use spectrum factorization to find $\Psi_+(s)$ and $\Psi_-(s)$.
(c) Find $\Phi_+(s)$.
(d) Find $W(y)$.
(e) Find the average waiting time W.
(f) We solved for $W(y)$ by the method of spectrum factorization. Can you describe another way to find $W(y)$?

8.10. Consider the system M/G/1. Using the spectral solution method for Lindley's integral equation, find
(a) $\Psi_+(s)$. {HINT: Interpret $[1 - B^*(s)]/s\bar{x}$.}
(b) $\Psi_-(s)$.
(c) $s\Phi_+(s)$.

8.11. Consider the queue $E_q/E_r/1$.

(a) Show that

$$\frac{\Psi_+(s)}{\Psi_-(s)} = \frac{F(s)}{1 - F(s)}$$

where $F(s) = 1 - (1 - s/\lambda q)^q(1 + s/\mu r)^r$.

(b) For $\rho < 1$, show that $F(s)$ has one zero at the origin, zeroes $s_1, s_2, \ldots, s_r$ in Re $(s) < 0$, and zeroes $s_{r+1}, s_{r+2}, \ldots, s_{r+q-1}$ in Re $(s) > 0$.

(c) Express $\Psi_+(s)$ and $\Psi_-(s)$ in terms of s_i.

(d) Express $W^*(s)$ in terms of s_i $(i = 1, 2, \ldots, r + q - 1)$.

8.12. Show that Eq. (8.71) is equivalent to Eq. (6.30).

8.13. Consider a D/D/1 queue with $\rho < 1$. Assume $w_0 = 4\bar{t}(1 - \rho)$.

(a) Calculate $w_n(y)$ using the procedure defined by Eq. (8.78) for $n = 0, 1, 2, \ldots$.

(b) Show that the known solution for

$$w(y) = \lim_{n\to\infty} w_n(y)$$

satisfies Eq. (8.79).

8.14. Consider an M/M/1 queue with $\rho < 1$. Assume $w_0 = 0$.

(a) Calculate $w_1(y)$ using the procedure defined by Eq. (8.78).

(b) Repeat for $w_2(y)$.

(c) Show that our known solution for

$$w(y) = \lim_{n\to\infty} w_n(y)$$

satisfies Eq. (8.79).

(d) Compare $w_2(y)$ with $w(y)$.

8.15. By first cubing Eq. (8.91) and then forming expectations, express $\sigma_{\tilde{w}}^2$ (the variance of the waiting time) in terms of the first three moments of $\tilde{t}$, $\tilde{x}$, and I.

8.16. Show that $P[\tilde{w} = 0] = 1 - \sigma$ from Eq. (8.117) by finding the constant term in a power-series expansion of $W^*(s)$.

8.17. Consider a G/G/1 system.

(a) Express $\hat{I}^*(s)$ in terms of the transform of the pdf of idle time in the given system.

(b) Using (a) find $\hat{I}^*(s)$ when the original system is the ordinary M/M/1.

(c) Using (a), show that the transform of the idle-time pdf in a G/M/1 queue is given by

$$I^*(s) = \frac{1 - A^*(s)}{s\bar{t}}$$

thereby reaffirming Eq. (8.121).

(d) Since either the original or the dual queue must be unstable (except for D/D/1), discuss the existence of the transform of the idle-time pdf for the unstable queue.

Epilogue

We have invested eight chapters (and two appendices!) in studying the theory of queueing systems. Occasionally we have been overjoyed at the beauty and generality of the results, but more often we have been overcome (with frustration) at the lack of real progress in the theory. (No, we never promised you a rose garden.) However, we did seduce you into believing that this study would provide worthwhile methods for practical application to many of today's pressing congestion problems. We confirm that belief in Volume II.

In the next volume, after a brief review of this one, we begin by taking a more relaxed view of G/G/1. In Chapter 2, we enter a new world leaving behind the rigor (and pain) of exact solutions to exact problems. Here we are willing to accept the raw facts of life, which state that our models are not perfect pictures of the systems we wish to analyze so we should be willing to accept approximations and bounds in our problem solution. Upper and lower bounds are found for the average delay in G/G/1 and we find that these are related to a very useful heavy traffic approximation for such queues. This approximation, in fact, predicts that the long waiting times are exponentially distributed. A new class of models is then introduced whereby the discrete arrival and departure processes of queueing systems are replaced first by a fluid approximation (in which these stochastic processes are replaced by their mean values as a function of time), and then secondly by a diffusion approximation (in which we permit a variation about these means). We happily find that these approximations give quite reasonable results for rather general queueing systems. In fact they even permit us to study the transient behavior not only of stable queues but also of saturated queues, and this is the material in the final section of Chapter 2 whereby we give Newell's treatment of the rush-hour approximation—an effective method indeed.

Chapter 3 points the way to our applications in time-shared computer systems by presenting some of the principal results for priority queueing systems. We study general methods and apply them to a number of important queueing disciplines. The conservation law for priority systems is established, preventing the useless search for nonrealizable disciplines.

In the remainder, we choose applications principally from the computer field, since these applications are perhaps the most recent and successful for the theory of queues. In fact, the queueing analysis of allocation of resources and job flow through computer systems is perhaps the only tool available to computer scientists in understanding the behavior of the complex interaction of users, programs, processes, and resources. In Chapter 4 we emphasize multi-access computer systems in isolation, handling demands of a large collection of competing users. We look for throughput and response time as well as utilization of resources. The major portion of this chapter is devoted to a particular class of algorithms known as *processor-sharing* algorithms, since they are singularly suited to queueing analysis and capture the essence of more difficult and more complex algorithms seen in real scheduling problems. Chapter 5 addresses itself to computers in networks, a field that is perhaps the fastest growing in the young computer industry itself (most of the references there are drawn from the last three years—a tell-tale indicator indeed). The chapter is devoted to developing methods of analysis and design for computer-communication networks and identifies many unsolved important problems. A specific existing network, the ARPANET, is used throughout as an example to guide the reader through the motivation and evaluation of the various techniques developed.

Now it remains for you, the reader, to sharpen and apply your new set of tools. The world awaits and you must serve!

APPENDIX I

Transform Theory Refresher: z-Transform and Laplace Transform

In this appendix we develop some of the properties and expressions for the z-transform and the Laplace transform as they apply to our studies in queueing theory. We begin with the z-transform since it is easier to visualize its operation. The forms and properties of both transforms are very similar, and we compare them later under the discussion of Laplace transforms.

I.1. WHY TRANSFORMS?

So often as we progress through the study of interesting physical systems we find that transforms appear in one form or another. These transforms occur in many varieties (e.g., z-transform, Laplace transform, Fourier transform, Mellin transform, Hankel transform, Abel transform) and with a variety of names (e.g., transform, characteristic function, generating function). Why is it that they appear so often? The answer has two parts; first, because they arise *naturally* in the formulation and the solution of systems problems; and second, because when we observe or introduce them into our solution method, they greatly *simplify* the calculations. Moreover, oftentimes they are the only tools we have available for proceeding with the solution at all.

Since transforms do appear naturally, we should inquire as to what gives rise to their appearance. The answer lies in the consideration of *linear time-invariant systems.* A system, in the sense that we use it here, is merely a transformation, or mapping, or input–output relationship between two functions. Let us represent a general system as a "black" box with an input f and an output g, as shown in Figure I.1. Thus the system operates on the function f to produce the function g. In what follows we will assume that these functions depend upon an independent time parameter t; this arbitrary choice results in no loss of generality but is convenient so that we may discuss certain notions more explicitly. Thus we assume that $f = f(t)$. In order to represent the input–output relationship between the functions $f(t)$ and $g(t)$

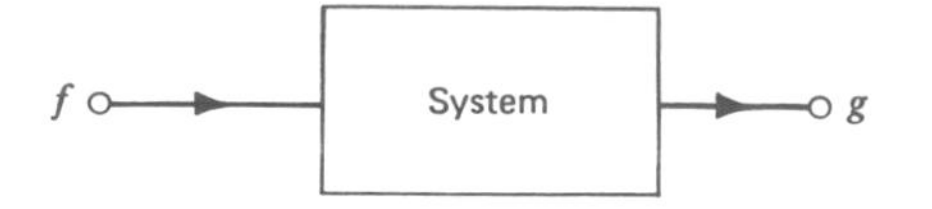

Figure I.1 A general system.

we use the notation

$$f(t) \to g(t) \tag{I.1}$$

to denote the fact that $g(t)$ is the output of our system when $f(t)$ is applied as input. A system is said to be *linear* if, when

$$f_1(t) \to g_1(t)$$

and

$$f_2(t) \to g_2(t)$$

then also the following is true:

$$af_1(t) + bf_2(t) \to ag_1(t) + bg_2(t) \tag{I.2}$$

where a and b are independent of the time variable t. Further, a system is said to be *time-invariant* if, when Eq. (I.1) holds, then the following is also true:

$$f(t + \tau) \to g(t + \tau) \tag{I.3}$$

for any τ. If the above two properties both hold, then our system is said to be a *linear time-invariant* system, and it is these with which we concern ourselves for the moment.

Whenever one studies such systems, one finds that *complex exponential functions of time* appear throughout the solution. Further, as we shall see, the transforms of interest merely represent ways of decomposing functions of time into sums (or integrals) of complex exponentials. That is, complex exponentials form the *building blocks* of our transforms, and so, we must inquire further to discover why these complex exponentials pervade our thinking with such systems. Let us now pose the fundamental question, namely, which functions of time $f(t)$ may pass through our linear time-invariant systems with no change in form; that is, for which $f(t)$ will $g(t) = Hf(t)$, where H is some scalar multiplier (with respect to t)? If we can discover such functions $f(t)$ we will then have found the "eigenfunctions," or "characteristic functions," or "invariants" of our system. Denoting these eigenfunctions by $f_e(t)$ it will be shown that they must be of the following form (to within an arbitrary scalar multiplier):

$$f_e(t) = e^{st} \tag{I.4}$$

where s is, in general, a complex variable. That is, the complex exponentials given in (I.4) form the set of eigenfunctions for *all* linear time-invariant systems. This result is so fundamental that it is worthwhile devoting a few lines to its derivation. Thus let us assume when we apply $f_e(t)$ that the output is of the form $g_e(t)$, that is,

$$f_e(t) = e^{st} \rightarrow g_e(t)$$

But, by the linearity property we have

$$e^{s\tau}f_e(t) = e^{s(t+\tau)} \rightarrow e^{s\tau}g_e(t)$$

where τ and therefore $e^{s\tau}$ are both constants. Moreover, from the time-invariance property we must have

$$f_e(t + \tau) = e^{s(t+\tau)} \rightarrow g_e(t + \tau)$$

From these last two, it must be that

$$e^{s\tau}g_e(t) = g_e(t + \tau)$$

The unique solution to this equation is given by

$$g_e(t) = He^{st}$$

which confirms our earlier hypothesis that the complex exponentials pass through our linear time-invariant systems unchanged except for the scalar multiplier H. H is independent of t but may certainly be a function of s and so we choose to write it as $H = H(s)$. Therefore, we have the final conclusion that

$$e^{st} \rightarrow H(s)e^{st} \tag{I.5}$$

and this fundamental result exposes the eigenfunctions of our systems.

In this way the complex exponentials are seen to be the basic functions in the study of linear time-invariant systems. Moreover, if it is true that the input to such a system is a complex exponential, then it is a trivial computation to evaluate the output of that system from Eq. (I.5) if we are given the function $H(s)$. Thus it is natural to ask that for any input $f(t)$ we would hope to be able to decompose $f(t)$ into a sum (or integral) of complex exponentials, each of which contributes to the overall output $g(t)$ through a computation of the form given in Eq. (I.5). Then the overall output may be found by summing (integrating) these individual components of the output. (The fact that the sum of the individual outputs is the same as the output of the sum of the individual inputs—that is, the complex exponential decomposition—is due to the linearity of our system.) The process of decomposing our input into sums of exponentials, computing the response to each from Eq. (I.5), and then reconstituting the output from sums of exponentials is referred to as the transform method of analysis. This approach, as we can see, arises very naturally from our foregoing statements. In this sense, transforms arise in a perfectly natural way. Moreover, we know that such systems are described by constant-coefficient linear differential equations, and so the common use of transforms in the solution of such equations is not surprising.

We still have not given a precise definition of the transform itself; be patient, for we are attempting to answer the question "why transforms?" If we were to pursue the line of reasoning that follows from Eq. (I.5), we would quickly encounter Laplace transforms. However, it is convenient at this point to consider only functions of discrete time rather than functions of continuous time, as we have so far been discussing. This change in direction brings us to a consideration of z-transforms and we will return to Laplace transforms later in this appendix. The reason for this switch is that it is easier to visualize operations on a discrete-time axis as compared to a continuous-time axis (and it also delays the introduction of the unit impulse function temporarily).

Thus we consider functions f that are defined only at discrete instants in time, which, let us say, are multiples of some basic time unit T. That is, $f(t) = f(t = nT)$, where $n = \ldots, -2, -1, 0, 1, 2, \ldots$. In order to incorporate this discrete-time axis into our notation we will denote the function $f(t = nT)$ by f_n. We assume further that our systems are also discrete in time. Thus we are led to consider linear time-invariant systems with inputs f and outputs g (also functions of discrete time) for which we obtain the following three equations corresponding to Eqs. (I.1)–(I.3):

$$f_n \rightarrow g_n \tag{I.6}$$

$$af_n^{(1)} + bf_n^{(2)} \rightarrow ag_n^{(1)} + bg_n^{(2)} \tag{I.7}$$

$$f_{n+m} \rightarrow g_{n+m} \tag{I.8}$$

where m is some integer constant. Here Eq. (I.7) is the expression of linearity whereas Eq. (I.8) is the expression of time-invariance for our discrete systems. We may ask the same fundamental question for these discrete systems and, of course, the answer will be essentially the same, namely, that the eigenfunctions are given by

$$f_n^{(e)} = e^{st} = e^{snT}$$

Once again the complex exponentials are the eigenfunctions. At this point it is convenient to introduce the definition

$$z \triangleq e^{-sT} \tag{I.9}$$

and so the eigenfunctions $f_n^{(e)}$ take the form

$$f_n^{(e)} = z^{-n}$$

Since s is a complex variable, so, too, is z. Following through steps essentially identical to those which led from Eq. (I.4) to Eq. (I.5) we find that

$$z^{-n} \rightarrow H(z)z^{-n} \tag{I.10}$$

where $H(z)$ is a function independent of n. This merely expresses the fact that the set of functions $\{z^{-n}\}$ for any value of z form the set of eigenfunctions for discrete linear time-invariant systems. Moreover, the function (constant) H either in Eq. (I.5) or (I.10) tells us precisely how much of a given complex exponential we get out of our linear system when we insert a unit amount of that exponential at the input. That is, H really describes the effect of the system on these exponentials; for this reason it is usually referred to as the *system* (*or transfer*) *function.*

Let us pursue this line of reasoning somewhat further. As we all know, a common way to discover what is inside a system is to kick it—hard and quickly. For our systems this corresponds to providing an input only at time $t = 0$ and then observing the subsequent output. Thus let us define the *Kronecker delta function* (also known as the *unit function*) as

$$u_n = \begin{cases} 1 & n = 0 \\ 0 & n \neq 0 \end{cases} \tag{I.11}$$

When we apply u_n to our system it is common to refer to the output as the *unit response*, and this is usually denoted by h_n. That is,

$$u_n \rightarrow h_n$$

From Eq. (I.8) we may therefore also write

$$u_{n+m} \rightarrow h_{n+m}$$

From the linearity property in Eq. (I.7) we have therefore

$$z^m u_{n+m} \rightarrow z^m h_{n+m}$$

Certainly we may multiply both expressions by unity, and so

$$z^{-n}z^n z^m u_{n+m} \rightarrow z^{-n}z^n z^m h_{n+m} \tag{I.12}$$

Furthermore, if we consider a set of inputs $\{f_n^{(i)}\}$, and if we define the output for each of these by

$$f_n^{(i)} \rightarrow g_n^{(i)}$$

then by the linearity of our system we must have

$$\sum_i f_n^{(i)} \rightarrow \sum_i g_n^{(i)} \tag{I.13}$$

If we now apply this last observation to Eq. (I.12) we have

$$z^{-n} \sum_m u_{n+m} z^{n+m} \to z^{-n} \sum_m h_{n+m} z^{n+m}$$

where the sum ranges over all integer values of m. From the definition in Eq. (I.11) it is clear that the sum on the left-hand side of this equation has only one nonzero term, namely, for $m = -n$, and this term is equal to unity; moreover, let us make a change of variable for the sum on the right-hand side of this expression, giving

$$z^{-n} \to z^{-n} \sum_k h_k z^k$$

This last equation is now in the same form as Eq. (I.10); it is obvious then that we have the relationship

$$H(z) = \sum_k h_k z^k \tag{I.14}$$

This last equation relates the system function $H(z)$ to the unit response h_k. Recall that our linear time-invariant system was completely* specified by knowledge of $H(z)$, since we could then determine the output for any of our eigenfunctions; similarly, knowledge of the unit response also completely* determines the operation of our linear time-invariant system. Thus it is no surprise that some explicit relationship must exist between the two, and, of course, this is given in Eq. (I.14).

Finally, we are in a position to answer the question—why transforms? The key lies in the expression (I.14), which is, itself, a transform (in this case a z-transform), which converts† the time function h_k into a function of a complex variable $H(z)$. This transform arose naturally in our study of linear time-invariant systems and was not introduced into the analysis in an artificial way. We shall see later that a similar relationship exists for continuous-time systems, as well, and this gives rise to the Laplace transform. Recalling that continuous-time systems may be described by constant-coefficient linear differential equations and that the use of transforms greatly simplifies the solution of these equations, we are not surprised that discrete-time systems lead to sets of constant-coefficient linear *difference* equations whose solution is simplified by the use of z-transforms. Lastly, we comment that the inputs f

* Completely specified in the sense that the only additional required information is the initial state of the system (e.g., the initial conditions of all the energy storage elements). Usually, the system is assumed to be in the zero-energy state, in which case we truly have a complete specification.

† Transforms not only change the form in which the information describing a given function is presented, but they also present this information in a simplified form which is convenient for mathematical manipulation.

and the outputs g are easily decomposed into weighted sums of complex exponentials by means of transforms, and of course, once this is done, then results such as (I.5) or (I.10) immediately give us the component-by-component output of our system for each of these inputs; the total output is then formed by summing the output components as in Eq. (I.13).

The fact that these transforms arise naturally in our system studies is really only a partial answer to our basic question regarding their use in analysis. The other and more pragmatic reason is that they greatly simplify the analysis itself; most often, in fact, the analysis can only proceed with the use of transforms leading us to a partial solution from which properties of the system behavior may be derived.

The remainder of this appendix is devoted to giving examples and properties of these two principal transforms which are so useful in queueing theory.

I.2. THE z-TRANSFORM [JURY 64, CADZ 73]

Let us consider a function of discrete time f_n, which takes on nonzero values only for the nonnegative integers, that is, for $n = 0, 1, 2, \ldots$ (i.e., for convenience we assume that $f_n = 0$ for $n < 0$). We now wish to compress this semi-infinite sequence into a single function in a way such that we can expand the compressed form back into the original sequence when we so desire. In order to do this, we must place a "tag" on each of the terms in the sequence f_n. We choose to tag the term f_n by multiplying it by z^n; since n is then unique for each term in the sequence, each tag is also unique. z will be chosen as some complex variable whose permitted range of values will be discussed shortly. Once we tag each term, we may then sum over all tagged terms to form our compressed function, which represents the original sequence. Thus we define the z-transform (also known as the generating function or geometric transform) for f_n as follows:

$$F(z) \triangleq \sum_{n=0}^{\infty} f_n z^n \tag{I.15}$$

$F(z)$ is clearly only a function of our complex variable z since we have summed over the index n; the notation we adopt for the z-transform is to use a capital letter that corresponds to the lower-case letter describing the sequence, as in Eq. (I.15). We recognize that Eq. (I.14) is, of course, in exactly this form. The z-transform for a sequence will exist so long as the terms in that sequence grow no faster than geometrically, that is, so long as there is some $a > 0$ such that

$$\lim_{n \to \infty} \frac{|f_n|}{a^n} = 0$$

Furthermore, given a sequence f_n its z-transform $F(z)$ is unique.

If the sum over all terms in the sequence f_n is finite, then certainly the unit disk $|z| \leq 1$ represents a range of analyticity for $F(z)$.* In such a case we have

$$F(1) = \sum_{n=0}^{\infty} f_n \tag{I.16}$$

We now consider some important *examples* of z-transforms. It is convenient to denote the relationship between a sequence and its transform by means of a double-barred, double-headed arrow†; thus Eq. (I.15) may be written as

$$f_n \Leftrightarrow F(z) \tag{I.17}$$

For our first example, let us consider the unit function as defined in Eq. (I.11). For this function and from the definition given in Eq. (I.15) we see that exactly one term in the infinite summation is nonzero, and so we immediately have the transform pair

$$u_n \Leftrightarrow 1 \tag{I.18}$$

For a related example, let us consider the unit function shifted to the right by k units, that is,

$$u_{n-k} = \begin{cases} 1 & n = k \\ 0 & n \neq k \end{cases}$$

From Eq. (I.15) again, exactly one term will be nonzero, giving

$$u_{n-k} \Leftrightarrow z^k$$

As a third example, let us consider the *unit step function* defined by

$$\delta_n = 1 \qquad \text{for} \qquad n = 0, 1, 2, \ldots$$

(recall that all functions are zero for $n < 0$). In this case we have a geometric series, that is,

$$\delta_n \Leftrightarrow \sum_{n=0}^{\infty} 1 z^n = \frac{1}{1 - z} \tag{I.19}$$

We note in this case that $|z| < 1$ in order for the z-transform to exist. An extremely important sequence often encountered is the geometric series

$$f_n = A\alpha^n \qquad n = 0, 1, 2, \ldots$$

* A function of a complex variable is said to be analytic at a point in the complex plane if that function has a unique derivative at that point. The Cauchy–Riemann necessary and sufficient condition for analyticity of such functions may be found in any text on functions of a complex variable [AHLF 66].

† The double bar denotes the transform relationship whereas the double heads on the arrow indicate that the journey may be made in either direction, $f \Rightarrow F$ and $F \Rightarrow f$.

Its z-transform may be calculated as

$$F(z) = \sum_{n=0}^{\infty} A\alpha^n z^n$$

$$= A\sum_{n=0}^{\infty} (\alpha z)^n$$

$$= \frac{A}{1 - \alpha z}$$

And so

$$A\alpha^n \Leftrightarrow \frac{A}{1 - \alpha z} \tag{I.20}$$

where, of course, the region of analyticity for this function is $|z| < 1/\alpha$; note that α may be greater or less than unity.

Linear transformations such as the z-transform enjoy a number of important *properties*. Many of these are listed in Table I.1. However, it is instructive for us to derive the *convolution* property which is most important in queueing systems. Let us consider two functions of discrete time f_n and g_n, which may take on nonzero values only for the nonnegative integers. Their respective z-transforms are, of course, $F(z)$ and $G(z)$. Let $\circledast$ denote the convolution operator, which is defined for f_n and g_n as follows:

$$f_n \circledast g_n \overset{\Delta}{=} \sum_{k=0}^{n} f_{n-k} g_k$$

We are interested in deriving the z-transform of the convolution for f_n and g_n, and this we do as follows:

$$f_n \circledast g_n \Leftrightarrow \sum_{n=0}^{\infty} (f_n \circledast g_n) z^n$$

$$= \sum_{n=0}^{\infty} \sum_{k=0}^{n} f_{n-k} g_k z^{n-k} z^k$$

However, since

$$\sum_{n=0}^{\infty} \sum_{k=0}^{n} = \sum_{k=0}^{\infty} \sum_{n=k}^{\infty}$$

we have

$$f_n \circledast g_n \Leftrightarrow \sum_{k=0}^{\infty} g_k z^k \sum_{n-k}^{\infty} f_{n-k} z^{n-k}$$

$$= \left(\sum_{k=0}^{\infty} g_k z^k\right)\left(\sum_{m=0}^{\infty} f_m z^m\right)$$

$$= G(z)F(z)$$

Table I.1
Some Properties of the z-Transform

	SEQUENCE		z-TRANSFORM
1.	$f_n \quad n = 0, 1, 2, \ldots$	$\Leftrightarrow$	$F(z) = \sum_{n=0}^{\infty} f_n z^n$
2.	$af_n + bg_n$		$aF(z) + bG(z)$
3.	$a^n f_n$		$F(az)$
4.	$f_{n/k} \quad n = 0, k, 2k, \ldots$		$F(z^k)$
5.	f_{n+1}		$\frac{1}{z}[F(z) - f_0]$
6.	$f_{n+k} \quad k > 0$		$\frac{F(z)}{z^k} - \sum_{i=1}^{k} z^{i-k-1} f_{i-1}$
7.	f_{n-1}		$zF(z)$
8.	$f_{n-k} \quad k > 0$		$z^k F(z)$
9.	nf_n		$z \frac{d}{dz} F(z)$
10.	$n(n-1)(n-2), \ldots, (n-m+1)f_n$		$z^m \frac{d^m}{dz^m} F(z)$
11.	$f_n \circledast g_n$		$F(z)G(z)$
12.	$f_n - f_{n-1}$		$(1 - z)F(z)$
13.	$\sum_{k=0}^{n} f_k \quad n = 0, 1, 2, \ldots$		$\frac{F(z)}{1 - z}$
14.	$\frac{\partial}{\partial a} f_n$ (a is a parameter of f_n)		$\frac{\partial}{\partial a} F(z)$
15.	Series sum property		$F(1) = \sum_{n=0}^{\infty} f_n$
16.	Alternating sum property		$F(-1) = \sum_{n=0}^{\infty} (-1)^n f_n$
17.	Initial value theorem		$F(0) = f_0$
18.	Intermediate value theorem		$\frac{1}{n!} \frac{d^n F(z)}{dz^n} \Big\vert_{z=0} = f_n$
19.	Final value theorem		$\lim_{z \to 1} (1 - z)F(z) = f_\infty$

Table I.2
Some z-Transform Pairs

SEQUENCE		z-TRANSFORM
1. $f_n \quad n = 0, 1, 2, \ldots$	$\Leftrightarrow$	$F(z) = \sum_{n=0}^{\infty} f_n z^n$
2. $u_n = \begin{cases} 1 & n = 0 \\ 0 & n \neq 0 \end{cases}$		1
3. u_{n-k}		z^k
4. $\delta_n = 1 \quad n = 0, 1, 2, \ldots$		$\dfrac{1}{1 - z}$
5. δ_{n-k}		$\dfrac{z^k}{1 - z}$
6. $A\alpha^n$		$\dfrac{A}{1 - \alpha z}$
7. $n\alpha^n$		$\dfrac{\alpha z}{(1 - \alpha z)^2}$
8. n		$\dfrac{z}{(1 - z)^2}$
9. $n^2\alpha^n$		$\dfrac{\alpha z(1 + \alpha z)}{(1 - \alpha z)^3}$
10. n^2		$\dfrac{z(1 + z)}{(1 - z)^3}$
11. $(n + 1)\alpha^n$		$\dfrac{1}{(1 - \alpha z)^2}$
12. $(n + 1)$		$\dfrac{1}{(1 - z)^2}$
13. $\dfrac{1}{m!}(n + m)(n + m - 1) \cdots (n + 1)\alpha^n$		$\dfrac{1}{(1 - \alpha z)^{m+1}}$
14. $\dfrac{1}{n!}$		e^z

In words, then, we have that the z-transform of the convolution of two sequences is equal to the product of the z-transform of each of the sequences themselves.

In Table I.1 we list a number of important properties of the z-transform, and following that in Table I.2 we provide a list of important common z-transforms.

Some comments regarding these tables are in order. First, in the property table we note that Property 2 is a statement of linearity, and Properties 3 and 4 are statements regarding scale change in the transform and time domain, respectively. Properties 5–8 regard translation in time and are most useful. In particular, note from Property 7 that the unit delay (delay by one unit of time) results in multiplication of the transform by the factor z whereas Property 5 states that a unit advance involves division by the factor z. Properties 9 and 10 show multiplication of the sequence by terms of the form $n(n-1)\cdots(n-m)$. Combinations of these may be used in order to find, for example, the transform of $n^2 f_n$; this may be done by recognizing that $n^2 = n(n-1) + n$, and so the transform of $n^2 f_n$ is merely $z^2\, d^2F(z)/dz^2 + z dF(z)/dz$. This shows the simple differentiation technique of obtaining more complex transforms. Perhaps the most important, however, is Property 11 showing that the convolution of two time sequences has a transform that is the product of the transform of each time sequence separately. Properties 12 and 13 refer to the difference and summation of various terms in the sequence. Property 14 shows if a is an independent parameter of f_n, differentiating the sequence with respect to this parameter is equivalent to differentiating the transform. Property 15 is also important and shows that the transform expression may be evaluated at $z = 1$ directly to give the sum of all terms in the sequence. Property 16 merely shows how to calculate the alternating sum. From the definition of the z-transform, the initial value theorem given in Property 17 is obvious and shows how to calculate the initial term of the sequence directly from the transform. Property 18, on the other hand, shows how to calculate *any* term in the original sequence directly from its z-transform by successive differentiation; this then corresponds to one method for calculating the sequence given its transform. It can be seen from Property 18 that the sequence f_n forms the coefficients in the Taylor-series expansion of $F(z)$ about the point 0. Since this power-series expansion is unique, then it is clear that the inversion process is also unique. Property 19 gives a direct method for calculating the final value of a sequence from its z-transform.

Table I.2 lists some useful transform pairs. This table can be extended considerably by making use of the properties listed in Table I.1; in some cases this has already been done. For example, Pair 5 is derived from Pair 4 by use of the delay theorem given as entry 8 in Table I.1. One of the more useful relationships is given in Pair 6 considered earlier.

Thus we see the effect of compressing a time sequence f_n into a single function of the complex variable z. Recall that the use of the variable z was to tag the terms in the sequence f_n so that they could be recovered from the compressed function; that is, f_n was tagged with the factor z^n. We have seen how to form the z-transform of the sequence [through Eq. (I.15)]. The problem confronting us now is to find the sequence f_n given the z-transform $F(z)$. There are basically three methods for carrying out this inversion. The *first* is the *power-series method*, which attempts to take the given function $F(z)$ and express it as a power series in z; once this is done the terms in the sequence f_n may be picked off by inspection since the tagging is now explicitly exposed. The power series may be obtained in one of two ways: the first way we have already seen through our intermediate value theorem expressed as Item 18 in Table I.1, that is,

$$f_n = \frac{1}{n!} \frac{d^n F(z)}{dz^n} \bigg|_{z=0}$$

(this method is useful if one is only interested in a few terms but is rather tedious if many terms are required); the second way is useful if $F(z)$ is expressible as a rational function of z (that is, as the ratio of a polynomial in z over a polynomial in z) and in this case one may divide the denominator into the numerator to pick off the sequence of leading terms in the power series directly. The power-series expansion method is usually difficult when many terms are required.

The *second* and most useful method for inverting z-transforms [that is, to calculate f_n from $F(z)$] is the *inspection method*. That is, one attempts to express $F(z)$ in a fashion such that it consists of terms that are recognizable as transform pairs, for example, from Table I.2. The standard approach for placing $F(z)$ in this form is to carry out a *partial-fraction expansion*,* which we now discuss. The partial-fraction expansion is merely an algebraic technique for expressing rational functions of z as sums of simple terms, each of which is easily inverted. In particular, we will attempt to express a rational $F(z)$ as a sum of terms, each of which looks either like a simple pole (see entry 6 in Table I.2) or as a multiple pole (see entry 13). Since the sum of the transforms equals the transform of the sum we may apply Property 2 from Table I.1 to invert each of these now recognizable forms separately, thereby carrying out the required inversion. To carry out the partial-fraction expansion we proceed as follows. We assume that $F(z)$ is in rational form, that is

$$F(z) = \frac{N(z)}{D(z)}$$

where both the numerator $N(z)$ and the denominator $D(z)$ are each

* This procedure is related to the Laurent expansion of $F(z)$ around each pole [GUIL 49].

polynomials in z.* Furthermore we will assume that $D(z)$ is already in factored form, that is,

$$D(z) = \prod_{i=1}^{k} (1 - \alpha_i z)^{m_i} \tag{I.21}$$

The product notation used in this last equation is defined as

$$\prod_{i=1}^{k} a_i \overset{\Delta}{=} a_1 a_2 \cdots a_k$$

Equation (I.21) implies that the ith root at $z = 1/\alpha_i$ occurs with multiplicity m_i. [We note here that in most problems of interest, the difficult part of the solution is to take an arbitrary polynomial such as $D(z)$ and to find its roots α_i so that it may be put in the factored form given in Eq. (I.21). At this point we assume that that difficult task has been accomplished.] If $F(z)$ is in this form then it is possible to express it as follows [GUIL 49]:

$$\begin{aligned} F(z) = {} & \frac{A_{11}}{(1 - \alpha_1 z)^{m_1}} + \frac{A_{12}}{(1 - \alpha_1 z)^{m_1 - 1}} + \cdots + \frac{A_{1m_1}}{(1 - \alpha_1 z)} \\ & + \frac{A_{21}}{(1 - \alpha_2 z)^{m_2}} + \frac{A_{22}}{(1 - \alpha_2 z)^{m_2 - 1}} + \cdots + \frac{A_{2m_2}}{1 - \alpha_2 z} + \cdots \\ & + \frac{A_{k1}}{(1 - \alpha_k z)^{m_k}} + \frac{A_{k2}}{(1 - \alpha_k z)^{m_k - 1}} + \cdots + \frac{A_{km_k}}{(1 - \alpha_k z)} \end{aligned} \tag{I.22}$$

This last form is exactly what we were looking for, since each term in this sum may be found in our table of transform pairs; in particular it is Pair 13 (and in the simplest case it is Pair 6). Thus if we succeed in carrying out the partial-fraction expansion, then by inspection we have our time sequence f_n. It remains now to describe the method for calculating the coefficients A_{ij}. The general expression for such a term is given by

$$A_{ij} = \frac{1}{(j-1)!}\left(-\frac{1}{\alpha_i}\right)^{j-1} \frac{d^{j-1}}{dz^{j-1}}\left[(1 - \alpha_i z)^{m_i} \frac{N(z)}{D(z)}\right]\Bigg|_{z=1/\alpha_i} \tag{I.23}$$

This rather formidable procedure is, in fact, rather straightforward as long as the function $F(z)$ is not terribly complex.

* We note here that a partial-fraction expansion may be carried out only if the degree of the numerator polynomial is strictly less than the degree of the denominator polynomial; if this is not the case, then it is necessary to divide the denominator into the numerator until the remainder is of lower degree than the denominator. This remainder divided by the original denominator may then be expanded in partial fractions by the method shown; the terms generated from the division also may be inverted by inspection making use of transform pair 3 in Table I.2. An alternative way of satisfying the degree condition is to attempt to factor out enough powers of z from the numerator if possible.

It is worthwhile at this point to carry out an example in order to demonstrate the method. Let us assume $F(z)$ is given by

$$F(z) = \frac{4z^2(1 - 8z)}{(1 - 4z)(1 - 2z)^2} \tag{I.24}$$

In this example the numerator and denominator both have the same degree and so it is necessary to bring the expression into proper form (numerator degree less than denominator degree). In this case our task is simple since we may factor out two powers of z (we are required to factor out only one power of z in order to bring the numerator degree below that of the denominator, but obviously in this case we may as well factor out both and simplify our calculations). Thus we have

$$F(z) = z^2\left[\frac{4(1 - 8z)}{(1 - 4z)(1 - 2z)^2}\right]$$

Let us define the term in square brackets as $G(z)$. We note in this example that the denominator has three poles: one at $z = 1/4$; and two (that is a double pole) at $z = 1/2$. Thus in terms of the variables defined in Eq. (I.21) we have $k = 2$, $\alpha_1 = 4$, $m_1 = 1$, $\alpha_2 = 2$, $m_2 = 2$. From Eq. (I.22) we are therefore seeking the following expansion:

$$\begin{aligned} G(z) &\triangleq \frac{4(1 - 8z)}{(1 - 4z)(1 - 2z)^2} \\ &= \frac{A_{11}}{1 - 4z} + \frac{A_{21}}{(1 - 2z)^2} + \frac{A_{22}}{(1 - 2z)} \end{aligned}$$

Terms such as A_{11} (that is, coefficients of simple poles) are easily obtained from Eq. (I.23) by multiplying the original function by the factor corresponding to the pole and then evaluating the result at the pole itself (that is, when z takes on a value that drives the factor to 0). Thus in our example we have

$$A_{11} = (1 - 4z)G(z)\big|_{z=1/4} = \frac{4[1 - (8/4)]}{[1 - (2/4)]^2} = -16$$

A_{21} may be evaluated in a similar way from Eq. (I.23) as follows:

$$A_{21} = (1 - 2z)^2G(z)\big|_{z=1/2} = \frac{4[1 - (8/2)]}{[1 - (4/2)]} = 12$$

Finally, in order to evaluate A_{22} we must apply the differentiation formula given Eq. (I.23) once, that is,

$$\begin{aligned} A_{22} &= -\frac{1}{2}\frac{d}{dz}[(1-2z)^2G(z)]\bigg|_{z=1/2} \\ &= -\frac{1}{2}\frac{d}{dz}\frac{4(1-8z)}{(1-4z)}\bigg|_{z=1/2} \\ &= -\frac{1}{2}\frac{(1-4z)(-32)-4(1-8z)(-4)}{(1-4z)^2}\bigg|_{z=1/2} \\ &= 8 \end{aligned}$$

Thus we conclude that

$$G(z) = \frac{-16}{1-4z} + \frac{12}{(1-2z)^2} + \frac{8}{1-2z}$$

This is easily shown to be equal to the original factored form of $G(z)$ by placing these terms over a common denominator. Our next step is to invert $G(z)$ by inspection. This we do by observing that the first and third terms are of the form given by transform pair 6 in Table I.2 and that the second term is given by transform pair 13. This, coupled with the linearity property 2 in Table I.1 gives immediately that

$$G(z) \Leftrightarrow g_n = \begin{cases} 0 & n < 0 \\ -16(4)^n + 12(n+1)(2)^n + 8(2)^n & n = 0, 1, 2, \ldots \end{cases} \tag{I.25}$$

Of course, we must now account for the factor z^2 to give the expression for f_n. As mentioned above we do this by taking advantage of Property 8 in Table I.1 and so we have (for $n = 2, 3, \ldots$)

$$f_n = -16(4)^{n-2} + 12(n-1)(2)^{n-2} + 8(2)^{n-2}$$

and so

$$f_n = \begin{cases} 0 & n < 2 \\ (3n-1)2^n - 4^n & n = 2, 3, 4, \ldots \end{cases}$$

This completes our example.

The *third* method for carrying out the inversion process is to use the *inversion formula*. This involves evaluating the following integral:

$$f_n = \frac{-1}{2\pi j}\oint_C F(z)z^{-1-n}\,dz \tag{I.26}$$

where $j = \sqrt{-1}$ and the integral is evaluated in the complex z-plane around a closed circular contour C, which is large enough* to surround all poles of $F(z)$. This method of evaluation works properly when factors of the form z^k are removed from the expression; the reduced expression is then evaluated and the final solution is obtained by taking advantage of Property 9 in Table I.1 as we shall see below. This contour integration is most easily performed by making use of the Cauchy residue theorem [GUIL 49]. This theorem may be stated as follows:

Cauchy Residue Theorem *The integral of $g(z)$ over a closed contour C containing within it only isolated singular points of $g(z)$ is equal to $2\pi j$ times the sum of the residues at these points, whenever $g(z)$ is analytic† on and within the closed contour C.*

An isolated singular point of an analytic function is a singular point whose neighborhood contains no other singular points; a simple pole (i.e., a pole of order one—see below) is the classical example. If $z = a$ is an isolated singular point of $g(z)$ and if $\gamma(z) = (z - a)^m g(z)$ is analytic at $z = a$ and $\gamma(a) \neq 0$, then $g(z)$ is said to have a *pole of order m* at $z = a$ with the *residue* r_a given by

$$r_a = \frac{1}{(m-1)!} \frac{d^{m-1}}{dz^{m-1}} (z - a)^m g(z) \bigg|_{z=a} \tag{I.27}$$

We note that the residue given in this last equation is almost the same as A_{ij} given in Eq. (I.23), the main difference being the form in which we write the pole. Thus we have now defined all that we need to apply the Cauchy residue theorem in order to evaluate the integral in Eq. (I.26) and thereby to recover the time function f_n from our z-transform. By way of illustration we carry out the calculation of our previous example given in Eq. (I.24). Making use of Eq. (I.26) we have

$$g_n = \frac{-1}{2\pi j} \oint_C \frac{4z^{-1-n}(1 - 8z)}{(1 - 4z)(1 - 2z)^2} \, dz$$

* Since Jordan's lemma (see p. 353) requires that $F(z) \to 0$ as $z \to \infty$ if we are to let the contour grow, then we require that any function $F(z)$ that we consider have this property; thus for rational functions of z if the numerator degree is not less than the denominator degree, then we must divide the numerator by the denominator until the remainder is of lower degree than the denominator, as we have seen earlier. The terms generated by this division are easily transformed by inspection, as discussed earlier, and it is only the remaining function which we now consider in this inversion method for the z-transform.

† A function $F(z)$ of a complex variable z is said to be analytic in a region of the complex plane if it is single-valued and differentiable at every point in that region.

where C is a circle large enough to enclose the poles of $F(z)$ at $z = 1/4$ and $z = 1/2$. Using the residue theorem and Eq. (I.27) we find that the residue at $z = 1/4$ is given by

$$r_{1/4} = \left(z - \frac{1}{4}\right)\frac{4z^{-1-n}(1-8z)}{(1-4z)(1-2z)^2}\bigg|_{z=1/4}$$

$$= \frac{-(1/4)^{-1-n}[1-(8/4)]}{[1-(2/4)]^2} = 16(4)^n$$

whereas the residue at $z = 1/2$ is calculated as

$$r_{1/2} = \frac{d}{dz}\left[\left(z-\frac{1}{2}\right)^2\frac{4z^{-1-n}(1-8z)}{(1-4z)(1-2z)^2}\right]\bigg|_{z=1/2}$$

$$= \frac{d}{dz}\left[\frac{z^{-1-n}(1-8z)}{1-4z}\right]\bigg|_{z=1/2}$$

$$= \frac{(1-4z)[(-1-n)z^{-2-n}(1-8z)+z^{-1-n}(-8)]-z^{-1-n}(1-8z)(-4)}{(1-4z)^2}\bigg|_{z=1/2}$$

$$= (-1)\left[(-1-n)\left(\frac{1}{2}\right)^{-2-n}(-3)+\left(\frac{1}{2}\right)^{-1-n}(-8)\right]-\left(\frac{1}{2}\right)^{-1-n}(-3)(-4)$$

$$= -12(n+1)2^n + 16(2)^n - 24(2)^n$$

Now we must take $2\pi j$ times the sum of the residues and then multiply by the factor preceding the integral in Eq. (I.26) (thus we must take -1 times the sum of the residues) to yield

$$g_n = -16(4)^n + 12(n+1)2^n + 8(2)^n \qquad n = 0, 1, 2, \ldots$$

But this last is exactly equal to the form for g_n in Eq. (I.25) found by the method of partial-fraction expansions. From here the solution proceeds as in that method, thus confirming the consistency of these two approaches.

Thus we have reviewed some of the techniques for applying and inverting the z-transform in the handling of discrete-time functions. The application of these methods in the solution of difference equations is carefully described in Sect. I.4 below.

I.3. THE LAPLACE TRANSFORM [WIDD 46]

The Laplace transform, defined below, enjoys many of the same properties as the z-transform. As a result, the following discussion very closely parallels that given in the previous section.

We now consider functions of continuous time $f(t)$, which take on nonzero values only for nonnegative values of the continuous parameter t. [Again for convenience we are assuming that $f(t) = 0$ for $t < 0$. For the more general case, most of these techniques apply as discussed in the paragraph containing Eq. (I.38) below.] As with discrete-time functions, we wish to take our continuous-time function and transform it from a function of t to a function of a new complex variable (say, s). At the same time we would like to be able to "untransform" back into the t domain, and in order to do this it is clear we must somehow "tag" $f(t)$ at each value of t. For reasons related to those described in Section I.1 the tag we choose to use is e^{-st}. The complex variable s may be written in terms of its real and complex parts as $s = \sigma + j\omega$ where, again, $j = \sqrt{-1}$. Having multiplied by this tag, we then integrate over all nonzero values in order to obtain our transform function defined as follows:

$$F^*(s) \triangleq \int_{-\infty}^{\infty} f(t)e^{-st}\,dt \tag{I.28}$$

Again, we have adopted the notation for general Laplace transforms in which we use a capital letter for the transform of a function of time, which is described in terms of a lower case letter. This is usually referred to as the "two-sided," or "bilateral" Laplace transform since it operates on both the negative and positive time axes. We have assumed that $f(t) = 0$ for $t < 0$, and in this case the lower limit of integration may be replaced by 0^-, which is defined as the limit of $0 - \epsilon$ as $\epsilon(> 0)$ goes to zero; further, we often denote this lower limit merely by 0 with the understanding that it is meant as 0^- (usually this will cause no confusion). There also exists what is known as the "one-sided" Laplace transform in which the lower limit is replaced by 0^+, which is defined as the limit of $0 + \epsilon$ as $\epsilon(> 0)$ goes to zero; this one-sided transform has application in the solution of transient problems in linear systems. It is important that the reader distinguish between these two transforms with zero as their lower limit since in the former case (the bilateral transform) any accumulation at the origin (as, for example, the unit impulse defined below) will be included in the transform, whereas in the latter case (the one-sided transform) it will be omitted.

For our assumed case in which $f(t) = 0$ for $t < 0$ we may write our transform as

$$F^*(s) = \int_{0}^{\infty} f(t)e^{-st}\,dt \tag{I.29}$$

where, we repeat, the lower limit is to be interpreted as 0^-. This Laplace transform will exist so long as $f(t)$ grows no faster than an exponential, that is, so long as there is some real number σ_a such that

$$\lim_{T\to\infty} \int_{0}^{T} |f(t)|\, e^{\sigma_a t}\,dt < \infty$$

The smallest possible value for σ_a is referred to as the abscissa of absolute convergence. Again we state that the Laplace transform $F^*(s)$ for a given function $f(t)$ is unique.

If the integral of $f(t)$ is finite, then certainly the right-half plane Re $(s) \geq 0$ represents a region of analyticity for $F^*(s)$; the notation Re ( ) reads as "the real part of the complex function within the parentheses." In such a case we have, corresponding to Eq. (I.16),

$$F^*(0) = \int_0^\infty f(t)\, dt \tag{I.30}$$

From our earlier definition in Eq. (I.9) we see that properties for the z-transform when $z = 1$ will correspond to properties for the Laplace transform when $s = 0$ as, for example, in Eqs. (I.16) and (I.30).

Let us now consider some important examples of Laplace transforms. We use notation here identical to that used in Eq. (I.17) for z-transforms, namely, we use a double-barred, double-headed arrow to denote the relationship between a function and its transform; thus, Eq. (I.29) may be written as

$$f(t) \Leftrightarrow F^*(s) \tag{I.31}$$

The use of the double arrow is a statement of the uniqueness of the transform as earlier.

As in the case of z-transforms, the most useful method for finding the inverse [that is, calculating $f(t)$ from $F^*(s)$] is the *inspection method*, namely, looking up the inverse in a table. Let us, therefore, concentrate on the calculation of some Laplace transform pairs. By far the most important Laplace transform pair to consider is for the one-sided exponential function, namely,

$$f(t) = \begin{cases} Ae^{-at} & t \geq 0 \\ 0 & t < 0 \end{cases}$$

Let us carry out the computation of this transform, as follows:

$$\begin{aligned} f(t) \Leftrightarrow F^*(s) &= \int_0^\infty Ae^{-at}e^{-st}\, dt \\ &= A\int_0^\infty e^{-(s+a)t}\, dt \\ &= \frac{A}{s+a} \end{aligned}$$

And so we have the fundamental relationship

$$Ae^{-at}\, \delta(t) \Leftrightarrow \frac{A}{s+a} \tag{I.32}$$

where we have defined the unit step function in continuous time as

$$\delta(t) = \begin{cases} 1 & t \geq 0 \\ 0 & t < 0 \end{cases} \tag{I.33}$$

In fact, we observe that the unit step function is a special case of our one-sided exponential function when $A = 1$, $a = 0$, and so we have immediately the additional pair

$$\delta(t) \Leftrightarrow \frac{1}{s} \tag{I.34}$$

We note that the transform in Eq. (I.32) has an abscissa of convergence $\sigma_a = -a$.

Thus we have calculated analogous z-transform and Laplace-transform pairs: the geometric series given in Eq. (I.20) with the exponential function given in Eq. (I.32) and also the unit step function in Eqs. (I.19) and (I.34), respectively. It remains to find the continuous analog of the unit function defined in Eq. (I.11) and whose z-transform is given in Eq. (I.18). This brings us face to face with the *unit impulse function.* The unit impulse function plays an important part in transform theory, linear system theory, as well as in probability and queueing theory. It therefore behoves us to learn to work with this function. Let us adopt the following notation

$$u_0(t) \triangleq \text{unit impulse function occurring at } t = 0$$

$u_0(t)$ corresponds to highly concentrated unit-area pulses that are of such short duration that they cannot be distinguished by available measurement instruments from other perhaps briefer pulses. Therefore, as one might expect, the exact shape of the pulse is unimportant, rather only its time of occurrence and its area matter. This function has been studied and utilized by scientists for many years [GUIL 49], among them Dirac, and so the unit impulse function is often referred to as the *Dirac delta function.* For a long time pure mathematicians have refrained from using $u_0(t)$ since it is a highly improper function, but years ago Schwartz's theory of distributions [SCHW 59] put the concept of a unit impulse function on firm mathematical ground. Part of the difficulty lies with the fact that the unit impulse function is not a function at all, but merely provides a notational way for handling discontinuities and their derivatives. In this regard we will introduce the unit impulse as the limit of a sequence without appealing to the more sophisticated generalized functions that place much of what we do in a more rigorous framework.

As we mentioned earlier, the exact shape of the pulse is unimportant. Let us therefore choose the following representative pulse shape for our discussion of impulses:

$$f_\alpha(t) = \begin{cases} \alpha & |t| \leq \dfrac{1}{2\alpha} \\ 0 & |t| > \dfrac{1}{2\alpha} \end{cases}$$

This rectangular wave form has a height and width dependent upon the parameter α as shown in Figure I.2. Note that this function has a constant area of unity (hence the name *unit* impulse function). As α increases, we note that the pulse gets taller and narrower. The limit of this sequence as $\alpha \rightarrow \infty$ (or the limit of any one of an infinite number of other sequences with similar properties, i.e., increasing height, decreasing width, unit area) is what we mean by the unit impulse "function." Thus we are led to the following description of the unit impulse function.

$$u_0(t) = \begin{cases} \infty & t = 0 \\ 0 & t \neq 0 \end{cases}$$

$$\int_{-\infty}^{\infty} u_0(t)\, dt = 1$$

This function is represented graphically by a vertical arrow located at the instant of the impulse and with a number adjacent to the head of the arrow

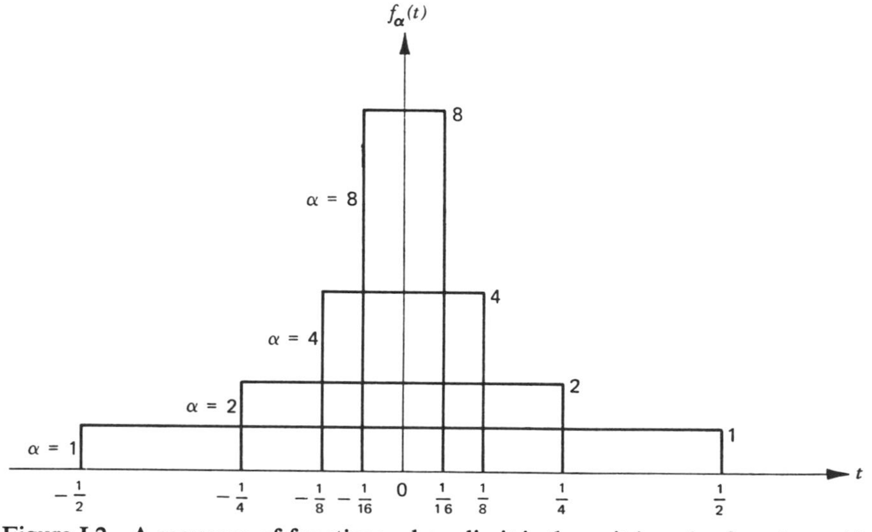

Figure I.2 A sequence of functions whose limit is the unit impulse function $u_0(t)$.

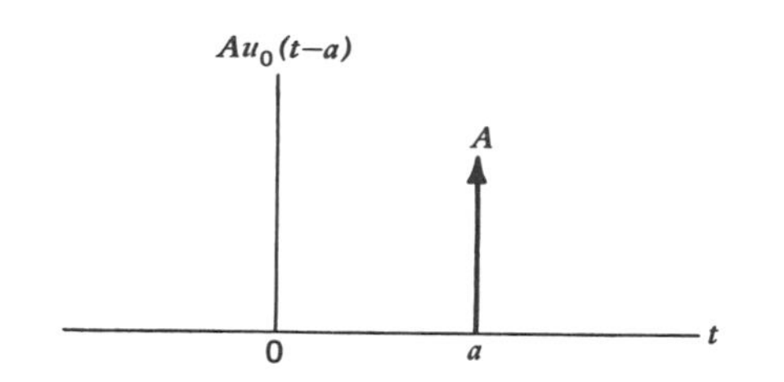

Figure I.3 Graphical representation of $Au_0(t - a)$.

indicating the area of the impulse; that is, A times a unit impulse function located at the point $t = a$ is denoted as $Au_0(t - a)$ and is depicted as in Figure I.3.

Let us now consider the integral of the unit impulse function. It is clear that if we integrate from $-\infty$ to a point t where $t < 0$ then the total integral must be 0 whereas if $t > 0$ then we will have successfully integrated past the unit impulse and thereby will have accumulated a total area of unity. Thus we conclude

$$\int_{-\infty}^{t} u_0(x)\,dx = \begin{cases} 1 & t \geq 0 \\ 0 & t < 0 \end{cases}$$

But we note immediately that the right-hand side is the same as the definition of the unit step function given in Eq. (I.33). Therefore, we conclude that the unit step function is the integral of the unit impulse function, and so the "derivative" of the unit step function must therefore be a unit impulse function. However, we recognize that the derivative of this discontinuous function (the step function) is not properly defined; once again we appeal to the theory of distributions to place this operation on a firm mathematical foundation. We will therefore assume this is a proper operation and proceed to use the unit impulse function as if it were an ordinary function.

One of the very important properties of the unit impulse function is its *sifting* property; that is, for an arbitrary differentiable function $g(t)$ we have

$$\int_{-\infty}^{\infty} u_0(t - x)g(x)\,dx = g(t)$$

This last equation merely says that the integral of the product of our function $g(x)$ with an impulse located at $x = t$ "sifts" the function $g(x)$ to produce its value at t, $g(t)$. We note that it is possible also to define the derivative of the unit impulse which we denote by $u_1(t) = du_0(t)/dt$; this is known as the *unit doublet* and has the property that it is everywhere 0 except in the vicinity of the origin where it runs off to ∞ just to the left of the origin and off to $-\infty$

just to the right of the origin, and, in addition, has a total area equal to zero. Such functions correspond to electrostatic dipoles, for example, used in physics. In fact, an impulse function may be likened to the force placed on a piece of paper when it is laid over the edge of a knife and pressed down, whereas a unit doublet is similar to the force the paper experiences when cut with scissors. Higher-order derivatives are possible and in general we may have $u_n(t) = du_{n-1}(t)/dt$. In fact, as we have seen, we may also go back down the sequence by integrating these functions as, for example, by generating the unit step function as the integral of the unit impulse function; the obvious notation for the unit step function, therefore, would be $u_{-1}(t)$ and so we may write $u_0(t) = du_{-1}(t)/dt$. [Note, from Eq. (I.33), that we have also reserved the notation $\delta(t)$ to represent the unit step function.] Thus we have defined an infinite sequence of specialized functions beginning with the unit impulse and proceeding to higher-order derivatives such as the doublet, and so on, as well as integrating the unit impulse and thereby generating the unit step function, the ramp, namely,

$$u_{-2}(t) \triangleq \int_{-\infty}^{t} u_{-1}(x)\,dx = \begin{cases} t & t \geq 0 \\ 0 & t < 0 \end{cases}$$

the parabola, namely,

$$u_{-3}(t) \triangleq \int_{-\infty}^{t} u_{-2}(x)\,dx = \begin{cases} \dfrac{t^2}{2} & t \geq 0 \\ 0 & t < 0 \end{cases}$$

and in general

$$u_{-n} \triangleq \int_{-\infty}^{t} u_{-(n-1)}(x)\,dx = \begin{cases} \dfrac{t^{n-1}}{(n-1)!} & t \geq 0 \\ 0 & t < 0 \end{cases} \tag{I.35}$$

This entire family is called the *family of singularity functions*, and the most important members are the unit step function and the unit impulse function.

Let us now return to our main discussion and consider the Laplace transform of $u_0(t)$. We proceed directly from Eq. (I.28) to obtain

$$u_0(t) \Leftrightarrow \int_0^{\infty} u_0(t)e^{-st}\,dt = 1$$

(Note that the lower limit is interpreted as 0^-.) Thus we see that the unit impulse has a Laplace transform equal to the constant unity.

Let us now consider some of the important properties of the transformation. As with the z-transform, the convolution property is the most important and we proceed to derive it here in the continuous time case. Thus consider two functions of continuous time $f(t)$ and $g(t)$, which take on nonzero values only for $t \geq 0$; we denote their Laplace transforms by $F^*(s)$ and $G^*(s)$, respectively. Defining ⊛ once again as the convolution operator, that is,

$$f(t) \circledast g(t) \triangleq \int_{-\infty}^{\infty} f(t - x)g(x)\, dx \tag{I.36}$$

which in our case reduces to

$$f(t) \circledast g(t) = \int_{0}^{t} f(t - x)g(x)\, dx$$

we may then ask for the Laplace transform of this convolution. We obtain this formally by plugging into Eq. (I.28) as follows:

$$\begin{aligned} f(t) \circledast g(t) &\Leftrightarrow \int_{t=0}^{\infty} (f(t) \circledast g(t))e^{-st}\, dt \\ &= \int_{t=0}^{\infty} \int_{x=0}^{t} f(t - x)g(x)\, dx\, e^{-st}\, dt \\ &= \int_{x=0}^{\infty} \int_{t=x}^{\infty} f(t - x)e^{-s(t-x)}\, dt\, g(x)e^{-sx}\, dx \\ &= \int_{x=0}^{\infty} g(x)e^{-sx}\, dx \int_{v=0}^{\infty} f(v)e^{-sv}\, dv \end{aligned}$$

And so we have

$$f(t) \circledast g(t) \Leftrightarrow F^*(s)G^*(s)$$

Once again we see that the transform of the convolution of two functions equals the product of the transforms of each. In Table I.3, we list a number of important properties of the Laplace transform, and in Table I.4 we list some of the important transforms themselves. In these tables we adopt the usual notation as follows:

$$\frac{d^n f(t)}{dt^n} \triangleq f^{(n)}(t) \tag{I.37}$$

$$\underbrace{\int \cdots \int}_{n \text{ times}} f(x)\, dx \triangleq f^{(-n)}(t)$$

For example, $f^{(-1)}(t) = \int_{-\infty}^{t} f(x)\, dx$; when we deal with functions which are zero for $t < 0$, then $f^{(-1)}(0^-) = 0$. We comment here that the one-sided transform that uses 0^+ as a lower limit in its definition is quite commonly used in transient analysis, but we prefer 0^- so as to include impulses at the origin.

The table of properties permits one to compute many transform pairs from a given pair. Property 2 is the statement of linearity and Property 3 describes the effect of a scale change. Property 4 gives the effect of a translation in time,

Table I.3
Some Properties of the Laplace Transform

FUNCTION		TRANSFORM
1. $f(t) \qquad t \geq 0$	$\Leftrightarrow$	$F^*(s) = \int_{0^-}^{\infty} f(t)e^{-st}\,dt$
2. $af(t) + bg(t)$		$aF^*(s) + bG^*(s)$
3. $f\left(\frac{t}{a}\right) \qquad (a > 0)$		$aF^*(as)$
4. $f(t - a)$		$e^{-as}F^*(s)$
5. $e^{-at}f(t)$		$F^*(s + a)$
6. $tf(t)$		$-\frac{dF^*(s)}{ds}$
7. $t^nf(t)$		$(-1)^n \frac{d^nF^*(s)}{ds^n}$
8. $\frac{f(t)}{t}$		$\int_{s_1=s}^{\infty} F^*(s_1)\,ds_1$
9. $\frac{f(t)}{t^n}$		$\int_{s_1=s}^{\infty} ds_1 \int_{s_2=s_1}^{\infty} ds_2 \cdots \int_{s_n=s_{n-1}}^{\infty} ds_n F^*(s_n)$
10. $f(t) \circledast g(t)$		$F^*(s)G^*(s)$
11.† $\frac{df(t)}{dt}$		$sF^*(s)$
12.† $\frac{d^nf(t)}{dt^n}$		$s^nF^*(s)$
13.† $\int_{-\infty}^{t} f(t)\,dt$		$\frac{F^*(s)}{s}$
14.† $\underbrace{\int_{-\infty}^{t} \cdots \int_{-\infty}^{t}}_{n \text{ times}} f(t)(dt)^n$		$\frac{F^*(s)}{s^n}$
15. $\frac{\partial}{\partial a} f(t) \qquad$ [a is a parameter]		$\frac{\partial}{\partial a} F(s)$
16. Integral property		$F^*(0) = \int_{0^-}^{\infty} f(t)\,dt$
17. Initial value theorem		$\lim_{s \to \infty} sF^*(s) = \lim_{t \to 0} f(t)$
18. Final value theorem		$\lim_{s \to 0} sF^*(s) = \lim_{t \to \infty} f(t)$ if $sF^*(s)$ is analytic for Re $(s) \geq 0$

† To be complete, we wish to show the form of the transform for entries 11–14 in the case when $f(t)$ may have nonzero values for $t < 0$ also:

$$\frac{d^nf(t)}{dt^n} \Leftrightarrow s^nF^*(s) - s^{n-1}f(0^-) - s^{n-2}f^{(1)}(0^-) - \cdots - f^{(n-1)}(0^-)$$

$$\underbrace{\int_{-\infty}^{t} \cdots \int_{-\infty}^{t}}_{n \text{ times}} f(t)(dt)^n \Leftrightarrow \frac{F^*(s)}{s^n} + \frac{f^{(-1)}(0^-)}{s^n} + \frac{f^{(-2)}(0^-)}{s^{n-1}} + \cdots + \frac{f^{(-n)}(0^-)}{s}$$

Table I.4
Some Laplace Transform Pairs

FUNCTION		TRANSFORM
1. $f(t) \quad t \geq 0$	$\Leftrightarrow$	$F^*(s) = \int_{0-}^{\infty} f(t)e^{-st}\,dt$
2. $u_0(t)$ (unit impulse)		1
3. $u_0(t-a)$		e^{-as}
4. $u_n(t) \triangleq \frac{d}{dt} u_{n-1}(t)$		s^n
5. $u_{-1}(t) \triangleq \delta(t)$ (unit step)		$\frac{1}{s}$
6. $u_{-1}(t-a)$		$\frac{e^{-as}}{s}$
7. $u_{-n}(t) \triangleq \frac{t^{n-1}}{(n-1)!}$		$\frac{1}{s^n}$
8. $Ae^{-at}\,\delta(t)$		$\frac{A}{s+a}$
9. $te^{-at}\,\delta(t)$		$\frac{1}{(s+a)^2}$
10. $\frac{t^n}{n!} e^{-at}\,\delta(t)$		$\frac{1}{(s+a)^{n+1}}$

whereas Property 5, its dual, gives the effect of a parameter shift in the transform domain. Properties 6 and 7 show the effect of multiplication by t (to some power), which corresponds to differentiation in the transform domain; similarly, Properties 8 and 9 show the effect of division by t (to some power), which corresponds to integration. Property 10, a most important property (derived earlier), shows the effect of convolution in the time domain going over to simple multiplication in the transform domain. Properties 11 and 12 give the effect of time differentiation; it should be noted that this corresponds to multiplication by s (to a power equal to the number of differentiations in time) times the original transform. In a similar way Properties 13 and 14 show the effect of time integration going over to division by s in the transform domain. Property 15 shows that differentiation with respect to a parameter of $f(t)$ corresponds to differentiation in the transform domain as well. Property 16, the integral property, shows the simple way in which the transform may be evaluated at the origin to give the total integral of $f(t)$. Properties 17 and 18, the initial and final value theorems, show how to compute the values for $f(t)$ at $t = 0$ and $t = \infty$ directly from the transform.

In Table I.4 we have a rather short list of important Laplace transform pairs. Much more extensive tables exist and may be found elsewhere [DOET 61]. Of course, as we said earlier, the table shown can be extended considerably by making use of the properties listed in Table I.3. We note, for example, that the transform pair 3 in Table I.4 is obtained from transform pair 2 by application of Property 4 in Table I.3. We point out again that this table is limited in length since we have included only those functions that find relevance to the material contained in this text.

So far in this discussion of Laplace transforms we have been considering only functions $f(t)$ for which $f(t) = 0$ for $t < 0$. This will be satisfactory for most of the work we consider in this text. However, there is an occasional need for transforming a function of time which may be nonzero anywhere on the real-time axis. For this purpose we must once again consider the lower limit of integration to be $-\infty$, that is,

$$F^*(s) = \int_{-\infty}^{\infty} f(t)e^{-st}\,dt \tag{I.38}$$

One can easily show that this (bilateral) Laplace transform may be calculated in terms of one-sided time functions and their transforms as follows. First we define

$$f_-(t) = \begin{cases} f(t) & t < 0 \\ 0 & t \geq 0 \end{cases}$$

$$f_+(t) = \begin{cases} 0 & t < 0 \\ f(t) & t \geq 0 \end{cases}$$

and so it immediately follows that

$$f(t) = f_-(t) + f_+(t)$$

We now observe that $f_-(-t)$ is a function that is nonzero only for positive values of t, and $f_+(t)$ is nonzero only for nonnegative values of t. Thus we have

$$f_+(t) \Leftrightarrow F_+{}^*(s)$$
$$f_-(-t) \Leftrightarrow F_-{}^*(s)$$

where these transforms are defined as in Eq. (I.29). However, we need the transform of $f_-(t)$ which is easily shown to be

$$f_-(t) \Leftrightarrow F_-{}^*(-s)$$

Thus, by the linearity of transforms, we may finally write the bilateral transform in terms of one-sided transforms:

$$F^*(s) = F_-{}^*(-s) + F_+{}^*(s)$$

As always, these Laplace transforms have abscissas of absolute convergence. Let us therefore define σ_+ as the convergence abscissa for $F_+^*(s)$; this implies that the region of convergence for $F_+^*(s)$ is Re $(s) > \sigma_+$. Similarly, $F_-^*(s)$ will have some abscissa of absolute convergence, which we will denote by σ_-, which implies that $F_-^*(s)$ converges for Re $(s) > \sigma_-$. It then follows directly that $F_-^*(-s)$ will have the same convergence abscissa (σ_-) but will converge for Re $(s) < \sigma_-$. Thus we have a situation where $F^*(s)$ converges for $\sigma_+ <$ Re $(s) < \sigma_-$ and therefore we will have a "convergence strip" if and only if $\sigma_+ < \sigma_-$; if such is not the case, then it is not useful to define $F^*(s)$. Of course, a similar argument can be made in the case of z-transforms for functions that take on nonzero values for negative time indices.

So far we have seen the effect of tagging our time function $f(t)$ with the complex exponential e^{-st} and then compressing (integrating) over all such tagged functions to form a new function, namely, the transform $F^*(s)$. The purpose of the tagging was so that we could later "untransform" or, if you will, "unwind" the transform in order to obtain $f(t)$ once again. In principle we know this is possible since a transform and its time function are uniquely related. So far, we have specified how to go in the one direction from $f(t)$ to $F^*(s)$. Let us now discuss the problem of inverting the Laplace transform $F^*(s)$ to recover $f(t)$. There are basically two methods for conducting this inversion: The *inspection method* and the *formal inversion integral method.* These two methods are very similar.

First let us discuss the *inspection* method, which is perhaps the most useful scheme for inverting transforms. Here, as with z-transforms, the approach is to rewrite $F^*(s)$ as a sum of terms, each of which can be recognized from the table of Laplace transform pairs. Then, making use of the linearity property, we may invert the transform term by term, and then sum the result to recover $f(t)$. Once again, the basic method for writing $F^*(s)$ as a sum of recognizable terms is that of the partial-fraction expansion. Our description of that method will be somewhat shortened here since we have discussed it at some length in the z-transform section. First, we will assume that $F^*(s)$ is a rational function of s, namely,

$$F^*(s) = \frac{N(s)}{D(s)}$$

where both the numerator $N(s)$ and denominator $D(s)$ are each polynomials in s. Again, we assume that the degree of $N(s)$ is less than the degree of $D(s)$; if this is not the case, $N(s)$ must be divided by $D(s)$ until the remainder is of degree less than the degree of $D(s)$, and then the partial-fraction expansion is carried out for this remainder, whereas the terms of the quotient resulting from the division will be simple powers of s, which may be inverted by appealing to Transform 4 in Table I.4. In addition, we will assume that the "hard" part of the problem has been done, namely, that $D(s)$ has been put in factored form

$$D(s) = \prod_{i=1}^{k} (s + a_i)^{m_i} \tag{I.39}$$

Once $F^*(s)$ is in this form we may then express it as the following sum:

$$\begin{aligned} F^*(s) = {} & \frac{B_{11}}{(s + a_1)^{m_1}} + \frac{B_{12}}{(s + a_1)^{m_1 - 1}} + \cdots + \frac{B_{1m_1}}{(s + a_1)} \\ & + \frac{B_{21}}{(s + a_2)^{m_2}} + \frac{B_{22}}{(s + a_2)^{m_2 - 1}} + \cdots + \frac{B_{2m_2}}{(s + a_2)} + \cdots \\ & + \frac{B_{k1}}{(s + a_k)^{m_k}} + \frac{B_{k2}}{(s + a_k)^{m_k - 1}} + \cdots + \frac{B_{km_k}}{(s + a_k)} \end{aligned} \tag{I.40}$$

Once we have expressed $F^*(s)$ as above we are then in a position to invert each term in this sum by inspection from Table I.4. In particular, Pairs 8 (for simple poles) and 10 (for multiple poles) give us the answer directly. As before, the method for calculating the coefficients B_{ij} is given in general by

$$B_{ij} = \frac{1}{(j-1)!} \frac{d^{j-1}}{ds^{j-1}} \left[(s + a_i)^{m_i} \frac{N(s)}{D(s)} \right] \bigg|_{s=-a_i} \tag{I.41}$$

Thus we have a complete prescription for finding $f(t)$ from $F^*(s)$ by inspection in those cases where $F^*(s)$ is rational and where $D(s)$ has been factored as in Eq. (I.39). This method works very well in those cases where $F^*(s)$ is not overly complex.

To elucidate some of these principles let us carry out a simple example. Assume that $F^*(s)$ is given by

$$F^*(s) = \frac{8(s^2 + 3s + 1)}{(s + 3)(s + 1)^3} \tag{I.42}$$

We have already written the denominator in factored form, and so we may proceed directly to expand $F^*(s)$ as in Eq. (I.40). Note that we have $k = 2$, $a_1 = 3$, $m_1 = 1$, $a_2 = 1$, $m_2 = 3$. Since the denominator degree (4) is greater than the numerator degree (2), we may immediately expand $F^*(s)$ as a partial fraction as given by Eq. (I.40), namely,

$$F^*(s) = \frac{B_{11}}{s + 3} + \frac{B_{21}}{(s + 1)^3} + \frac{B_{22}}{(s + 1)^2} + \frac{B_{23}}{(s + 1)}$$

Evaluation of the coefficients B_{ij} proceeds as follows. B_{11} is especially simple since no differentiations are required, and we obtain

$$B_{11} = (s + 3)F^*(s)\big|_{s=-3} = 8\,\frac{(9 - 9 + 1)}{(-2)^3} = -1$$

B_{21} is also easy to evaluate:

$$B_{21} = (s + 1)^3F^*(s)\big|_{s=-1} = 8\,\frac{(1 - 3 + 1)}{2} = -4$$

For B_{22} we must differentiate once, namely,

$$\begin{aligned} B_{22} &= \frac{d}{ds}\left[\frac{8(s^2 + 3s + 1)}{s + 3}\right]\bigg|_{s=-1} \\ &= 8\,\frac{(s + 3)(2s + 3) - (s^2 + 3s + 1)(1)}{(s + 3)^2}\bigg|_{s=-1} \\ &= 8\,\frac{s^2 + 6s + 8}{(s + 3)^2}\bigg|_{s=-1} = 8\,\frac{1 - 6 + 8}{(2)^2} \\ &= 6 \end{aligned}$$

Lastly, the calculation of B_{23} involves two differentiations; however, we have already carried out the first differentiation, and so we take advantage of the form we have derived in B_{22} just prior to evaluation at $s = -1$; furthermore, we note that since $j = 3$, we have for the first time an effect due to the term $(j - 1)!$ from Eq. (I.41). Thus

$$\begin{aligned} B_{23} &= \frac{1}{2!}\,\frac{d^2}{ds^2}\left[8\,\frac{(s^2 + 3s + 1)}{s + 3}\right]\bigg|_{s=-1} \\ &= \frac{1}{2}\,(8)\,\frac{d}{ds}\left[\frac{s^2 + 6s + 8}{(s + 3)^2}\right]\bigg|_{s=-1} \\ &= 4\,\frac{(s + 3)^2(2s + 6) - (s^2 + 6s + 8)(2)(s + 3)}{(s + 3)^4}\bigg|_{s=-1} \\ &= 4\,\frac{(2)^2(4) - (1 - 6 + 8)(2)(2)}{(2)^4} \\ &= 1 \end{aligned}$$

This completes the evaluation of the constants B_{ij} to give the partial-fraction expansion

$$F^*(s) = \frac{-1}{s+3} + \frac{-4}{(s+1)^3} + \frac{6}{(s+1)^2} + \frac{1}{(s+1)} \tag{I.43}$$

This last form lends itself to inversion by inspection as we had promised. In particular, we observe that the first and last terms invert directly according to transform pair 8 from Table I.4, whereas the second and third terms invert directly from Pair 10 of that table; thus we have for $t \geq 0$ the following:

$$f(t) = -e^{-3t} - 2t^2e^{-t} + 6te^{-t} + e^{-t} \tag{I.44}$$

and of course, $f(t) = 0$ for $t < 0$.

In the course of carrying out an inversion by partial-fraction expansions there are two natural points at which one can conduct a test to see if any errors have been made: first, once we have the partial-fraction expansion [as in our example, the result given in Eq. (I.43)], then one can combine this sum of terms into a single term over a common denominator and check that this single term corresponds to the original given $F^*(s)$; the other check is to take the final form for $f(t)$ and carry out the forward transformation and confirm that it gives the original $F^*(s)$ [of course, one then gets $F^*(s)$ expanded directly as a partial fraction].

The *second* method for finding $f(t)$ from $F^*(s)$ is to use the *inversion integral*

$$f(t) = \frac{1}{2\pi j}\int_{\sigma_c - j\infty}^{\sigma_c + j\infty} F^*(s)e^{st}\,ds \tag{I.45}$$

for $t \geq 0$ and $\sigma_c > \sigma_a$. The integration in the complex s-plane is taken to be a straight-line integration parallel to the imaginary axis and lying to the right of σ_a, the abscissa of absolute convergence for $F^*(s)$. The usual means for carrying out this integration is to make use of the Cauchy residue theorem as applied to the integral in the complex domain around a closed contour. The closed contour we choose for this purpose is a semicircle of infinite radius as shown in Figure I.4. In this figure we see the path of integration required for Eq. (I.45) is $s_3 - s_1$ and the semicircle of infinite radius closing this contour is given as $s_1 - s_2 - s_3$. If the integral along the path $s_1 - s_2 - s_3$ is 0, then the integral along the entire closed contour will in fact give us $f(t)$ from Eq. (I.45). To establish that this contribution is 0, we need

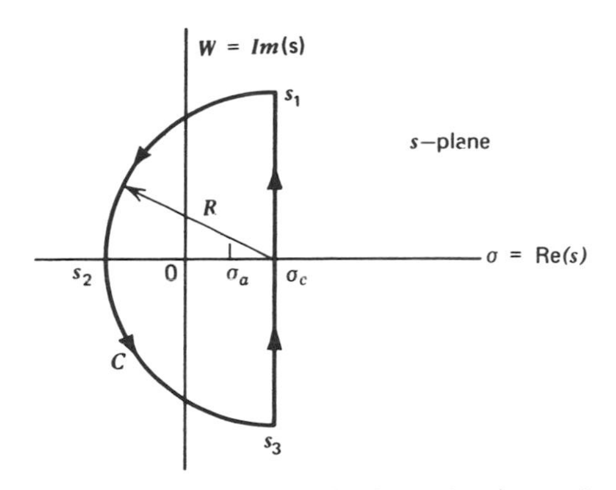

Figure I.4 Closed contour for inversion integral.

Jordan's Lemma If $|F^*(s)| \to 0$ as $R \to \infty$ on $s_1 - s_2 - s_3$, then

$$\int_{s_1-s_2-s_3} F^*(s)e^{st}\,ds = 0 \qquad \text{for } t > 0$$

Thus, in order to carry out the complex inversion integral shown in Eq. (I.45), we must first express $F^*(s)$ in a form for which Jordan's lemma applies. Having done this we may then evaluate the integral around the closed contour C by calculating residues and using Cauchy's residue theorem. This is most easily carried out if $F^*(s)$ is in rational form with a factored denominator as in Eq. (I.39). In order for Jordan's lemma to apply, we will require, as we did before, that the degree of the numerator be strictly less than the degree of the denominator, and if this is not so, we must divide the rational function until the remainder has this property. That is all there is to the method. Let us carry this out on our previous example, namely that given in Eq. (I.42). We note this is already in a form for which Jordan's lemma applies, and so we may proceed directly with Cauchy's residue theorem. Our poles are located at $s = -3$ and $s = -1$. We begin by calculating the residue at $s = -3$, thus

$$\begin{aligned} r_{-3} &= (s+3)F^*(s)e^{st}\big|_{s=-3} \\ &= \left.\frac{8(s^2+3s+1)e^{s}{}_{t}}{(s+1)^3}\right|_{s=-3} \\ &= \frac{8(9-9+1)e^{-3t}}{(-2)^3} \\ &= -e^{-3t} \end{aligned}$$

Similarly, we must calculate the residue at $s = -1$, which requires the differentiations indicated in our residue formula Eq. (I.27):

$$r_{-1} = \frac{1}{2!}\frac{d^2}{ds^2}(s+1)^3F^*(s)e^{st}\Big|_{s=-1}$$

$$= \frac{1}{2}\frac{d^2}{ds^2}\frac{8(s^2+3s+1)e^{st}}{(s+3)}\Big|_{s=-1}$$

$$= \frac{1}{2}\frac{d}{ds}\left[\frac{(s+3)[8(2s+3)e^{st}+8(s^2+3s+1)te^{st}]-8(s^2+3s+1)e^{st}}{(s+3)^2}\right]\Big|_{s=-1}$$

$$= \frac{1}{2}\frac{1}{(s+3)^4}\{(s+3)^2[8(2s+3)e^{st}+8(s^2+3s+1)te^{st}$$

$$+(s+3)8[2e^{st}+(2s+3)te^{st}]$$

$$+(s+3)8[(2s+3)te^{st}+(s^2+3s+1)t^2e^{st}]$$

$$-8(2s+3)e^{st}-8(s^2+3s+1)te^{st}]$$

$$-[(s+3)8[(2s+3)e^{st}+(s^2+3s+1)te^{st}]$$

$$-8(s^2+3s+1)e^{st}]2(s+3)\}\,|_{s=-1}$$

$$= e^{-t}+6te^{-t}-2t^2e^{-t}$$

Combining these residues we have

$$f(t) = -e^{-3t}+e^{-t}+6te^{-t}-2t^2e^{-t} \qquad t \geq 0$$

Thus we see that our solution here is the same as in Eq. (I.44), as it must be; we have once again that $f(t) = 0$ for $t < 0$.

In our earlier discussion of the (bilateral) Laplace transform we discussed functions of time $f_-(t)$ and $f_+(t)$ defined for the negative and positive real-time axis, respectively. We also observed that the transform for each of these functions was analytic in a left half-plane and a right half-plane, respectively, as measured from their appropriate abscissas of absolute convergence. Moreover, in our last inversion method [the application of Eq. (I.45)] we observed that closing the contour by a semicircle of infinite radius in a counterclockwise direction gave a result for $t > 0$. We comment now that had we closed the contour in a clockwise fashion to the right, we would have obtained the result that would have been applicable for $t < 0$, assuming that the contribution of this contour could be shown to be 0 by Jordan's lemma. In order to invert a bilateral transform, we proceed by obtaining first $f(t)$ for positive values of t and then for negative values of t. For the first we take a path of integration within the convergence strip defined by $\sigma_- < \sigma_c < \sigma_+$ and then closing the contour with a counterclockwise semicircle; for $t < 0$, we take the same vertical contour but close it with a semicircle to the right.

As may be anticipated from our contour integration methods, it is sometimes necessary to determine exactly how many singularities of a function exist within a closed region. A very powerful and convenient theorem which aids us in this determination is given as follows:

Rouché's Theorem [GUIL 49] *If $f(s)$ and $g(s)$ are analytic functions of s inside and on a closed contour C, and also if $|g(s)| < |f(s)|$ on C, then $f(s)$ and $f(s) + g(s)$ have the same number of zeroes inside C.*

I.4. USE OF TRANSFORMS IN THE SOLUTION OF DIFFERENCE AND DIFFERENTIAL EQUATIONS

As we have already mentioned, transforms are extremely useful in the solution of both differential and difference equations with constant coefficients. In this section we illustrate that technique; we begin with difference equations using z-transforms and then move on to differential equations using Laplace transforms, preparing us for the more complicated differential-difference equations encountered in the text for which we need both methods simultaneously.

Let us consider the following general Nth-order linear difference equation with constant coefficients

$$a_N g_{n-N} + a_{N-1} g_{n-N+1} + \cdots + a_0 g_n = e_n \tag{I.46}$$

where the a_i are known constant coefficients, g_i are unknown functions to be found, and e_n is a given function of n. In addition, we assume we are given N boundary equations (e.g., initial conditions). As always with such equations, the solution which we are seeking consists of both a homogeneous and a particular solution, namely,

$$g_n = g_n^{(h)} + g_n^{(p)}$$

just as with differential equations. We know that the homogeneous solution must satisfy the homogeneous equation

$$a_N g_{n-N} + \cdots + a_0 g_n = 0 \tag{I.47}$$

The general form of solution to Eq. (I.47) is

$$g_n^{(h)} = A\alpha^n$$

where A and α are yet to be determined. If we substitute the proposed solution into Eq. (I.47), we find

$$a_N A\alpha^{n-N} + a_{N-1} A\alpha^{n-N+1} + \cdots + a_0 A\alpha^n = 0 \tag{I.48}$$

This Nth-order polynomial clearly has N solutions, which we will denote by $\alpha_1, \alpha_2, \ldots, \alpha_N$, assuming for the moment that all the α_i are distinct. Associated with each such solution is an arbitrary constant A_i which will be determined from the initial conditions for the difference equation (of which there must be N). By cancelling the common term $A\alpha^{n-N}$ from Eq. (I.48) we finally arrive at the *characteristic equation* which determines the values α_i

$$a_N + a_{N-1}\alpha + a_{N-2}\alpha^2 + \cdots + a_0\alpha^N = 0 \qquad \text{(I.49)}$$

Thus the search for the homogeneous solution is now reduced to finding the N roots of our characteristic equation (I.49). If all N of the α_i are distinct, then the homogeneous solution is

$$g_n^{(h)} = A_1\alpha_1^{\ n} + A_2\alpha_2^{\ n} + \cdots + A_N\alpha_N^{\ n}$$

In the case of nondistinct roots, we have a slightly different situation. In particular, let α_1 be a multiple root of order k; in this case the k equal roots will contribute to the homogeneous solution in the following form:

$$(A_{11}n^{k-1} + A_{12}n^{k-2} + \cdots + A_{1,k-1}n + A_{1k})\alpha_1^{\ n}$$

and similarly for any other multiple roots. As far as the particular solution $g_n^{(p)}$ is concerned, we know that it must be found by an appropriate *guess* from the form of e_n.

Let us illustrate some of these principles by means of an example. Consider the second-order difference equation

$$6g_n - 5g_{n-1} + g_{n-2} = 6\left(\frac{1}{5}\right)^n \qquad n = 2, 3, 4, \ldots \qquad \text{(I.50)}$$

This equation gives the relationship among the unknown functions g_n for $n = 2, 3, 4, \ldots$. Of course, we must give two initial conditions (since the order is 2) and we choose these to be $g_0 = 0$, $g_1 = 6/5$. In order to find the homogeneous solution we must form Eq. (I.49), which in this case becomes

$$6\alpha^2 - 5\alpha + 1 = 0$$

and so the two values of α which solve this equation are

$$\alpha_1 = \frac{1}{2}$$

$$\alpha_2 = \frac{1}{3}$$

and thus we have the homogeneous solution

$$g_n^{(h)} = A_1\left(\frac{1}{2}\right)^n + A_2\left(\frac{1}{3}\right)^n$$

The particular solution must be guessed at, and the correct guess in this case is

$$g_n^{(p)} = B\left(\frac{1}{5}\right)^n$$

If we plug $g_n^{(p)}$ as given back into our basic equation, namely, Eq. (I.50), we find that $B = 1$ and so we are convinced that the particular solution is correct. Thus our *complete* solution is given by

$$g_n = g_n^{(h)} + g_n^{(p)} = A_1\left(\frac{1}{2}\right)^n + A_2\left(\frac{1}{3}\right)^n + \left(\frac{1}{5}\right)^n$$

We use the initial conditions to solve for A_1 and A_2 and find $A_1 = 8$ and $A_2 = -9$. Thus our final solution is

$$g_n = \left(\frac{1}{2}\right)^{n-3} - \left(\frac{1}{3}\right)^{n-2} + \left(\frac{1}{5}\right)^n \qquad n = 0, 1, 2, \ldots \tag{I.51}$$

This completes the standard way for solving our difference equation.

Let us now describe the method of z-transforms for solving difference equations. Assume once again that we are given Eq. (I.46) and that it is good in the range $n = k, k + 1, \ldots$. Our approach begins by defining the following z-transform:

$$G(z) = \sum_{n=0}^{\infty} g_n z^n \tag{I.52}$$

From our earlier discussion we know that once we have found $G(z)$ we may then apply our inversion techniques to find the desired solution g_n. Our next step is to multiply the nth equation from Eq. (I.46) by z^n and then form the sum of all such multiplied equations from k to infinity; that is, we form

$$\sum_{n=k}^{\infty} \sum_{i=0}^{N} a_i g_{n-i} z^n = \sum_{n=k}^{\infty} e_n z^n$$

We then carry out the summations and attempt to recognize $G(z)$ in this single equation. Next we solve for $G(z)$ algebraically and then proceed with our inversion techniques to obtain the solution. This method does not require that we guess at the particular solution, and so in that sense is simpler than the direct method; however, as we shall see, it still has the basic difficulty that we must solve the characteristic equation [Eq. (I.49)] and in general this is the difficult part of the solution. However, even if we cannot solve for the roots α_i it is possible to obtain meaningful properties of the solution g_n from the perhaps unfactored form for $G(z)$.

Let us solve our earlier example using the method of z-transforms. Accordingly we begin with Eq. (I.50), multiply by z^n and then sum; the sum will go from 2 to infinity since this is the applicable range for that equation. Thus

$$\sum_{n=2}^{\infty} 6g_n z^n - \sum_{n=2}^{\infty} 5g_{n-1} z^n + \sum_{n=2}^{\infty} g_{n-2} z^n = \sum_{n=2}^{\infty} 6\left(\frac{1}{5}\right)^n z^n$$

We now factor out enough powers of z from each sum so that these powers match the subscript on g thusly:

$$6\sum_{n=2}^{\infty} g_n z^n - 5z\sum_{n=2}^{\infty} g_{n-1} z^{n-1} + z^2\sum_{n=2}^{\infty} g_{n-2} z^{n-2} = \sum_{n=2}^{\infty} 6\left(\frac{1}{5}\right)^n z^n$$

Focusing on the first summation we see that it is almost of the form $G(z)$ except that it is missing the terms for $n = 0$ and $n = 1$ [see Eq. (I.52)]; applying this observation to each of the sums on the left-hand side and carrying out the summation on the right-hand side directly, we find

$$6[G(z) - g_0 - g_1 z] - 5z[G(z) - g_0] + z^2 G(z) = \frac{6(1/5)^2 z^2}{1 - (1/5)z}$$

Observe how the first term in this last equation reflects the fact that our summation was missing the first two terms for $G(z)$. Solving for $G(z)$ algebraically we find

$$G(z) = \frac{6g_0 + 6g_1 z - 5g_0 z + (6/25)z^2/[1 - (1/5)z]}{6 - 5z + z^2}$$

If we now use our given values for g_0 and g_1, we have

$$G(z) = \left(\frac{1}{5}\right)\frac{z(6 - z)}{[1 - (1/3)z][1 - (1/2)z][1 - (1/5)z]}$$

Proceeding with a partial-fraction expansion of this last form we obtain

$$G(z) = \frac{-9}{1 - (1/3)z} + \frac{8}{1 - (1/2)z} + \frac{1}{1 - (1/5)z}$$

which by our usual inversion methods yields the final solution

$$g_n = -9\left(\frac{1}{3}\right)^n + 8\left(\frac{1}{2}\right)^n + \left(\frac{1}{5}\right)^n \qquad n = 0, 1, 2, \ldots$$

Note that this is exactly the same as Eq. (I.51) and so our method checks. We comment here that even were we not able to invert the given form for $G(z)$ we could still have found certain of its properties; for example, we could find that the sum of all terms is given immediately by $G(1)$, that is,

$$G(1) = \sum_{n=0}^{\infty} g_n = \frac{15}{4}$$

Let us now consider the application of the Laplace transform to the solution of constant-coefficient linear differential equations. Consider an Nth-order equation of the following form:

$$a_N \frac{d^N f(t)}{dt^N} + a_{N-1} \frac{d^{N-1} f(t)}{dt^{N-1}} + \cdots + a_1 \frac{df(t)}{dt} + a_0 f(t) = e(t) \qquad \text{(I.53)}$$

Here the coefficients a_i are given constants, and $e(t)$ is a given driving function. Along with this equation we must also be given N initial conditions in order to carry out a complete solution; these conditions typically are the values of the first N derivatives at some instant, usually at time zero. It is required to find the function $f(t)$. As usual, we will have a homogeneous solution $f^{(h)}(t)$, which solves the homogeneous equation [when $e(t) = 0$] as well as a particular solution $f^{(p)}(t)$ that corresponds to the nonhomogeneous equation. The form for the homogeneous solution will be

$$f^{(h)}(t) = Ae^{\alpha t}$$

If we substitute this into Eq. (I.53) we obtain

$$a_N A\alpha^N e^{\alpha t} + a_{N-1} A\alpha^{N-1} e^{\alpha t} + \cdots + a_1 A\alpha e^{\alpha t} + a_0 A e^{\alpha t} = 0$$

This equation will have N solutions $\alpha_1, \alpha_2, \ldots, \alpha_n$, which must solve the *characteristic equation*

$$a_N \alpha^N + a_{N-1} \alpha^{N-1} + \cdots + a_1 \alpha + a_0 = 0$$

which is equivalent to Eq. (I.49) with a change in subscripts. If all of the α_i are distinct, then the general form for our homogeneous solution will be

$$f^{(h)}(t) = A_1 e^{\alpha_1 t} + A_2 e^{\alpha_2 t} + \cdots + A_N e^{\alpha_N t}$$

The evaluation of the coefficients A_i is carried out making use of the initial conditions. In the case of multiple roots we have the following modification. Let us assume that α_1 is a repeated root of order k; this multiple root will contribute to the homogeneous solution in the following way:

$$(A_{11} t^{k-1} + A_{12} t^{k-2} + \cdots + A_{1\,k-1} t + A_{1k}) e^{\alpha_1 t}$$

and in the case of more than one multiple root the modification is obvious. As usual, one must guess in order to find the particular solution $f^{(p)}(t)$. The complete solution then is, of course, the sum of the homogeneous and particular solutions, namely,

$$f(t) = f^{(h)}(t) + f^{(p)}(t)$$

Let us apply this method to the solution of the following differential equation for illustrative purposes:

$$\frac{d^2f(t)}{dt^2} - 6\frac{df(t)}{dt} + 9f(t) = 2t \tag{I.54}$$

with the two initial conditions $f(0^-) = 0$ and $df(0^-)/dt = 0$. Forming the characteristic equation

$$\alpha^2 - 6\alpha + 9 = 0 \tag{I.55}$$

we find the following multiple root:

$$\alpha_1 = \alpha_2 = 3$$

and so the homogeneous solution must be of the form

$$f^{(h)}(t) = (A_{11}t + A_{12})e^{3t}$$

Making an appropriate guess for the particular solution we try

$$f^{(p)}(t) = B_1 + B_2t$$

Substituting this back into the basic equation (I.54) we find that $B_1 = 4/27$ and $B_2 = 2/9$. Thus our complete solution takes the form

$$f(t) = (A_{11}t + A_{12})e^{3t} + \frac{4}{27} + \frac{2}{9}t$$

Since our initial conditions state that both $f(t)$ and its first derivative must be zero at $t = 0^-$, we find that $A_{11} = 2/9$ and $A_{12} = -4/27$, which gives for our final and complete solution

$$f(t) = \frac{2}{9}\left(t - \frac{2}{3}\right)e^{3t} + \frac{2}{9}\left(t + \frac{2}{3}\right) \qquad t \geq 0 \tag{I.56}$$

The Laplace transform provides an alternative method for solving constant-coefficient linear differential equations. The method is based upon Properties 11 and 12 of Table I.3, which relate the derivative of a time function to its Laplace transform. The approach is to make use of these properties to transform both sides of the given differential equation into an equation involving the Laplace transform of the unknown function $f(t)$ itself, which we denote as usual by $F^*(s)$. This algebraic equation is then solved for $F^*(s)$, and is then inverted by any of our methods in order to immediately yield the complete solution for $f(t)$. No guess is required in order to find the particular solution, since it comes out of the inversion procedure directly.

Let us apply this technique to our previous example. We begin by transforming both sides of Eq. (I.54), which will require that we take advantage of our initial conditions as follows:

$$s^2F^*(s) - sf(0^-) - f^{(1)}(0^-) - 6sF^*(s) + 6f(0^-) + 9F^*(s) = \frac{2}{s^2}$$

In carrying out this last operation we have taken advantage of Laplace transform pair 7 from Table I.4. Since our initial conditions are both zero, we may eliminate certain terms in this last equation and proceed directly to solve for $F^*(s)$ thusly:

$$F^*(s) = \frac{2/s^2}{s^2 - 6s + 9}$$

We must now factor this last equation, which is the same problem we faced in finding the roots of Eq. (I.55) in the direct method, and as usual forms the basically difficult part of all direct and indirect methods. Carrying this out we have

$$F^*(s) = \frac{2}{s^2(s-3)^2}$$

We are now in the position to make a partial-fraction expansion yielding

$$F^*(s) = \frac{2/9}{s^2} + \frac{4/27}{s} + \frac{2/9}{(s-3)^2} + \frac{-4/27}{s-3}$$

Inverting as usual we then obtain, for $t \geq 0$,

$$f(t) = \frac{2}{9}t + \frac{4}{27} + \frac{2}{9}te^{3t} - \frac{4}{27}e^{3t}$$

which is identical to our former solution given in Eq. (I.56).

In our study of queueing systems we often encounter not only difference equations and differential equations but also the combination in the form of differential-difference equations. That is, if we refer back to Eq. (I.53) and replace the time functions by time functions that depend upon an index, say n, and if we then display a set of differential equations for various values of n, then we have an infinite set of differential-difference equations. The solution to such equations often requires that we take both the z-transform on the discrete index n and the Laplace transform on the continuous time parameter t. Examples of this type of analysis are to be found in the text itself.

REFERENCES

AHLF 66 Ahlfors, L. V., *Complex Analysis*, 2nd Edition, McGraw-Hill (New York), 1966.

CADZ 73 Cadzow, J. A., *Discrete-Time Systems*, Prentice-Hall (Englewood Cliffs, N.J.), 1973.

DOET 61 Doetsch, G., *Guide to the Applications of Laplace Transforms*, Van Nostrand (Princeton), 1961.

GUIL 49 Guillemin, E. A., *The Mathematics of Circuit Analysis*, Wiley (New York), 1949.

JURY 64 Jury, E. I., *Theory and Application of the z-Transform Method*, Wiley (New York), 1964.

SCHW 59 Schwartz, L., *Théorie des Distributions*, 2nd printing, Actualitiés scientifiques et industrielles Nos. 1245 and 1122, Hermann et Cie. (Paris), Vol. 1 (1957), Vol. 2 (1959).

WIDD 46 Widder, D. V., *The Laplace Transform*, Princeton University Press (Princeton), 1946.

APPENDIX II

Probability Theory Refresher

In this appendix we review selected topics from probability theory, which are relevant to our discussion of queueing systems. Mostly, we merely list the important definitions and results with an occasional example. The reader is expected to be familiar with this material, which corresponds to a good first course in probability theory. Such a course would typically use one of the following texts that contain additional details and derivations: Feller, Volume I [FELL 68]; Papoulis [PAPO 65]; Parzen [PARZ 60]; or Davenport [DAVE 70].

Probability theory concerns itself with describing random events. A typical dictionary definition of a random event is an event lacking aim, purpose, or regularity. Nothing could be further from the truth! In fact, it is the *extreme regularity* that manifests itself in collections of random events, that makes probability theory interesting and useful. The notion of statistical regularity is central to our studies. For example, if one were to toss a fair coin four times, one expects on the average two heads and two tails. Of course, there is one chance in sixteen that no heads will occur. As a consequence, if an unusual sequence came up (that is, no heads), we would not be terribly surprised nor would we suspect the coin was unfair. On the other hand, if we tossed the coin a million times, then once again we expect approximately half heads and half tails, but in this case, if no heads occurred, we would be more than surprised, we would be indignant and with overwhelming assurance could state that this coin was clearly unfair. In fact, the odds are better than 10^{88} to 1 that at least 490,000 heads will occur! This is what we mean by statistical regularity, namely, that we can make some very precise statements about large collections of random events.

II.1. RULES OF THE GAME

We now describe the rules of the game for creating a mathematical model for probabilistic situations, which is to correspond to real-world experiments. Typically one examines three features of such experiments:

1. A set of possible experimental *outcomes*.

2. A grouping of these outcomes into classes called *results*.
3. The *relative frequency* of these classes in many independent trials of the experiment.

The relative frequency f_c of a class c is merely the number of times the experimental outcome falls into that class, divided by the number of times the experiment is performed; as the number of experimental trials increases, we expect f_c to reach a limit due to our notion of statistical regularity.

The *mathematical model* we create also has three quantities of interest that are in one-to-one relation with the three quantities listed above in the experimental world. They are, respectively:

1. A *sample space* which is a collection of objects which we denote by S. S corresponds to the set of mutually exclusive exhaustive outcomes of the model of an experiment. Each object (i.e., possible outcome) ω in the set S is referred to as a *sample point*.
2. A family of *events* $\mathscr{E}$ denoted $\{A, B, C, \ldots\}$ in which each event is a set of sample points $\{\omega\}$. An event corresponds to a class or result in the real world.
3. A *probability measure* P which is an assignment (mapping) of the events defined on S into the set of real numbers. P corresponds to the relative frequency in the experimental situation. The notation $P[A]$ is used to denote the real number associated with the event A. This assignment must satisfy the following properties (axioms):

   (a) For any event A, $0 \leq P[A] \leq 1$. (II.1)

   (b) $P[S] = 1$. (II.2)

   (c) If A and B are "mutually exclusive" events [see (II.4) below], then $P[A \cup B] = P[A] + P[B]$. (II.3)

It is appropriate at this point to define some set theoretic notation [for example, the use of the symbol $\cup$ in property (c)]. Typically, we describe an event A as follows: $A = \{\omega\colon \omega$ satisfies the membership property for the event $A\}$; this is read as "A is the set of sample points ω such that ω satisfies the membership property for the event A." We further define

$$A^c = \{\omega\colon \omega \text{ not in } A\} = \textit{complement} \text{ of } A$$

$$A \cup B = \{\omega\colon \omega \text{ in } A \text{ or } B \text{ or both}\} = \textit{union} \text{ of } A \text{ and } B$$

$$A \cap B = AB = \{\omega\colon \omega \text{ in } A \text{ and } B\} = \textit{intersection} \text{ of } A \text{ and } B$$

$$\varphi = S^c = \text{null event (contains no sample points since } S \text{ contains all the points)}$$

If $AB = \varphi$, then A and B are said to be *mutually exclusive* (or disjoint). A set of events whose union forms the sample space S is said to be an *exhaustive* set of events. We are therefore led to the definition of a set of *mutually exclusive exhaustive* events $\{A_1, A_2, \ldots, A_n\}$, which have the properties

$$\begin{aligned} A_iA_j &= \varphi \qquad \text{for all } i \neq j \\ A_1 \cup A_2 \cup \cdots \cup A_n &= S \end{aligned} \tag{II.4}$$

We note further that $A \cup A^c = S$, $AA^c = \varphi$, $AS = A$, $A\varphi = \varphi$, $A \cup S = S$, $A \cup \varphi = A$, $S^c = \varphi$, and $\varphi^c = S$. Also, we comment that the union and intersection operators are commutative, associative, and distributive.

The triplet $(S, \mathscr{E}, P)$ along with Axioms (II.1)–(II.3) form a *probability system*. These three axioms are all that one needs in order to develop an axiomatic theory of probability whenever the number of events that can be defined on the sample space S is finite. [When the number of such events is infinite it is necessary to include an additional axiom which extends Axiom (II.3) to include the infinite union of disjoint events. This leads us to the notion of a Borel field and of infinite additivity of probability measures. We do not discuss the details further in this refresher.] Lastly, we comment that Axiom (II.2) is nothing more than a normalization statement and the choice of unity for this normalization is quite arbitrary (but also very natural).

Two other definitions are now in order. The first is that of *conditional probability*. The conditional probability of the event A given that the event B occurred (denoted as $P[A \mid B]$) is defined as

$$P[A \mid B] \triangleq \frac{P[AB]}{P[B]}$$

whenever $P[B] \neq 0$. The introduction of the conditional event B forces us to restrict attention from the original sample space S to a new sample space defined by the event B; since this new constrained sample space must now have a total probability of unity, we magnify the probabilities associated with conditional events by dividing by the term $P[B]$ as given above.

The second additional notion we need is that of *statistical independence* of events. Two events A, B are said to be statistically independent if and only if

$$P[AB] = P[A]P[B] \tag{II.5}$$

For three events A, B, C we require that each pair of events satisfies Eq. (II.5) and in addition

$$P[ABC] = P[A]P[B]P[C]$$

This definition extends of course to n events requiring the n-fold factoring of the probability expression as well as all the $(n - 1)$-fold factorings all the way down to all the pairwise factorings. It is easy to see for two independent events A, B that $P[A \mid B] = P[A]$, which merely says that knowledge of the occurrence of the event B in no way affects the probability of the occurrence of the independent event A.

The *theorem of total probability* is especially simple and important. It relates the probability of an event B and a set of mutually exclusive exhaustive events $\{A_i\}$ as defined in Eq. (II.4). The theorem is

$$P[B] = \sum_{i=1}^{n} P[A_iB]$$

which merely says that if the event B is to occur it must occur in conjunction with exactly one of the mutually exclusive exhaustive events A_i. However from the definition of conditional probability we may always write

$$P[A_iB] = P[A_i \mid B]\ P[B] = P[B \mid A_i]\ P[A_i]$$

Thus we have the second important form of the theorem of total probability, namely,

$$P[B] = \sum_{i=1}^{n} P[B \mid A_i]\ P[A_i]$$

This last equation is perhaps one of the most useful for us in studying queueing theory. It suggests the following approach for finding the probability of some complex event B, namely, first to condition the event B on some event A_i in such a way that the calculation of the occurrence of event B given this condition is less complex, and then of course to multiply by the probability of the conditional event A_i to yield the joint probability $P[A_iB]$; this having been done for a set of mutually exclusive exhaustive events $\{A_i\}$ we may then sum these probabilities to find the probability of the event B. Of course, this approach can be extended and we may wish to condition the event B on more than one event then uncondition each of these events suitably (by multiplying by the probability of the appropriate condition) and then sum all possible forms of all conditions. We will use this approach many times in the text.

We now come to the well-known *Bayes' theorem.* Once again we consider a set of events $\{A_i\}$, which are mutually exclusive and exhaustive. The theorem says

$$P[A_i \mid B] = \frac{P[B \mid A_i]\ P[A_i]}{\sum_{j=1}^{n} P[B \mid A_j]\ P[A_j]}$$

This theorem permits us to calculate the probability of one event conditioned on a second by calculating the probability of the second conditioned on the first and other similar terms.

A simple example is in order here to illustrate some of these ideas. Consider that you have just entered a gambling casino in Las Vegas. You approach a dealer who is known to have an identical twin brother; the twins cannot be distinguished. It is further known that one of the twins is an honest dealer whereas the second twin is a cheating dealer in the sense that when you play with the honest dealer you lose with probability one-half, whereas when you play with the cheating dealer you lose with probability p (if p is greater than one-half, he is cheating against you whereas if p is less than one-half he is cheating for you). Furthermore, it is equally likely that upon entering the casino you will find one or the other of these two dealers. Consider that you now play one game with the particular twin whom you encounter and further that you lose. Of course you are disappointed and you would now like to calculate the probability that the dealer you faced was in fact the cheat, for if you can establish that this probability is close to unity, you have a case for suing the casino. Let D_H be the event that you play with the honest dealer and let D_C be the event that you play with the cheating dealer; further let L be the event that you lose. What we are then asking for is $P[D_C \mid L]$. It is not immediately obvious how to make this calculation; however, if we apply Bayes' theorem the calculation itself is trivial, for

$$P[D_C \mid L] = \frac{P[L \mid D_C]\ P[D_C]}{P[L \mid D_C]\ P[D_C] + P[L \mid D_H]\ P[D_H]}$$

In this application of Bayes' theorem the collection of mutually exclusive exhaustive events is the set $\{D_H, D_C\}$, for one of these two events must occur and both cannot occur simultaneously. Our problem is now trivial since each term on the right-hand side is easily calculated and leads us to

$$P[D_C \mid L] = \frac{p(\frac{1}{2})}{p(\frac{1}{2}) + (\frac{1}{2})(\frac{1}{2})} = \frac{2p}{2p + 1}$$

This is the answer we were seeking and we find that the probability of having faced a cheating dealer, given that we lost in one play, ranges from 0 ($p = 0$) to 2/3 ($p = 1$). Thus, even if we know that the cheating dealer is completely dishonest ($p = 1$), we can only say that with probability 2/3 we faced this cheat, given that we lost one play.

As a final word on elementary topics, let us remind the reader that the number of *permutations* of N objects taken K at a time is

$$\frac{N!}{(N - K)!} = N(N - 1) \cdots (N - K + 1)$$

whereas the number of *combinations* of N things taken K at a time is denoted by $\binom{N}{K}$ and is given by

$$\binom{N}{K} = \frac{N!}{K!\,(N-K)!}$$

II.2. RANDOM VARIABLES

So far we have described a probability system which consists of the triplet $(S, \mathscr{E}, P)$, that is, a sample space, a set of events, and a probability assignment to the events of that sample space. We are now in a position to define the important concept of a *random variable*. A random variable is a variable whose value depends upon the outcome of a random experiment. Since the outcomes of our random experiments are represented as points $\omega \in S$ then to each such outcome ω, we associate a real number $X(\omega)$, which is in fact the value the random variable takes on when the experimental outcome is ω. Thus our (real) random variable $X(\omega)$ is nothing more than a function defined on the sample space, or if you will, a mapping from the points of the sample space into the (real) line.

As an example, let us consider the random experiment which consists of one play of a game of blackjack in Las Vegas. The sample space consists of all possible pairs of scores that can be obtained by the dealer and the player. Let us assume that we have grouped all such sample points into three (mutually exclusive) events of interest: lose (L), draw (D), or win (W). In order to complete the probability system we must assign probabilities to each of these events as follows*: $P[L] = 3/8$, $P[D] = 1/4$, $P[W] = 3/8$. Thus our probability system may be represented as in the Venn diagram of Figure II.1. The numbers in parentheses are of course the probabilities. Now for the random variable $X(\omega)$. Let us assume that if we win the game we win \$5, if we draw we win \$0, and if we lose we win $-$\$5 (that is, we lose \$5). Let our winnings on this single play of blackjack be the random variable $X(\omega)$. We may therefore define this variable as follows:

$$X(\omega) = \begin{cases} +5 & \omega \in W \\ 0 & \omega \in D \\ -5 & \omega \in L \end{cases}$$

Similarly, we may represent this random variable as the mapping shown in Figure II.2.

* This is the most difficult step in practice, that is, determining appropriate numbers to use in our model of the real world.

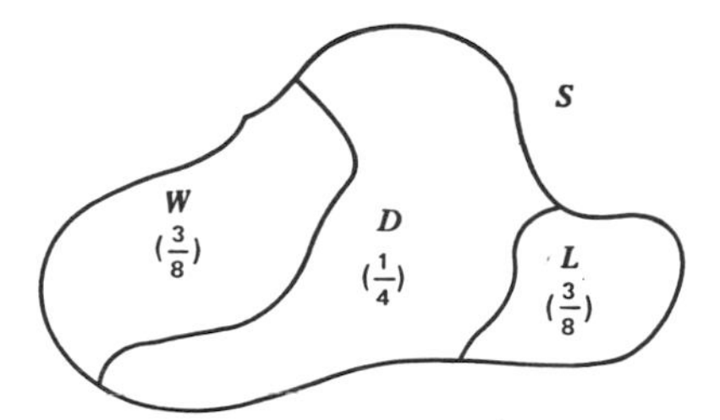

Figure II.1 The probability system for the blackjack example.

The *domain* of the random variable $X(\omega)$ is the set of events $\mathscr{E}$ and the values it takes on form its *range*. We note in passing that the probability assignment P may itself be thought of as a random variable since it satisfies the definition; this particular assignment P, however, has further restrictions on it, namely, those given in Axioms (II.1)–(II.3).

We are mainly interested in describing the *probability* that the random variable $X(\omega)$ takes on certain values. To this end we define the following shorthand notation for events:

$$[X = x] \triangleq \{\omega: X(\omega) = x\} \tag{II.6}$$

We may discuss the probability of this event which we define as

$$P[X = x] = \text{probability that } X(\omega) \text{ is equal to } x$$

which is merely the sum of the probabilities associated with each point ω for which $X(\omega) = x$. For our example we have

$$\begin{aligned} P[X = -5] &= 3/8 \\ P[X = 0] &= 1/4 \\ P[X = 5] &= 3/8 \end{aligned} \tag{II.7}$$

Another convenient form for expressing the probabilities associated with the random variable is the *probability distribution function* (PDF) also known

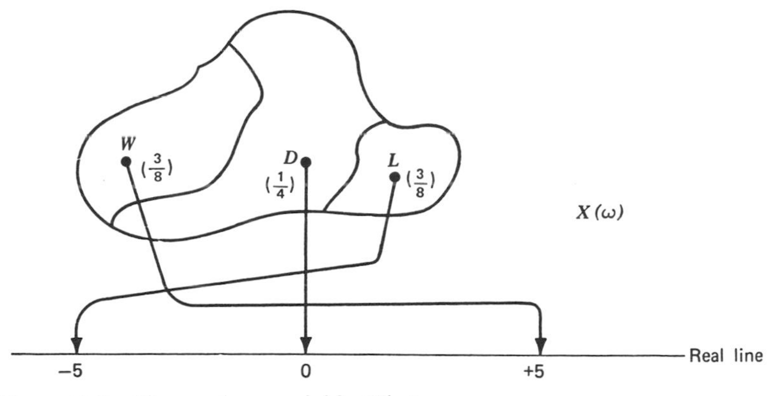

Figure II.2 The random variable $X(\omega)$.

as the cumulative distribution function. For this purpose we define notation similar to that given in Eq. (II.6), namely,

$$[X \leq x] = \{\omega \colon X(\omega) \leq x\}$$

We then have that the PDF is defined as

$$F_X(x) \triangleq P[X \leq x]$$

which expresses the probability that the random variable X takes on a value less than or equal to x. The important properties of this function are

$$F_X(x) \geq 0 \tag{II.8}$$

$$F_X(\infty) = 1$$

$$F_X(-\infty) = 0$$

$$F_X(b) - F_X(a) = P[a < X \leq b] \qquad \text{for} \qquad a < b \tag{II.9}$$

$$F_X(b) \geq F_X(a) \qquad \text{for} \qquad a \leq b$$

Thus $F_X(x)$ is a nonnegative monotonically nondecreasing function with limits 0 and 1 at $-\infty$ and $+\infty$, respectively. In addition $F_X(x)$ is assumed to be continuous from the right. For our blackjack example we then have the function given in Figure II.3. We note that at points of discontinuity the PDF takes on the upper value (as indicated by the dot) since the function is piecewise continuous from the right. From Property (II.9) we may easily calculate the probability that our random variable lies in a given interval. Thus for our blackjack example, we may write $P[-2 < x \leq 6] = 5/8$, $P[1 < x \leq 4] = 0$, and so on.

For purposes of calculation it is much more convenient to work with a function closely related to the PDF rather than with the PDF itself. Thus we

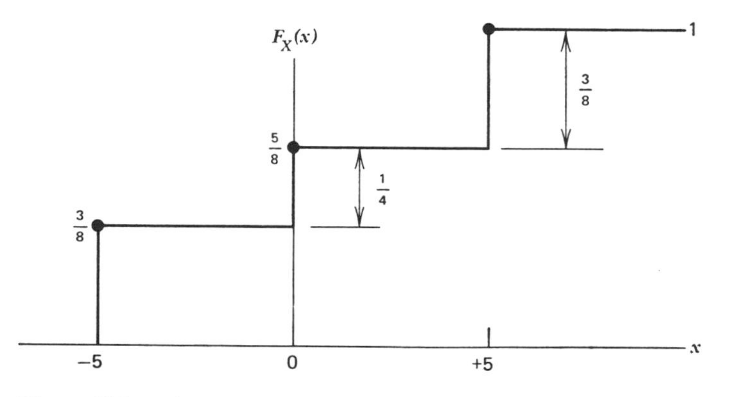

Figure II.3 The PDF for the blackjack example.

are led to the definition of the *probability density function* (pdf) defined as follows:

$$f_X(x) \triangleq \frac{dF_X(x)}{dx} \tag{II.10}$$

Of course, we are immediately faced with the question of whether or not such a derivative exists and if so over what interval. We temporarily avoid that question and assume that $F_X(x)$ possesses a continuous derivative everywhere (which is false for our blackjack example). As we shall see later, it is possible to define the pdf even when the PDF contains jumps. We may "invert" Eq. (II.10) to yield

$$F_X(x) = \int_{-\infty}^{x} f_X(y)\,dy \tag{II.11}$$

From this and Eq. (II.8) we have

$$f_X(x) \geq 0$$

Since $F_X(\infty) = 1$, we have from Eq. (II.11)

$$\int_{-\infty}^{\infty} f_X(x)\,dx = 1$$

Thus the pdf is a function which when integrated over an interval gives the probability that the random variable X lies in that interval, namely, for $a < b$ we have

$$P[a < X \leq b] = \int_{a}^{b} f_X(x)\,dx$$

For $a \to b$, and the axiom stated in Eq. (II.1) we see that this last equation also implies

$$f_X(x) \geq 0$$

As an example, let us consider an exponentially distributed random variable defined as one for which

$$F_X(x) = \begin{cases} 1 - e^{-\lambda x} & 0 \leq x \\ 0 & x < 0 \end{cases} \tag{II.12}$$

where $\lambda > 0$.
The corresponding pdf is given by

$$f_X(x) = \begin{cases} \lambda e^{-\lambda x} & 0 \leq x \\ 0 & x < 0 \end{cases} \tag{II.13}$$

For this example, the probability that the random variable lies between the values $a(>0)$ and $b(>a)$ may be calculated in either of the two following ways:

$$P[a < x \leq b] = F_X(b) - F_X(a)$$
$$= e^{-\lambda a} - e^{-\lambda b}$$
$$P[a < x \leq b] = \int_a^b f_X(x)\, dx$$
$$= e^{-\lambda a} - e^{-\lambda b}$$

From our blackjack example we notice that the PDF has a derivative which is everywhere 0 except at the three critical points ($x = -5, 0, +5$). In order to complete the definition for the pdf when the PDF is discontinuous we recognize that we must introduce a function such that when it is integrated over the region of the discontinuity it yields a value equal to the size of the discontinuous jump; that is, in the blackjack example the probability density function must be such that when integrated from $-5 - \epsilon$ to $-5 + \epsilon$ (for small $\epsilon > 0$) it should yield a probability equal to 3/8. Such a function has already been studied in Appendix I and is, of course, the impulse function (or Dirac delta function). Recall that such a function $u_0(x)$ is given by

$$u_0(x) = \begin{cases} \infty & x = 0 \\ 0 & x \neq 0 \end{cases}$$

$$\int_{-\infty}^{\infty} u_0(x)\, dx = 1$$

and also that it is merely the derivative of the unit step function as can be seen from

$$\int_{-\infty}^{x} u_0(y)\, dy = \begin{cases} 0 & x < 0 \\ 1 & x \geq 0 \end{cases}$$

Using the graphical notation in Figure I.3, we may properly describe the pdf for our blackjack example as in Figure II.4. We note immediately that this representation gives exactly the information we had in Eq. (II.7), and therefore the use of impulse functions is overly cumbersome for such problems. In particular if we define a discrete random variable as one that takes on values over a discrete set (finite or countable) then the use of the pdf* is a bit heavy and unnecessary although it does fit into our general definition in the obvious way. On the other hand, in the case of a random variable that takes on values over a continuum it is perfectly natural to use the pdf and in the

* In the discrete case, the function $P[X = x_k]$ is often referred to as the *probability mass function*. The generalization to the pdf leads one to the notion of a *mass density* function.

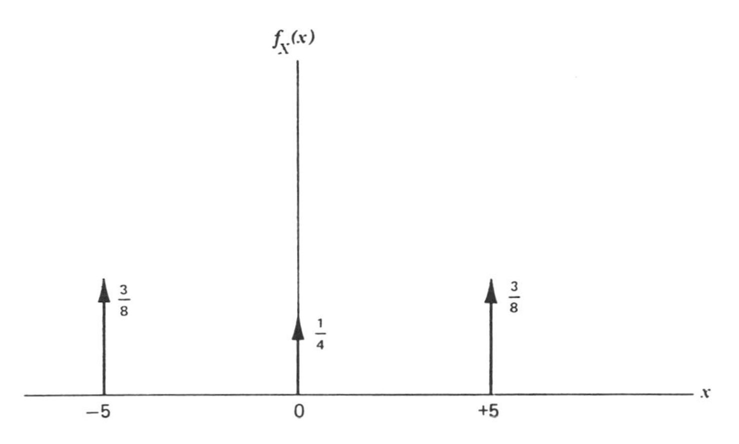

Figure II.4 The pdf for the blackjack example.

case where there is also a nonzero probability that the random variable takes on a specific value (i.e., that the PDF contains jumps) then the use of the pdf is necessary as well as is the introduction of the impulse function to account for these points of accumulation. We are thus led to distinguish between a *discrete* random variable, a purely *continuous* random variable (one whose PDF is continuous and everywhere differentiable), and the third case of a *mixed* random variable which contains some discrete as well as continuous portions.* So, for example, let us consider a random variable that represents the lifetime of an automobile. We will assume that there is a finite probability, say of value p, that the automobile will be inoperable immediately upon delivery, and therefore will have a lifetime of length zero. On the other hand, if the automobile is operable upon delivery then we will assume that the remainder of its lifetime is exponentially distributed as given in Eqs. (II.12) and (II.13). Thus for this automobile lifetime we have a PDF and a pdf as given in Figure II.5. Thus we clearly see the need for impulse functions in describing interesting random variables.

We have now discussed the notion of a probability system $(S, \mathscr{E}, P)$ and the notion of a random variable $X(\omega)$ defined upon the sample space S. There is, of course, no reason why we cannot define *many* random variables on the same sample space. Let us consider the case of two random variables X and Y defined for some probability system $(S, \mathscr{E}, P)$. In this case we have

* It can be shown that any PDF may be decomposed into a sum of three parts, namely, a pure jump function (containing only discontinuous jumps), a purely continuous portion, and a singular portion (which rarely occurs in distribution functions of interest and which will be considered no further in this text).

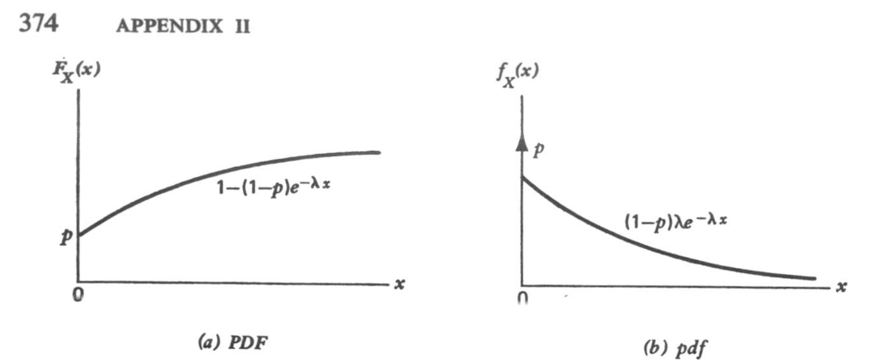

Figure II.5 PDF and pdf for automobile lifetime.

the natural extension of the PDF for two random variables, namely,

$$F_{XY}(x, y) \triangleq P[X \leq x, Y \leq y]$$

which is merely the probability that X takes on a value less than or equal to x at the same time Y takes on a value less than or equal to y; that is, it is the sum of the probabilities associated with all sample points in the intersection of the two events $\{\omega: X(\omega) \leq x\}$, $\{\omega: Y(\omega) \leq y\}$. $F_{XY}(x, y)$ is referred to as the *joint* PDF. Of course, associated with this function is a joint probability density function defined as

$$f_{XY}(x, y) \triangleq \frac{d^2 F_{XY}(x, y)}{dx\, dy}$$

Given a joint pdf, one naturally inquires as to the "marginal" density function for one of the variables and this is clearly given by integrating over all possible values of the second variable, thus

$$f_X(x) = \int_{y=-\infty}^{\infty} f_{XY}(x, y)\, dy \tag{II.14}$$

We are now in a position to define the notion of *independence* between random variables. Two random variables X and Y are said to be independent if and only if

$$f_{XY}(x, y) = f_X(x) f_Y(y)$$

that is, if their joint pdf factors into the product of the one-dimensional pdf's. This is very much like the definition for two independent events as given in Eq. (II.5). However, for three or more random variables, the definition is essentially the same as for two, namely, $X_1, X_2, \ldots, X_n$ are said to be independent random variables if and only if

$$f_{X_1 X_2 \cdots X_n}(x_1, x_2, \ldots, x_n) = f_{X_1}(x_1) f_{X_2}(x_2) \cdots f_{X_n}(x_n)$$

This last is a much simpler test than that required for multiple *events* to be independent.

With more than one random variable, we can now define conditional distributions and densities as follows. For example, we could ask for the PDF of the random variable X conditioned on some given value of the random variable Y, which would be expressed as $P[X \le x \mid Y = y]$. Similarly, the conditional pdf on X, given Y, is defined as

$$f_{X \mid Y}(x \mid y) \overset{\Delta}{=} \frac{d}{dx} P[X \le x \mid Y = y] = \frac{f_{XY}(x, y)}{f_Y(y)}$$

much as the definition for conditional probability of events.

To review again, we see that a random variable is defined as a mapping from the sample space for some probability system into the real line and from this mapping the PDF may easily be determined. Usually, however, a random variable is not given in terms of its sample space and the mapping, but rather directly in terms of its PDF or pdf.

It is possible to define one random variable Y in terms of a second random variable X, in which case Y would be referred to as a *function* of the random variable X. In its most general form we then have

$$Y = g(X) \tag{II.15}$$

where $g(\cdot)$ is some given function of its argument. Thus, once the value for X is determined, then the value for Y may be computed; however, the value for X depends upon the sample point ω, and therefore so does the value of Y which we may therefore write as $Y = Y(\omega) = g(X(\omega))$. Given the random variable X and its PDF, one should be able to calculate the PDF for the random variable Y, once the function $g(\cdot)$ is known. In principle, the computation takes the following form:

$$F_Y(y) = P[Y \le y] = P[\{\omega : g(X(\omega)) \le y\}]$$

In general, this computation is rather complex.

One random variable may be a function of *many* other random variables rather than just one. A particularly important form which often arises is in fact the *sum* of a collection of independent random variables $\{X_i\}$, namely,

$$Y = \sum_{i-1}^{n} X_i \tag{II.16}$$

Let us derive the distribution function of the sum of *two* independent random variables ($n = 2$). It is clear that this distribution is given by

$$F_Y(y) = P[Y \le y] = P[X_1 + X_2 \le y]$$

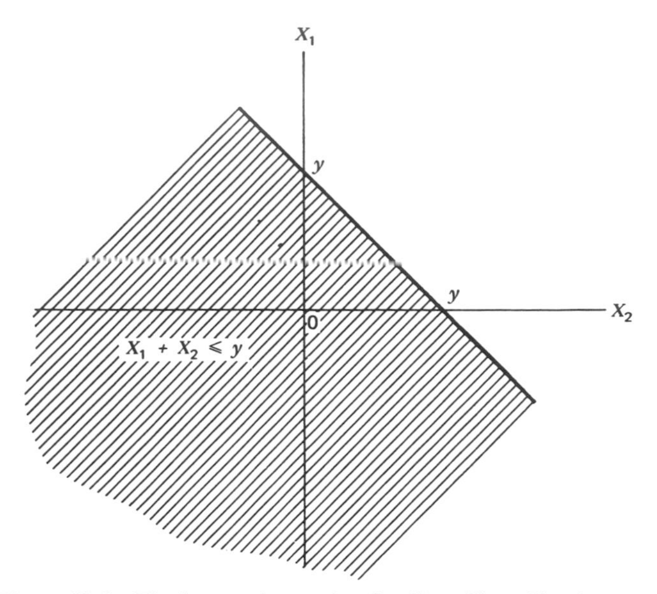

Figure II.6 The integration region for $Y = X_1 + X_2 \leq y$.

We have the situation shown in Figure II.6. Integrating over the indicated region we have

$$F_Y(y) = \int_{-\infty}^{\infty} \int_{-\infty}^{y-x_2} f_{X_1X_2}(x_1, x_2)\, dx_1\, dx_2$$

Due to the independence of X_1 and X_2 we then obtain the PDF for Y as

$$\begin{aligned} F_Y(y) &= \int_{-\infty}^{\infty} \left[\int_{-\infty}^{y-x_2} f_{X_1}(x_1)\, dx_1 \right] f_{X_2}(x_2)\, dx_2 \\ &= \int_{-\infty}^{\infty} F_{X_1}(y - x_2) f_{X_2}(x_2)\, dx_2 \end{aligned}$$

Finally, forming the pdf from this PDF, we have

$$f_Y(y) = \int_{-\infty}^{\infty} f_{X_1}(y - x_2) f_{X_2}(x_2)\, dx_2$$

This last equation is merely the *convolution* of the density functions for X_1 and X_2 and, as in Eq. (I.36), we denote this convolution operator (which is both associative and commutative) by an asterisk enclosed within a circle. Thus

$$f_Y(y) = f_{X_1}(y) \circledast f_{X_2}(y)$$

In a similar fashion, one easily shows for the case of arbitrary n that the pdf for Y as defined in Eq. (II.16) is given by the convolution of the pdf's for the X_i's, that is,

$$f_Y(y) = f_{X_1}(y) \circledast f_{X_2}(y) \circledast \cdots \circledast f_{X_n}(y) \tag{II.17}$$

II.3. EXPECTATION

In this section we discuss certain measures associated with the PDF and the pdf for a random variable. These measures will in general be called *expectations* and they deal with integrals of the pdf. As we saw in the last section, the pdf involves certain difficulties in its definition, and these difficulties were handily resolved by the use of impulse functions. However, in much of the literature on probability theory and in most of the literature on queueing theory the use of impulses is either not accepted, not understood or not known; as a result, special care and notation has been built up to get around the problem of differentiating discontinuous functions. The result is that many of the integrals encountered are Stieltjes integrals rather than the usual Riemann integrals with which we are most familiar. Let us take a moment to define the Stieltjes integral. A *Stieltjes* integral is defined in terms of a non-decreasing function $F(x)$ and a continuous function $\varphi(x)$; in addition, two sets of points $\{t_k\}$ and $\{\xi_k\}$ such that $t_{k-1} < \xi_k \le t_k$ are defined and a limit is considered where $\max |t_k - t_{k-1}| \to 0$. From these definitions, consider the sum

$$\sum_k \varphi(\xi_k)[F(t_k) - F(t_{k-1})]$$

This sum tends to a limit as the intervals shrink to zero independent of the sets $\{t_k\}$ and $\{\xi_k\}$ and the limit is referred to as the Stieltjes integral of φ with respect to F. This Stieltjes integral is written as

$$\int \varphi(x)\, dF(x)$$

Of course, we recognize that the PDF may be identified with the function F in this definition and that $dF(x)$ may be identified with the pdf [say, $f(x)$] through

$$dF(x) = f(x)\, dx$$

by definition. Without the use of impulses the pdf may not exist; however, the Stieltjes integral will always exist and therefore it avoids the issue of impulses. However, in this text we will feel free to incorporate impulse functions and therefore will work with both the Riemann and Stieltjes integrals; when impulses are permitted in the function $f(x)$ we then have the following identity:

$$\int \varphi(x)\,dF(x) = \int \varphi(x)f(x)\,dx$$

We will use both notations throughout the text in order to familiarize the student with the more common Stieltjes integral for queueing theory, as well as with the more easily manipulated Riemann integral with impulse functions. Having said all this we may now introduce the definition of expectation.

The *expectation* of a real random variable $X(\omega)$ denoted by $E[X]$ and also by $\bar{X}$ is given by the following:

$$E[X] \triangleq \bar{X} \triangleq \int_{-\infty}^{\infty} x\,dF_X(x) \tag{II.18}$$

This last is given in the form of a Stieltjes integral; in the form of a Riemann integral we have, of course,

$$E[X] = \bar{X} = \int_{-\infty}^{\infty} xf_X(x)\,dx$$

The expectation of X is also referred to as the *mean* or *average value* of X. We may also write

$$E[X] = \int_0^{\infty} [1 - F_X(x)]\,dx - \int_{-\infty}^{0} F_X(x)\,dx$$

which, upon integrating by parts, is easily shown to be equal to Eq. (II.18) so long as $E[X] < \infty$. Similarly, for X a nonnegative random variable, this form becomes

$$E[X] = \int_0^{\infty} [1 - F_X(x)]\,dx \qquad X \geq 0$$

In general, the expectation of a random variable is equal to the product of the value the random variable may take on and the probability it takes on this value, summed (integrated) over all possible values.

Now let us consider once again a new random variable Y, which is a function of our first random variable X, namely, as in Eq. (II.15)

$$Y = g(X)$$

We may define the expectation $E_Y[Y]$ for Y in terms of its PDF just as we did for X; the subscript Y on the expectation is there to distinguish expectation with respect to Y as opposed to any other random variables (in this case X). Thus we have

$$E_Y[Y] = \int_{-\infty}^{\infty} yf_Y(y)\,dy$$

This last computation requires that we find either $F_Y(y)$ or $f_Y(y)$, which as mentioned in the previous section, may be a rather complex computation. However, the *fundamental theorem of expectation* gives a much more straightforward calculation for this expectation in terms of distribution of the underlying random variable X, namely,

$$
\begin{aligned}
E_Y[y] &= E_X[g(X)] \\
&= \int_{-\infty}^{\infty} g(x) f_X(x)\, dx
\end{aligned}
$$

We may define the expectation of the *sum* of two random variables given by the following obvious generalization of the one-dimensional case:

$$
\begin{aligned}
E[X+Y] &= \int_{-\infty}^{\infty}\int_{-\infty}^{\infty} (x+y) f_{XY}(x, y)\, dx\, dy \\
&= \int_{-\infty}^{\infty}\int_{-\infty}^{\infty} x f_{XY}(x, y)\, dx\, dy + \int_{-\infty}^{\infty}\int_{-\infty}^{\infty} y f_{XY}(x, y)\, dx\, dy \\
&= \int_{-\infty}^{\infty} x f_X(x)\, dx + \int_{-\infty}^{\infty} y f_Y(y)\, dy \\
&= E[X] + E[Y] \qquad \text{(II.19)}
\end{aligned}
$$

This may also be written as $(\overline{X+Y} = \bar{X} + \bar{Y})$. In going from the second line to the third line we have taken advantage of Eq. (II.14) of the previous section in which the marginal density was defined from the joint density. We have shown the very important result, that *the expectation of the sum of two random variables is always equal to the sum of the expectations of each—this is true whether or not these random variables are independent*. This very nice property comes from the fact that the expectation operator is a linear operator. The more general statement of this property for any number of random variables, independent or not, is that *the expectation of the sum is always equal to the sum of the expectations*, that is,

$$
E[X_1 + X_2 + \cdots + X_n] = E[X_1] + E[X_2] + \cdots + E[X_n]
$$

A similar question may be asked about the *product* of two random variables, that is,

$$
E[XY] = \int_{-\infty}^{\infty}\int_{-\infty}^{\infty} xy f_{XY}(x, y)\, dx\, dy
$$

In the special case where the two random variables X and Y are *independent*, we may write the pdf for this joint random variable as the product of the pdf's for the individual random variables, thus obtaining

$$
E[XY] = \int_{-\infty}^{\infty}\int_{-\infty}^{\infty} xy f_X(x) f_Y(y)\, dx\, dy = E[X]E[Y] \qquad \text{(II.20)}
$$

This last equation (which may also be written as $\overline{XY} = \overline{X}\,\overline{Y}$) states that the expectation of the product is equal to the product of the expectations if the random variables are independent. A result similar to that expressed in Eq. (II.20) applies also to *functions* of independent random variables. That is, if we have two independent random variables X and Y and functions of each denoted by $g(X)$ and $h(Y)$, then by arguments exactly the same as those leading to Eq. (II.20) we may show

$$E[g(X)h(Y)] = E[g(X)]E[h(Y)] \tag{II.21}$$

Often we are interested in the expectation of the *power* of a random variable. In fact, this is so common that a special name has been coined so that the expected value of the nth power of a random variable is referred to as its *nth moment*. Thus, by definition (really this follows from the fundamental theorem of expectation), the nth moment of X is given by

$$E[X^n] \triangleq \overline{X^n} \triangleq \int_{-\infty}^{\infty} x^n f_X(x)\,dx$$

Furthermore, the *nth central moment* of this random variable is given as follows:

$$\overline{(X - \overline{X})^n} \triangleq \int_{-\infty}^{\infty} (x - \overline{X})^n f_X(x)\,dx$$

The nth central moment may be expressed in terms of the first n moments themselves; to show this we first write down the following identity making use of the binomial theorem

$$(X - \overline{X})^n = \sum_{k=0}^{n} \binom{n}{k} X^k (-\overline{X})^{n-k}$$

Taking expectations on both sides we then have

$$\begin{aligned} \overline{(X - \overline{X})^n} &= \overline{\sum_{k=0}^{n} \binom{n}{k} X^k (-\overline{X})^{n-k}} \\ &= \sum_{k=0}^{n} \binom{n}{k} \overline{X^k} (-\overline{X})^{n-k} \end{aligned} \tag{II.22}$$

In going from the first to the second line in this last equation we have taken advantage of the fact that the expectation of a sum is equal to the sum of the expectations and that the expectation of a constant is merely the constant itself. Now for a few observations. First we note that the 0th moment of a random variable is just unity. Also, the 0th central moment must be one. The first central moment must be 0 since

$$\overline{(X - \overline{X})} = \overline{X} - \overline{X} = 0$$

The second central moment is extremely important and is referred to as the *variance*; a special notation has been adopted for the variance and is given by

$$\sigma_X{}^2 \triangleq \overline{(X - \overline{X})^2}$$
$$\triangleq \overline{X^2} - (\overline{X})^2$$

In the second line of this last equation we have taken advantage of Eq. (II.22) and have expressed the variance (a central moment) in terms of the first two moments themselves. The square root of the variance σ_X is referred to as the *standard deviation.* The ratio of the standard deviation to the mean of a random variable is a most important quantity in statistics and also in queueing theory; this ratio is referred to as the *coefficient of variation* and is denoted by

$$C_X \triangleq \frac{\sigma_X}{\overline{X}} \qquad \text{(II.23)}$$

II.4. TRANSFORMS, GENERATING FUNCTIONS, AND CHARACTERISTIC FUNCTIONS

In probability theory one encounters a variety of functions (in particular, expectations) all of which are close relatives of each other. Included in this class is the *characteristic function* of a random variable, its *moment generating function*, the *Laplace transform of its probability density function*, and its *probability generating function.* In this section we wish to define and distinguish these various forms and to indicate a common central property that they share.

The *characteristic function* of a random variable X, denoted by $\phi_X(u)$, is given by

$$\phi_X(u) \triangleq E[e^{juX}]$$
$$= \int_{-\infty}^{\infty} e^{jux} f_X(x)\, dx$$

where $j = \sqrt{-1}$ and where u is an arbitrary real variable. (Note that except for the sign of the exponent, the characteristic function is the Fourier transform of the pdf for X). Clearly,

$$|\phi_X(u)| \le \int_{-\infty}^{\infty} |e^{jux}|\, |f_X(x)|\, dx$$

and since $|e^{jux}| = 1$, we have

$$|\phi_X(u)| \le \int_{-\infty}^{\infty} f_X(x)\, dx$$

which shows that

$$|\phi_X(u)| \le 1$$

An important property of the characteristic function may be seen by expanding the exponential in the integrand in terms of its power series and then integrating each term separately as follows:

$$\phi_X(u) = \int_{-\infty}^{\infty} f_X(x)\left[1 + jux + \frac{(jux)^2}{2!} + \cdots\right] dx$$
$$= 1 + ju\overline{X} + \frac{(ju)^2}{2!}\overline{X^2} + \cdots$$

From this expansion, we see that the characteristic function is expressed in terms of all the moments of X. Now, if we set $u = 0$ we find that $\phi_X(0) = 1$. Similarly, if we first form $d\phi_X(u)/du$ and then set $u = 0$, we obtain $j\overline{X}$. Thus, in general, we have

$$\left.\frac{d^n \phi_X(u)}{du^n}\right|_{u=0} = j^n \overline{X^n} \qquad \text{(II.24)}$$

This last important result gives a rather simple way for calculating a constant times the nth moment of the random variable X.

Since this property is frequently used, we find it convenient to adopt the following simplified notation (consistent with that in Eq. I.37) for the nth derivative of an arbitrary function $g(x)$, evaluated at some fixed value $x = x_0$:

$$g^{(n)}(x_0) \triangleq \left.\frac{d^n g(x)}{dx^n}\right|_{x=x_0} \qquad \text{(II.25)}$$

Thus the result in Eq. (II.24) may be rewritten as

$$\phi_X^{(n)}(0) = j^n \overline{X^n}.$$

The *moment generating function* denoted by $M_X(v)$ is given below along with the appropriate differential relationship that yields the nth moment of X directly.

$$M_X(v) \triangleq E[e^{vX}]$$
$$= \int_{-\infty}^{\infty} e^{vx} f_X(x)\, dx$$
$$M_X^{(n)}(0) = \overline{X^n}$$

where v is a real variable. From this last property it is easy to see where the name "moment generating function" comes from. The derivation of this moment relationship is the same as that for the characteristic function.

Another important and useful function is the *Laplace transform of the pdf* of a random variable X. We find it expedient to use a notation now in which the PDF for a random variable is labeled in a way that identifies the random variable without the use of subscripts. Thus, for example, if we have a random variable X, which represents, say, the interarrival time between adjacent customers to a system, then we define $A(x)$ to be the PDF for X;

$$A(x) = P[X \leq x]$$

where the symbol A is keyed to the word "*A*rrival." Further, the pdf for this example would be denoted $a(x)$. Finally, then, we denote the Laplace transform of $a(x)$ by $A^*(s)$ and it is given by the following:

$$\begin{aligned} A^*(s) &\triangleq E[e^{-sX}] \\ &\triangleq \int_{-\infty}^{\infty} e^{-sx} a(x)\, dx \end{aligned}$$

where s is a complex variable. Here we are using the "two-sided" transform; however, as mentioned in Section I.3, since most of the random variables we deal with are nonnegative, we often write

$$A^*(s) = \int_{0^-}^{\infty} e^{-sx} a(x)\, dx$$

The reader should take special note that the lower limit 0 is defined as 0^-; that is, the limit comes in from the left so that we specifically mean to include any impulse functions at the origin. In the fashion identical to that for the moment generating function and for the characteristic function, we may find the moments of X through the following formula:

$$A^{*(n)}(0) = (-1)^n \overline{X^n} \tag{II.26}$$

For nonnegative random variables

$$|A^*(s)| \leq \int_{0}^{\infty} |e^{-sx}|\, |a(x)|\, dx$$

But the complex variable s consists of a real part Re $(s) = \sigma$ and an imaginary part Im $(s) = \omega$ such that $s = \sigma + j\omega$. Then we have

$$\begin{aligned} |e^{-sx}| &= |e^{-\sigma x} e^{-j\omega x}| \\ &\leq |e^{-\sigma x}|\, |e^{-j\omega x}| \\ &= |e^{-\sigma x}| \end{aligned}$$

Moreover, for Re $(s) \geq 0$, $|e^{-\sigma x}| \leq 1$ and so we have from these last two equations and from $\int_0^\infty a(x)\, dx = 1$,

$$|A^*(s)| \leq 1 \qquad \text{Re}\,(s) \geq 0$$

It is clear that the three functions $\phi_X(u)$, $M_X(v)$, $A^*(s)$ are all close relatives of each other. In particular, we have the following relationship:

$$\phi_X(sj) = M_X(-s) = A^*(s)$$

Thus we are not surprised that the moment generating properties (by differentiation) are so similar for each; this property is the central property that we will take advantage of in our studies. Thus the nth moment of X is calculable from any of the following expressions:

$$\overline{X^n} = j^{-n}\phi_X^{(n)}(0)$$
$$\overline{X^n} = M_X^{(n)}(0)$$
$$\overline{X^n} = (-1)^n A^{*(n)}(0)$$

It is perhaps worthwhile to carry out an example demonstrating these properties. Consider the continuous random variable X, which represents, say, the interarrival time of customers to a system and which is exponentially distributed, that is,

$$f_X(x) = a(x) = \begin{cases} \lambda e^{-\lambda x} & x \geq 0 \\ 0 & x < 0 \end{cases}$$

By direct substitution into the defining integrals we find immediately that

$$\phi_X(u) = \frac{\lambda}{\lambda - ju}$$
$$M_X(v) = \frac{\lambda}{\lambda - v}$$
$$A^*(s) = \frac{\lambda}{\lambda + s}$$

It is always true that

$$\phi_X(0) = M_X(0) = A^*(0) = 1$$

and, of course, this checks out for our example as well. Using our expression for the first moment we find through any one of our three functions that

$$\bar{X} = \frac{1}{\lambda}$$

and we may also verify that the second moment may be calculated from any of the three to yield

$$\overline{X^2} = \frac{2}{\lambda^2}$$

and so it goes in calculating all of the moments.

In the case of a *discrete* random variable described, for example, by

$$g_k = P[X = k]$$

we make use of the *probability generating function* denoted by $G(z)$ as follows:

$$G(z) \triangleq E[z^X] = \sum_k z^k g_k \tag{II.27}$$

where z is a complex variable. It should be clear from our discussion in Appendix I that $G(z)$ is nothing more than the z-transform of the discrete sequence g_k. As with the continuous transforms, we have for $|z| \leq 1$

$$|G(z)| \leq \sum_k |z^k|\, |g_k| \leq \sum g_k$$

and so

$$|G(z)| \leq 1 \qquad \text{for } |z| \leq 1 \tag{II.28}$$

Note that the first derivative evaluated at $z = 1$ yields the first moment of X

$$G^{(1)}(1) = \overline{X} \tag{II.29}$$

and that the second derivative yields

$$G^{(2)}(1) = \overline{X^2} - \overline{X}$$

in a fashion similar to that for continuous random variables.* Note that in all cases

$$G(1) = 1$$

Let us apply these methods to the blackjack example considered earlier in this appendix. Working either with Eq. (II.7), which gives the probability of various winnings or with the impulsive pdf given in Figure II.4, we find that the probability generating function for the number of dollars won in a game of blackjack is given by

$$G(z) = \frac{3}{8} z^{-5} + \frac{1}{4} + \frac{3}{8} z^5$$

We note here that, of course, $G(1) = 1$ and further, that the mean winnings may be calculated as

$$\overline{X} = G^{(1)}(1) = 0$$

Let us now consider the sum of n independent variables X_i, namely, $Y = \sum_{i=1}^{n} X_i$, as defined in Eq. (II.16). If we form the characteristic function

* Thus we have that $\sigma_X{}^2 = G^{(2)}(1) + G^{(1)}(1) - [G^{(1)}(1)]^2$.

for Y, we have by definition

$$\begin{aligned}\phi_Y(u) &\triangleq E[e^{juY}] \\ &= E\left[e^{ju\sum_{i=1}^{n} X_i}\right] \\ &= E[e^{juX_1}e^{juX_2}\cdots e^{juX_n}]\end{aligned}$$

Now in Eq. (II.21) we showed that the expectation of the product of functions of independent random variables is equal to the product of the expectations of each function separately; applying this to the above we have

$$\phi_Y(u) = E[e^{juX_1}]E[e^{juX_2}]\cdots E[e^{juX_n}]$$

Of course the right-hand side of this equation is just a product of characteristic functions, and so

$$\phi_Y(u) = \phi_{X_1}(u)\phi_{X_2}(u)\cdots\phi_{X_n}(u) \tag{II.30}$$

In the case where each of the X_i is *identically distributed*, then, of course, the characteristic functions will all be the same, and so we may as well drop the subscript on X_i and conclude

$$\phi_Y(u) = [\phi_X(u)]^n \tag{II.31}$$

We have thus shown that the characteristic function of a sum of n identically distributed independent random variables is the nth power of the characteristic function of the individual random variable itself. This important result also applies to our other transforms, namely, the moment generating function, the Laplace transform and the z-transform. It is this significant property that accounts, in no small way, for the widespread use of transforms in probability theory and in the theory of stochastic processes.

Let us say a few more words now about sums of independent random variables. We have seen in Eq. (II.17) that the pdf of a sum of independent variables is equal to the convolution of the pdf for each; also, we have seen in Eq. (II.30) that the transform of the sum is equal to the product of the transforms for each. From Eq. (II.19) it is clear (regardless of the independence) that the expectation of the sum equals the sum of the expectations, namely,

$$\overline{Y} = \overline{X}_1 + \overline{X}_2 + \cdots + \overline{X}_n \tag{II.32}$$

For $n = 2$ we see that the second moment of Y must be

$$\overline{Y^2} = \overline{(X_1 + X_2)^2} = \overline{X_1^2} + 2\overline{X_1X_2} + \overline{X_2^2}$$

And also in this case

$$(\overline{Y})^2 = (\overline{X}_1 + \overline{X}_2)^2 = (\overline{X}_1)^2 + 2\overline{X}_1\overline{X}_2 + (\overline{X}_2)^2$$

Forming the variance of Y and then using these last two equations we have

$$\begin{aligned}\sigma_Y^{\,2} &= \overline{Y^2} - (\overline{Y})^2 \\ &= \overline{X_1^{\,2}} - (\overline{X_1})^2 + \overline{X_2^{\,2}} - (\overline{X_2})^2 + 2(\overline{X_1X_2} - \overline{X_1}\,\overline{X_2}) \\ &= \sigma_{X_1}^{\,2} + \sigma_{X_2}^{\,2} + 2(\overline{X_1X_2} - \overline{X_1}\,\overline{X_2})\end{aligned}$$

Now if X_1 and X_2 are also independent, then $\overline{X_1X_2} = \overline{X_1}\,\overline{X_2}$, giving the final result

$$\sigma_Y^{\,2} = \sigma_{X_1}^{\,2} + \sigma_{X_2}^{\,2}$$

In a similar fashion it is easy to show that the variance of the sum of n *independent* random variables is equal to the sum of the variances of each, that is,

$$\sigma_Y^{\,2} = \sigma_{X_1}^{\,2} + \sigma_{X_2}^{\,2} + \cdots + \sigma_{X_n}^{\,2}$$

Continuing with sums of independent random variables let us now assume that the number of these variables that are to be summed together is itself a random variable, that is, we define

$$Y = \sum_{i=1}^{N} X_i$$

where $\{X_i\}$ is a set of identically distributed independent random variables, each with mean $\overline{X}$ and variance $\sigma_X^{\,2}$, and where N is *also* a random variable with mean and variance $\overline{N}$ and $\sigma_N^{\,2}$, respectively; we assume that N is also independent of the X_i. In this case, $F_Y(y)$ is said to be a *compound* distribution. Let us now find $Y^*(s)$, which is the Laplace transform of the pdf for Y. By definition of the transform and due to the independence of all the random variables we may write down

$$\begin{aligned}Y^*(s) &= E\left[e^{-s\sum_{i=1}^{N} X_i}\right] \\ &= \sum_{n=0}^{\infty} E\left[e^{-s\sum_{i=1}^{n} X_i}\right] P[N = n] \\ &= \sum_{n=0}^{\infty} E[e^{-sX_1}] \cdots E[e^{-sX_n}] P[N = n]\end{aligned}$$

But since $\{X_i\}$ is a set of identically distributed random variables, we have

$$Y^*(s) = \sum_{n=0}^{\infty} [X^*(s)]^n P[N = n] \qquad \text{(II.33)}$$

where we have denoted the Laplace transform of the pdf for each of the X_i by $X^*(s)$. The final expression given in Eq. (II.33) is immediately recognized as the z-transform for the random variable N, which we choose to denote by $N(z)$ as defined in Eq. (II.27); in Eq. (II.33), z has been replaced by $X^*(s)$. Thus we finally conclude

$$Y^*(s) = N(X^*(s)) \tag{II.34}$$

Thus a *random* sum of identically distributed independent random variables has a transform that is related to the transforms of the sum's random variables and of the number of terms in the sum, as given above. Let us now find an expression similar to that in Eq. (II.32); in that equation for the case of identically distributed X_i we had $\bar{Y} = n\bar{X}$, where n was a given constant. Now, however, the number of terms in the sum is a random quantity and we must find the new mean $\bar{Y}$. We proceed by taking advantage of the moment generating properties of our transforms [Eq. (II.26)]. Thus differentiating Eq. (II.34), setting $s = 0$, and then taking the negative of the result we find

$$\bar{Y} = \bar{N}\bar{X}$$

which is a perfectly reasonable result. Similarly, one can find the variance of this random sum by differentiating twice and then subtracting off the mean squared to obtain

$$\sigma_Y^2 = \bar{N}\sigma_X^2 + (\bar{X})^2\sigma_N^2$$

This last result perhaps is not so intuitive.

II.5. INEQUALITIES AND LIMIT THEOREMS

In this section we present some of the classical inequalities and limit theorems in probability theory.

Let us first consider bounding the probability that a random variable exceeds some value. If we know only the mean value of the random variable, then the following *Markov inequality* can be established for a nonnegative random variable X:

$$P[X \geq x] \leq \frac{\bar{X}}{x}$$

Since only the mean value of the random variable is utilized, this inequality is rather weak. The *Chebyshev inequality* makes use of the mean and variance and is somewhat tighter; it states that for any $x > 0$,

$$P[|X - \bar{X}| \geq x] \leq \frac{\sigma_X^2}{x^2}$$

Other simple inequalities involve moments of two random variables, as follows: First we have the *Cauchy–Schwarz inequality*, which makes a statement about the expectation of a product of random variables in terms of the second moments of each.

$$(\overline{XY})^2 \leq \overline{X^2}\,\overline{Y^2} \tag{II.35}$$

A generalization of this last is *Hölder's inequality*, which states for $\alpha > 1$, $\beta > 1$, $\alpha^{-1} + \beta^{-1} = 1$, and $X > 0$, $Y > 0$ that

$$\overline{XY} \leq (\overline{X^\alpha})^{1/\alpha}(\overline{Y^\beta})^{1/\beta}$$

whenever the indicated expectations exist. Note that the Cauchy–Schwartz inequality is the (important) special case in which $\alpha = \beta = 2$. The *triangle inequality* relates the expectation of the absolute value of a sum to the sum of the expectations of the absolute values, namely,

$$\overline{|X + Y|} \leq \overline{|X|} + \overline{|Y|}$$

A generalization of the triangle inequality, which is known as the C_r*-inequality*, is

$$\overline{|X + Y|^r} \leq C_r[\overline{|X|^r} + \overline{|Y|^r}]$$

where

$$C_r = \begin{cases} 1 & 0 < r \leq 1 \\ 2^{r-1} & 1 < r \end{cases}$$

Next we bound the expectation of a *convex* function g of an arbitrary random variable X (whose first moment $\overline{X}$ is assumed to exist). A convex function $g(x)$ is one that lies on or below all of its chords, that is, for any $x_1 \leq x_2$, and $0 \leq \alpha \leq 1$

$$g(\alpha x_1 + (1 - \alpha)x_2) \leq \alpha g(x_1) + (1 - \alpha)g(x_2)$$

For such convex functions g and random variables X, we have *Jensen's inequality* as follows:

$$\overline{g(X)} \geq g(\overline{X})$$

When we deal with sums of random variables, we find that some very nice limiting properties exist. Let us once again consider the sum of n independent identically distributed random variables X_i, but let us now divide that sum by the number of terms n, thusly

$$W_n = \frac{1}{n}\sum_{i=1}^{n} X_i$$

This arithmetic mean is often referred to as the *sample mean*. We assume that each of the X_i has a mean given by $\overline{X}$ and a variance $\sigma_X{}^2$. From our earlier discussion regarding means and variances of sums of independent random variables we have

$$\overline{W}_n = \overline{X}$$

$$\sigma_{W_n}^2 = \frac{\sigma_X^2}{n}$$

If we now apply the Chebyshev inequality to the random variable W_n and make use of these last two observations, we may express our bound in terms of the mean and variance of the random variable X itself thusly

$$P[|W_n - \overline{X}| \geq x] \leq \frac{\sigma_X^2}{nx^2} \tag{II.36}$$

This very important result says that the arithmetic mean of the sum of n independent and identically distributed random variables will approach its expected value as n increases. This is due to the decreasing value of σ_X^2/nx^2 as n grows (σ_X^2/x^2 remains constant). In fact, this leads us directly to the *weak law of large numbers*, namely, that for any $\epsilon > 0$ we have

$$\lim_{n\to\infty} P[|W_n - \overline{X}| \geq \epsilon] = 0$$

The *strong law of large numbers* states that

$$\lim_{n\to\infty} W_n = \overline{X} \qquad \text{with probability one}$$

Once again, let us consider the sum of n independent identically distributed random variables X_i each with mean $\overline{X}$ and variance σ_X^2. The *central limit theorem* concerns itself with the normalized random variable Z_n defined by

$$Z_n = \frac{\sum_{i=1}^{n} X_i - n\overline{X}}{\sigma_X\sqrt{n}} \tag{II.37}$$

and states that the PDF for Z_n tends to the standard *normal distribution* as n increases; that is, for any real number x we have

$$\lim_{n\to\infty} P[Z_n \leq x] = \Phi(x)$$

where

$$\Phi(x) \triangleq \int_{-\infty}^{x} \frac{1}{(2\pi)^{1/2}} e^{-y^2/2}\, dy$$

That is, the appropriately normalized sum of a large number of independent random variables tends to a Gaussian, or a normal distribution. There are many other forms of the central limit theorem that deal, for example, with dependent random variables.

A rather sophisticated means for bounding the tail of the sum of a large number of independent random variables is available in the form of the *Chernoff bound*. It involves an inequality similar to the Markov and Chebyshev inequalities, but makes use of the entire distribution of the random variable itself (in particular, the moment generating function). Thus let us consider the sum of n independent identically distributed random variables X_i as given by

$$Y = \sum_{i=1}^{n} X_i$$

From Eq. (II.31) we know that the moment generating function for Y, $M_Y(v)$, is related to the moment generating function for each of the random variables X_i [namely, $M_X(v)$] through the relationship

$$M_Y(v) = [M_X(v)]^n \tag{II.38}$$

As with our earlier inequalities, we are interested in the probability that our sum exceeds a certain value, and this may be calculated as

$$P[Y \geq y] = \int_y^{\infty} f_Y(w)\, dw \tag{II.39}$$

Clearly, for $v \geq 0$ we have that the unit step function [see Eq. (I.33)] is bounded above by the following exponential:

$$u_{-1}(w - y) \leq e^{v(w-y)}$$

Applying this inequality to Eq. (II.39) we have

$$P[Y \geq y] \leq e^{-vy} \int_{-\infty}^{\infty} e^{vw} f_Y(w)\, dw \qquad \text{for } v \geq 0$$

However, the integral on the right-hand side of this equation is merely the moment generating function for Y, and so we have

$$P[Y \geq y] \leq e^{-vy} M_Y(v) \qquad v \geq 0 \tag{II.40}$$

Let us now define the "*semi-invariant*" generating function

$$\gamma(v) \triangleq \log M(v)$$

(Here we are considering natural logarithms.) Applying this definition to Eq. (II.38) we immediately have

$$\gamma_Y(v) = n\gamma_X(v)$$

and applying these last two to Eq. (II.40) we arrive at

$$P[Y \geq y] \leq e^{-vy+n\gamma_X(v)} \qquad v \geq 0$$

Since this last is good for any value of v (≥ 0), we should choose v to create the tightest possible bound; this is simply carried out by differentiating the exponent and setting it equal to zero. We thus find the optimum relationship between v and y as

$$y = n\gamma_X^{(1)}(v) \tag{II.41}$$

Thus the Chernoff bound for the tail of a density function takes the final form*

$$P[Y \geq n\gamma_X^{(1)}(v)] \leq e^{n[\gamma_X(v) - v\gamma_X^{(1)}(v)]} \qquad v \geq 0 \tag{II.42}$$

It is perhaps worthwhile to carry out an example demonstrating the use of this last bounding procedure. For this purpose, let us go back to the second paragraph in this appendix, in which we estimated the odds that at least 490,000 heads would occur in a million tosses of a fair coin. Of course, that calculation is the same as calculating the probability that no more than 510,000 heads will occur in the same experiment assuming the coin is fair. In this example the random variable X may be chosen as follows

$$X = \begin{cases} 1 & \text{heads} \\ 0 & \text{tails} \end{cases}$$

Since Y is the sum of a million trials of this experiment, we have that $n = 10^6$, and we now ask for the complementary probability that Y add up to 510,000 or more, namely, $P[Y \geq 510{,}000]$. The moment-generating function for X is

$$M_X(v) = \tfrac{1}{2} + \tfrac{1}{2}e^v$$

and so

$$\gamma_X(v) = \log \frac{1}{2}(1 + e^v)$$

Similarly

$$\gamma_X^{(1)}(v) = \frac{e^v}{1 + e^v}$$

From our formula (II.41) we then must have

$$n\gamma_X^{(1)}(v) = 10^6 \frac{e^v}{1 + e^v} = 510{,}000 = y$$

Thus we have

$$e^v = \frac{51}{49}$$

and

$$v = \log \frac{51}{49}$$

* The same derivation leads to a bound on the "lower tail" in which all three inequalities from Eq. (II.42) face thusly: $\leq$. For example $v \leq 0$.

Thus we see typically how v might be calculated. Plugging these values back into Eq. (II.42) we conclude

$$P[Y \geq 510{,}000] \leq e^{10^6[\log(50/49) - 0.51 \log(51/49)]}$$

This computation shows that the probability of exceeding 510,000 heads in a million tosses of a fair coin is less than 10^{-88} (this is where the number in our opening paragraphs comes from). An alternative way of carrying out this computation would be to make use of the central limit theorem. Let us do so as an example. For this we require the calculation of the mean and variance of X which are easily seen to be $\bar{X} = 1/2$, $\sigma_X^2 = 1/4$. Thus from Eq. (II.37) we have

$$Z_n = \frac{Y - 10^6(1/2)}{(1/2)10^3}$$

If we require Y to be greater than 510,000, then we are requiring that Z_n be greater than 20. If we now go to a table of the cumulative normal distribution, we find that

$$P[Z \geq 20] = 1 - \Phi(20) \cong 25 \times 10^{-90}$$

Again we see the extreme implausibility of such an event occurring. On the other hand, the Chebyshev inequality, as given in Eq. (II.36), yields the following;

$$P\left[\left|W_n - \frac{1}{2}\right| \geq 0.01\right] \leq \frac{0.25}{10^6 \cdot 10^{-4}} = 25 \times 10^{-4}$$

This result is twice as large as it should be for our calculation since we have effectively calculated both tails (namely, the probability that more than 510,000 or less than 490,000 heads would occur); thus the appropriate answer for the Chebyshev inequality would be that the probability of exceeding 510,000 heads is less than or equal to 12.5×10^{-4}. Note what a poor result this inequality gives compared to the central limit theorem approximation, which in this case is comparable to the Chernoff bound.

II.6. STOCHASTIC PROCESSES

It is often said that queueing theory is part of the theory of applied stochastic processes. As such, the main portion of this text is really the proper sequel to this section on stochastic processes; here we merely state some of the fundamental definitions and concepts.

We begin by considering a probability system $(S, \mathscr{E}, P)$, which consists of a sample space S, a set of events $\{A, B, \ldots\}$, and a probability measure P. In addition, we have already introduced the notion of a random variable

$X(\omega)$. A *stochastic process* may be defined as follows: For each sample point $\omega \in S$ we assign a time function $X(t, \omega)$. This family of functions forms a stochastic process; alternatively, we may say that for each t included in some appropriate parameter set, we choose a random variable $X(t, \omega)$. This is a collection of random variables depending upon t. Thus a stochastic process (or random function) is a function* $X(t)$ whose values are random variables. An example of a random process is the sequence of closing prices for a given security on the New York Stock Exchange; another example is the temperature at a given point on the earth as a function of time.

We are immediately confronted with the problem of completely specifying a random process $X(t)$. For this purpose we define, for each allowed t, a PDF, which we denote by $F_X(x, t)$ and which is given by

$$F_X(x, t) = P[X(t) \leq x]$$

Further we define for each of n allowable t, $\{t_1, t_2, \ldots, t_n\}$ a joint PDF, given by

$$F_{X_1X_2\cdots X_n}(x_1, x_2, \ldots, x_n; t_1, t_2, \ldots, t_n) \triangleq P[X(t_1) \leq x_1, X(t_2) \leq x_2, \ldots, X(t_n) \leq x_n]$$

and we use the vector notation $F_{\mathbf{X}}(\mathbf{x}; \mathbf{t})$ to denote this function.

A stochastic process $X(t)$ is said to be *stationary* if all $F_{\mathbf{X}}(\mathbf{x}, \mathbf{t})$ are invariant to shifts in time; that is, for any given constant τ the following holds:

$$F_{\mathbf{X}}(\mathbf{x}; \mathbf{t} + \tau) = F_{\mathbf{X}}(\mathbf{x}; \mathbf{t})$$

where the notation $\mathbf{t} + \tau$ implies the vector $(t_1 + \tau, t_2 + \tau, \ldots, t_n + \tau)$. Of most interest in the theory of stochastic processes are these stationary random functions.

In order to completely specify a stochastic process, then, one must give $F_{\mathbf{X}}(\mathbf{x}; \mathbf{t})$ for all possible subsets of $\{x_i\}$, $\{t_i\}$, and all n. This is a monstrous task in general! Fortunately, for many of the interesting stochastic processes, it is possible to provide this specification in very simple terms.

Some other definitions are in order. The first is the definition of the pdf for a stochastic process, and this is defined by

$$f_{\mathbf{X}}(\mathbf{x}; \mathbf{t}) \triangleq \frac{\partial F_{\mathbf{X}}(\mathbf{x}; \mathbf{t})}{\partial \mathbf{X}}$$

Second, we often discuss the mean value of a stochastic process given by

$$\overline{X(t)} = E[X(t)] = \int_{-\infty}^{\infty} x f_X(x; t)\, dx$$

* Usually we denote $X(t, \omega)$ by $X(t)$ for simplicity.

Next, we introduce the *autocorrelation* of $X(t)$ given by

$$R_{XX}(t_1, t_2) = E[X(t_1)X(t_2)]$$
$$= \int_{-\infty}^{\infty}\int_{-\infty}^{\infty} x_1 x_2 f_{X_1X_2}(x_1, x_2; t_1, t_2)\, dx_1\, dx_2$$

A large theory of stochastic process has been developed, known as *second-order* theory, in which these processes are classified and distinguished only on the basis of their mean $\overline{X(t)}$ and autocorrelation $R_{XX}(t_1, t_2)$. In the case of stationary random processes, we have

$$\overline{X(t)} = \bar{X} \tag{II.43}$$

and

$$R_{XX}(t_1, t_2) = R_{XX}(t_2 - t_1) \tag{II.44}$$

that is, R_{XX} is a function only of the time difference $\tau = t_2 - t_1$. In the stationary case, then, random processes are characterized in the second-order theory *only* by a constant (their mean $\bar{X}$) and a one-dimensional function $R_{XX}(\tau)$. A random process is said to be *wide-sense stationary* if Eqs. (II.43) and (II.44) hold. Note that all stationary processes are wide-sense stationary, but *not* conversely.

REFERENCES

DAVE 70 Davenport, W. B. Jr., *Probability and Random Processes*, McGraw-Hill (New York), 1970.

FELL 68 Feller, W., *An Introduction to Probability Theory and Its Applications*, 3rd Edition, Vol. 1, Wiley (New York), 1968.

PAPO 65 Papoulis, A., *Probability, Random Variables, and Stochastic Processes*, McGraw-Hill (New York), 1965.

PARZ 60 Parzen, E., *Modern Probability Theory and Its Applications*, Wiley (New York), 1960.

Glossary of Notation*

(Only the notation used often in this book is included below.)

NOTATION†	DEFINITION	TYPICAL PAGE REFERENCE
$A_n(t) = A(t)$	$P[t_n \leq t] = P[\tilde{t} \leq t]$	13
$A_n^*(s) = A^*(s)$	Laplace transform of $a(t)$	14
a_k	kth moment of $a(t)$	14
$a_n(t) = a(t)$	$dA_n(t)/dt = dA(t)/dt$	14
$B_n(x) = B(x)$	$P[x_n \leq x] = P[\tilde{x} \leq x]$	14
$B_n^*(s) = B^*(s)$	Laplace transform of $b(x)$	14
b_k	kth moment of $b(x)$	14
$b_n(x) = b(x)$	$dB_n(x)/dx = dB(x)/dx$	14
C_b^2	Coefficient of variation for service time	187
C_n	nth customer to enter the system	11
$C_n(u) = C(u)$	$P[u_n \leq u]$	281
$C_n^*(s) = C^*(s)$	Laplace transform of $c_n(u) = c(u)$	285
$c_n(u) = c(u)$	$dC_n(u)/du = dC(u)/du$	281
D	Denotes deterministic distribution	viii
d_k	$P[\tilde{q} = k]$	176
$E[X] = \bar{X}$	Expectation of the random variable X	378
E_i	System state i	27
E_r	Denotes r-stage Erlangian distribution	124
FCFS	First-come-first-served	8
$F_X(x)$	$P[X \leq x]$	370

* In those few cases where a symbol has more than one meaning, the context (or a specific statement) resolves the ambiguity.

† The use of the notation $y_n \to y$ is meant to indicate that $y = \lim y_n$, as $n \to \infty$ whereas $y(t) \to y$ indicates that $y = \lim y(t)$ as $t \to \infty$.

$f_X(x)$	$dF_X(x)/dx$	371
G	Denotes general distribution	viii
$G(y)$	Busy-period distribution	208
$G^*(s)$	Laplace transform of $g(y)$	211
g_k	kth moment of busy-period duration	213
$g(y)$	$dG(y)/dy$	215
H_R	Denotes R-stage hyperexponential distribution	141
$Im(s)$	Imaginary part of the complex variable s	293
$I_n \to I$	Duration of the (nth) idle period	206
$I^*(s)$	Laplace transform of idle-period density	307
$\hat{I}^*(s)$	Laplace transform of idle-time density in the dual system	310
K	Size of finite storage	viii
LCFS	Last-come-first-served	8
M	Denotes exponential distribution	viii
M	Size of finite population	viii
m	Number of servers	viii
$N_q(t) \to N_q$	Number of customers in queue at time t	17
$N(t) \to N$	Number of customers in system at time t	11
$O(x)$	$\lim_{x\to 0} O(x)/x = K < \infty$	284
$o(x)$	$\lim_{x\to 0} o(x)/x = 0$	48
P	Matrix of transition probabilities	31
$P[A]$	Probability of the event A	364
$P[A \mid B]$	Probability of the event A conditioned on the event B	365
PDF	Probability distribution function	369
pdf	Probability density function	371
$P_k(t)$	$P[N(t) = k]$	55
p_{ij}	P[next state is E_j | current state is E_i]	27
$p_{ij}(s, t)$	$P[X(t) = j \mid X(s) = i]$	46
p_k	P[k customers in system]	90
$Q(z)$	z-transform of $P[\tilde{q} = k]$	192
$q_{ij}(t)$	Transition rates at time t	48
$q_n \to \tilde{q}$	Number left behind by departure (of C_n)	177
$q_n' \to \tilde{q}'$	Number found by arrival (of C_n)	242

$Rc(s)$	Real part of the complex variable s	340
r_{ij}	P[next node is j | current node is i]	149
r_k	$P[\tilde{q}' = k]$	176
$S_n(y) \rightarrow S(y)$	$P[s_n \leq y] \rightarrow P[\tilde{s} \leq y]$	14
$S_n^*(s) \rightarrow S^*(s)$	Laplace transform of $s_n(y) \rightarrow s(y)$	14
s	Laplace transform variable	339
$s_n \rightarrow \tilde{s}$	Time in system (for C_n)	14
$s_n(y) \rightarrow s(y)$	$dS_n(y)/dy \rightarrow dS(y)/dy$	14
$\bar{s}_n \rightarrow \bar{s} = T$	Average time in system (for C_n)	14
$\overline{s_n^k} \rightarrow \overline{s^k}$	kth moment of $s_n(y)$	14
T	Average time in system	14
$t_n \rightarrow \tilde{t}$	Interarrival time (between C_{n-1} and C_n)	14
$\bar{t}_n = \bar{t} = 1/\lambda$	Average interarrival time	14
$\overline{t^k}$	kth moment of $a(t)$	14
$U(t)$	Unfinished work in system at time t	206
$u_0(t)$	Unit impulse function	341
$u_n \rightarrow \tilde{u}$	$u_n = x_n - t_{n+1} \rightarrow \tilde{u} = \tilde{x} - \tilde{t}$	277
$V(z)$	z-transform of $P[\tilde{v} = k]$	184
$v_n \rightarrow \tilde{v}$	Number of arrivals during service time (of C_n)	177
W	Average time in queue	14
W_0	Average remaining service time	190
$W_-(y)$	Complementary waiting time	284
$W_n(y) \rightarrow W(y)$	$P[w_n \leq y] \rightarrow P[\tilde{w} \leq y]$	14
$W_n^*(s) \rightarrow W^*(s)$	Laplace transform of $w_n(y)$	14
$w_n \rightarrow \tilde{w}$	Waiting time (for C_n) in queue	14
$w_n(y) \rightarrow w(y)$	$dW_n(y)/dy \rightarrow dW(y)/dy$	14
$\bar{w}_n \rightarrow \bar{w} = W$	Average waiting time (for C_n)	14
$\overline{w_n^k} \rightarrow \overline{w^k}$	kth moment of $w_n(y)$	14
$X(t)$	State of stochastic process $X(t)$ at time t	19
$x_n \rightarrow \tilde{x}$	Service time (of C_n)	14
$\overline{x^k}$	kth moment of $b(x)$	14
$\bar{x}_n = \bar{x} = 1/\mu$	Average service time	14
Y	Busy-period duration	206
z	z-transform variable	327

$\alpha(t)$	Number of arrivals in $(0, t)$	15
γ_i	(External) input rate to node i	149
$\delta(t)$	Number of departures in $(0, t)$	16
λ	Average arrival rate	14
λ_k	Birth (arrival) rate when $N = k$	53
μ	Average service rate	14
μ_k	Death (service) rate when $N = k$	54
$\boldsymbol{\pi}^{(n)} \to \boldsymbol{\pi}$	Vector of state probabilities $\pi_k^{(n)}$	31
$\pi_k^{(n)} \to \pi_k$	P[system state (at nth step) is E_k]	29
$\prod_{i=1}^{k} a_i$	$a_1 \ a_2 \cdots a_k$ (Product notation)	334
ρ	Utilization factor	18
σ	Root for G/M/m	249
σ_a^2	Variance of interarrival time	305
σ_b^2	Variance of service time	305
τ_n	Arrival time of C_n	12
$\Phi_+(s)$	Laplace transform of $W(y)$	285
$\Phi_-(s)$	Laplace transform of $W_-(y)$	285
$\triangleq$	Equals by definition	11
$(0, t)$	The interval from 0 to t	15
$\bar{X} = E[X]$	Expectation of the random variable X	378
$(y)^+$	max $[0, y]$	277
$\binom{n}{k}$	Binomial coefficient $= \dfrac{n!}{k!\,(n-k)!}$	368
A/B/m/K/M	m-Server queue with $A(t)$ and $B(x)$ identified by A and B, respectively, with storage capacity of size K, and with a customer population of size M (if any of the last two descriptors are missing they are assumed to be infinite)	viii
$F^{(n)}(a)$	$d^nF(y)/dy^n \big\|_{y=a}$	382
$f_{(k)}(x)$	$f(x) \circledast \cdots \circledast f(x)$ k-fold convolution	200
$\circledast$	Convolution operator	376
$f \to g$	Input f gives output g	322
$A \leftrightarrow B$	Statement A implies statement B and conversely	68
$f \Leftrightarrow F$	f and F form a transform pair	328

Summary of Important Results

Following is a collection of the basic results (those marked by ▬) from this text in the form of a list of equations. To the right of each equation is the page number where it first appears in a meaningful way; this is to aid the reader in locating the descriptive text and theory relevant to that equation.

GENERAL SYSTEMS

Equation	Page
$\rho = \lambda\bar{x}$ (G/G/1)	18
$\rho \triangleq \lambda\bar{x}/m$ (G/G/m)	18
$T = \bar{x} + W$	18
$\bar{N} = \lambda T$ (Little's result)	17
$\bar{N}_q = \lambda W$	17
$\bar{N}_q = \bar{N} - \rho$	188
$dP_k(t)/dt =$ flow rate into E_k—flow rate out of E_k	59
$p_k = r_k$ (for Poisson arrivals)	176
$r_k = d_k$ [$N(t)$ makes unit changes]	176

MARKOV PROCESSES

For a summary of discrete state Markov chains, see the table on pp. 402–403.

POISSON PROCESSES

Equation	Page
$P_k(t) = \dfrac{(\lambda t)^k}{k!} e^{-\lambda t} \qquad k \geq 0, t \geq 0$	60
$\overline{N(t)} = \lambda t$	62
$\sigma^2_{N(t)} = \lambda t$	62
$E[z^{N(t)}] = e^{\lambda t(z-1)}$	63

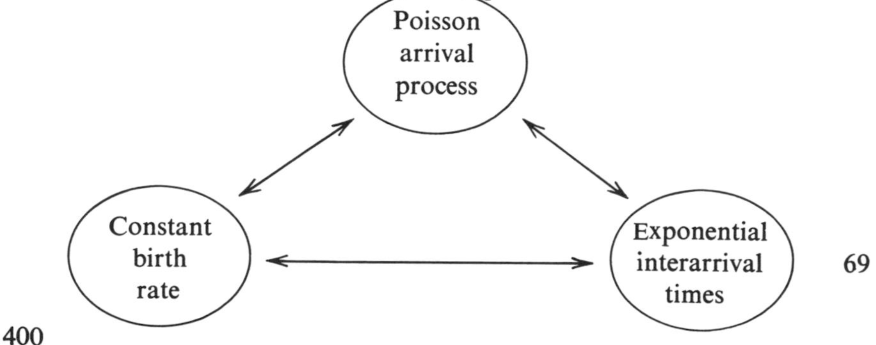

BIRTH-DEATH SYSTEMS

$$\frac{dP_k(t)}{dt} = -(\lambda_k + \mu_k)P_k(t) + \lambda_{k-1}P_{k-1}(t) + \mu_{k+1}P_{k+1}(t) \qquad k \geq 1 \qquad 57$$

$$\frac{dP_0(t)}{dt} = -\lambda_0 P_0(t) + \mu_1 P_1(t) \qquad k = 0 \qquad 57$$

$$p_k = p_0 \prod_{i=0}^{k-1} \frac{\lambda_i}{\mu_{i+1}} \quad \text{(equilibrium solution)} \qquad 92$$

$$p_0 = \frac{1}{1 + \sum_{k=1}^{\infty} \prod_{i=0}^{k-1} \frac{\lambda_i}{\mu_{i+1}}} \qquad 92$$

M/M/1

$$P_k(t) = e^{-(\lambda+\mu)t}\left[\rho^{(k-i)/2} I_{k-i}(at) + \rho^{(k-i-1)/2} I_{k+i+1}(at) + (1-\rho)\rho^k \sum_{j=k+i+2}^{\infty} \rho^{-j/2} I_j(at)\right] \qquad 77$$

$$p_k = (1-\rho)\rho^k \qquad 96$$

$$\bar{N} = \rho/(1-\rho) \qquad 96$$

$$\sigma_N^{\ 2} = \rho/(1-\rho)^2 \qquad 97$$

$$W = \frac{\rho/\mu}{1-\rho} \qquad 191$$

$$T = \frac{1/\mu}{1-\rho} \qquad 98$$

$$P[\geq k \text{ in system}] = \rho^k \qquad 99$$

$$s(y) = \mu(1-\rho)e^{-\mu(1-\rho)y} \qquad y \geq 0 \qquad 202$$

$$S(y) = 1 - e^{-\mu(1-\rho)y} \qquad y \geq 0 \qquad 202$$

$$w(y) = (1-\rho)u_0(y) + \lambda(1-\rho)e^{-\mu(1-\rho)y} \qquad y \geq 0 \qquad 203$$

$$W(y) = 1 - \rho e^{-\mu(1-\rho)y} \qquad y \geq 0 \qquad 203$$

Summary of Discrete-State Markov Chains

	DISCRETE-TIME		CONTINUOUS-TIME	
	HOMOGENEOUS	NONHOMOGENEOUS	HOMOGENEOUS	NONHOMOGENEOUS
One-step transition probability	$p_{ij} \triangleq P[X_{n+1} = j \mid X_n = i]$	$p_{ij}(n, n+1) \triangleq P[X_{n+1} = j \mid X_n = i]$	$p_{ij} \triangleq P[X(t + \Delta t) = j \mid X(t) = i]$	$p_{ij}(t, t + \Delta t) \triangleq P[X(t + \Delta t) = j \mid X(t) = i]$
Matrix of one-step transition probabilities	$\mathbf{P} \triangleq [p_{ij}]$	$\mathbf{P}(n) \triangleq [p_{ij}(n, n+1)]$	$\mathbf{P} \triangleq [p_{ij}]$	$\mathbf{P}(t) \triangleq [p_{ij}(t, t + \Delta t)]$
Multiple-step transition probabilities	$p_{ij}^{(m)} \triangleq p[X_{n+m} = j \mid X_n = i]$	$p_{ij}(m, n) \triangleq P[X_n = j \mid X_m = i]$	$p_{ij}(t) \triangleq P[X(s + t) = j \mid X(s) = i]$	$p_{ij}(s, t) \triangleq P[X(t) = j \mid X(s) = i]$
Matrix of multiple-step transition probabilities	$\mathbf{P}^{(m)} \triangleq [p_{ij}^{(m)}]$	$\mathbf{H}(m, n) \triangleq [p_{ij}(m, n)]$	$\mathbf{H}(t) \triangleq [p_{ij}(t)]$	$\mathbf{H}(s, t) \triangleq [p_{ij}(s, t)]$
Chapman–Kolmogorov equation	$p_{ij}^{(m)} = \sum_k p_{ik}^{(m-q)} p_{kj}^{(q)}$ $\mathbf{P}^{(m)} = \mathbf{P}^{(m-q)}\mathbf{P}^{(q)}$	$p_{ij}(m, n) = \sum_k p_{ik}(m, q) p_{kj}(q, n)$ $\mathbf{H}(m, n) = \mathbf{H}(m, q)\mathbf{H}(q, n)$	$p_{ij}(t) = \sum_k p_{ik}(t - s) p_{kj}(s)$ $\mathbf{H}(t) = \mathbf{H}(t - s)\mathbf{H}(s)$	$p_{ij}(s, t) = \sum_k p_{ik}(s, u) p_{kj}(u, t)$ $\mathbf{H}(s, t) = \mathbf{H}(s, u)\mathbf{H}(u, t)$

Table (continued)

Forward equation	$\mathbf{P}^{(m)} = \mathbf{P}^{(m-1)}\mathbf{P}$	$\mathbf{H}(m, n) = \mathbf{H}(m, n-1)\mathbf{P}(n-1)$	$d\mathbf{H}(t)/dt = \mathbf{H}(t)\mathbf{Q}$	$\partial\mathbf{H}(s, t)/\partial t = \mathbf{H}(s, t)\mathbf{Q}(t)$
Backward equation	$\mathbf{P}^{(m)} = \mathbf{P}\mathbf{P}^{(m-1)}$	$\mathbf{H}(m, n) = \mathbf{P}(m)\mathbf{H}(m+1, n)$	$d\mathbf{H}(t)/dt = \mathbf{Q}\mathbf{H}(t)$	$\partial\mathbf{H}(s, t)/\partial s = -\mathbf{Q}(s)\mathbf{H}(s, t)$
Solution	$\mathbf{P}^{(m)} = \mathbf{P}^m$	$\mathbf{H}(m, n) = \mathbf{P}(m)\mathbf{P}(m+1)\cdots\mathbf{P}(n-1)$	$\mathbf{H}(t) = e^{\mathbf{Q}t}$	$\mathbf{H}(s, t) = \exp\left[\int_s^t \mathbf{Q}(u)\,du\right]$
Transition-rate matrix	—	—	$\mathbf{Q} = \lim_{\Delta t \to 0} \frac{\mathbf{P} - \mathbf{I}}{\Delta t}$	$\mathbf{Q}(t) = \lim_{\Delta t \to 0} \frac{\mathbf{P}(t) - \mathbf{I}}{\Delta t}$
State probability	$\pi_j^{(n)} \triangleq P[X_n = j]$	$\pi_j^{(n)} \triangleq P[X_n = j]$	$\pi_j(t) \triangleq P[X(t) = j]$	$\pi_j(t) \triangleq P[X(t) = j]$
Matrix of state probabilities	$\boldsymbol{\pi}^{(n)} \triangleq [\pi_j^{(n)}]$	$\boldsymbol{\pi}^{(n)} \triangleq [\pi_j^{(n)}]$	$\boldsymbol{\pi}(t) \triangleq [\pi_j(t)]$	$\boldsymbol{\pi}(t) \triangleq [\pi_j(t)]$
Forward equation	$\boldsymbol{\pi}^{(n)} = \boldsymbol{\pi}^{(n-1)}\mathbf{P}$	$\boldsymbol{\pi}^{(n)} = \boldsymbol{\pi}^{(n-1)}\mathbf{P}(n-1)$	$d\boldsymbol{\pi}(t)/dt = \boldsymbol{\pi}(t)\mathbf{Q}$	$d\boldsymbol{\pi}(t)/dt = \boldsymbol{\pi}(t)\mathbf{Q}(t)$
solution	$\boldsymbol{\pi}^{(n)} = \boldsymbol{\pi}^{(0)}\mathbf{P}^n$	$\boldsymbol{\pi}^{(n)} = \boldsymbol{\pi}^{(0)}\mathbf{P}(0)\mathbf{P}(1)\cdots\mathbf{P}(n-1)$	$\boldsymbol{\pi}(t) = \boldsymbol{\pi}(0)e^{\mathbf{Q}t}$	$\boldsymbol{\pi}(t) = \boldsymbol{\pi}(0)\exp\left[\int_0^t \mathbf{Q}(u)\,du\right]$
Equilibrium solution	$\boldsymbol{\pi} = \boldsymbol{\pi}\mathbf{P}$	—	$\boldsymbol{\pi}\mathbf{Q} = \mathbf{0}$	—
Transform relationships	$[\mathbf{I} - z\mathbf{P}]^{-1} \Leftrightarrow \mathbf{P}^n$	—	$[s\mathbf{I} - \mathbf{Q}]^{-1} \Leftrightarrow \mathbf{H}(t)$	—

$P[\text{interdeparture time} \leq t] = 1 - e^{-\lambda t} \qquad t \geq 0$ 148

$$g(y) \triangleq \frac{1}{y(\rho)^{1/2}} e^{-(\lambda+\mu)y} I_1[2y(\lambda\mu)^{1/2}]$$ 215

$$f_n = \frac{1}{n}\binom{2n-2}{n-1}\rho^{n-1}(1+\rho)^{1-2n}$$ 218

$$p_k = \begin{cases} \dfrac{1-\lambda/\mu}{1-(\lambda/\mu)^{K+1}}\left(\dfrac{\lambda}{\mu}\right)^k & 0 \leq k \leq K \\ 0 & \text{otherwise} \end{cases} \qquad \text{(M/M/1/K)}$$ 104

$$p_k = \frac{\dfrac{M!}{(M-k)!}\left(\dfrac{\lambda}{\mu}\right)^k}{\displaystyle\sum_{i=0}^{M}\frac{M!}{(M-i)!}\left(\frac{\lambda}{\mu}\right)^i} \qquad \text{(M/M/1//M)}$$ 107

$$P(z) = \frac{\mu(1-\rho)(1-z)}{\mu(1-z) - \lambda z[1-G(z)]} \qquad \text{(M/M/1 bulk arrival)}$$ 136

$$p_k = \left(1 - \frac{1}{z_0}\right)\left(\frac{1}{z_0}\right)^k \qquad k = 0, 1, 2, \ldots \quad \text{(M/M/1 bulk service)}$$ 139

M/M/m

$$p_k = \begin{cases} p_0 \dfrac{(m\rho)^k}{k!} & k \leq m \\ p_0 \dfrac{(\rho)^k m^m}{m!} & k \geq m \end{cases}$$ 102

$$p_0 = \left[\sum_{k=0}^{m-1}\frac{(m\rho)^k}{k!} + \left(\frac{(m\rho)^m}{m!}\right)\left(\frac{1}{1-\rho}\right)\right]^{-1}$$ 103

$$P[\text{queueing}] = \frac{\left(\dfrac{(m\rho)^m}{m!}\right)\left(\dfrac{1}{1-\rho}\right)}{\left[\displaystyle\sum_{k=0}^{m-1}\frac{(m\rho)^k}{k!} + \left(\frac{(m\rho)^m}{m!}\right)\left(\frac{1}{1-\rho}\right)\right]} \qquad \text{(Erlang C formula)}$$ 103

$$p_k = \frac{(\lambda/\mu)^k}{k!} \Big/ \sum_{i=0}^{m}\frac{(\lambda/\mu)^i}{i!} \qquad \text{(M/M/m/m)}$$ 105

$$p_m = \frac{(\lambda/\mu)^m}{m!} \Big/ \sum_{i=0}^{m}\frac{(\lambda/\mu)^i}{i!} \qquad \text{(M/M/m/m)} \quad \text{(Erlang's loss formula)}$$ 106

M/D/1

$$\bar{q} = \frac{\rho}{1-\rho} - \frac{\rho^2}{2(1-\rho)} \qquad 188$$

$$W = \frac{\rho \bar{x}}{2(1-\rho)} \qquad 191$$

$$G(y) = \sum_{n=1}^{[y/\bar{x}]} \frac{(n\rho)^{n-1}}{n!} e^{-n\rho} \qquad 219$$

$$f_n = \frac{(n\rho)^{n-1}}{n!} e^{-n\rho} \qquad 219$$

E_r (r-stage Erlang Distribution)

$$b(x) = \frac{r\mu(r\mu x)^{r-1} e^{-r\mu x}}{(r-1)!} \qquad x \geq 0 \qquad 124$$

$$\sigma_b = \frac{1}{\mu(r)^{1/2}} \qquad 124$$

M/E$_r$/1

$$P_j = (1-\rho) \sum_{i=1}^{r} A_i (z_i)^{-j} \qquad j = 1, 2, \ldots, r \qquad 129$$

E$_r$/M/1

$$p_k = \begin{cases} 1-\rho & k = 0 \\ \rho(z_0^{\,r} - 1) z_0^{\,-rk} & k > 0 \end{cases} \qquad 133$$

H$_R$ (R-stage Hyperexponential Distribution)

$$b(x) = \sum_{i=1}^{R} \alpha_i \mu_i e^{-\mu_i x} \qquad x \geq 0 \qquad 141$$

$$C_b^{\,2} \geq 1 \qquad 143$$

MARKOVIAN NETWORKS

$$\lambda_i = \gamma_i + \sum_{j=1}^{N} \lambda_j r_{ji} \qquad 149$$

$$p(k_1, k_2, \ldots, k_N) = p_1(k_1) p_2(k_2) \cdots p_N(k_N) \qquad 150$$

(open) where $p_i(k_i)$ is solution to isolated M/M/m$_i$

$$p(k_1, k_2, \ldots, k_N) = \frac{1}{G(K)} \prod_{i=1}^{N} \frac{x_i^{k_i}}{\beta(k_i)} \qquad \text{(closed)} \qquad 152$$

LIFE AND RESIDUAL LIFE

$$f_X(x) = \frac{xf(x)}{m_1} \quad \text{(lifetime density of sampled interval)} \qquad 171$$

$$\hat{f}(y) = \frac{1 - F(y)}{m_1} \quad \text{(residual life density)} \qquad 172$$

$$\hat{F}^*(s) = \frac{1 - F^*(s)}{sm_1} \quad \text{(residual life transform)} \qquad 172$$

$$r_n = \frac{m_{n+1}}{(n+1)m_1} \quad \text{(nth moment of residual life)} \qquad 173$$

$$r_1 = \frac{m_2}{2m_1} \quad \text{(mean residual life)} \qquad 173$$

$$r(x) = \frac{f(x)}{1 - F(x)} \quad \text{(failure rate)} \qquad 173$$

M/G/1

$$r_k = p_k = d_k \qquad 176$$

$$q_{n+1} = q_n - \Delta_{q_n} + v_{n+1} \qquad 181$$

$$\bar{v} = \rho \qquad 183$$

$$\overline{v^2} - \bar{v} = \lambda^2\overline{x^2} = \rho^2(1 + C_b^2) \qquad 187$$

$$V(z) = B^*(\lambda - \lambda z) \qquad 184$$

$$\bar{q} = \rho + \rho^2\frac{(1 + C_b^2)}{2(1 - \rho)} \quad \text{(P–K mean value formula)} \qquad 187$$

$$\frac{T}{\bar{x}} = 1 + \rho\frac{(1 + C_b^2)}{2(1 - \rho)} \quad \text{(P–K mean value formula)} \qquad 191$$

$$\frac{W}{\bar{x}} = \rho\frac{(1 + C_b^2)}{2(1 - \rho)} \quad \text{(P–K mean value formula)} \qquad 191$$

$$W = \frac{W_0}{1 - \rho} \quad \text{(P–K mean value formula)} \qquad 190$$

$$W_0 \triangleq \frac{\lambda\overline{x^2}}{2} \qquad 190$$

$$Q(z) = B^*(\lambda - \lambda z)\frac{(1 - \rho)(1 - z)}{B^*(\lambda - \lambda z) - z} \quad \text{(P–K transform equation)} \qquad 194$$

$$W^*(s) = \frac{s(1-\rho)}{s - \lambda + \lambda B^*(s)} \qquad \text{(P–K transform equation)} \qquad 200$$

$$S^*(s) = B^*(s)\frac{s(1-\rho)}{s - \lambda + \lambda B^*(s)} \qquad 199$$

$$P[I \le y] = 1 - e^{-\lambda y} \qquad y \ge 0 \qquad 208$$

$$G^*(s) = B^*(s + \lambda - \lambda G^*(s)) \qquad 212$$

$$G(y) = \int_0^y \sum_{n=1}^{\infty} e^{-\lambda x}\frac{(\lambda x)^{n-1}}{n!}\, b_{(n)}(x)\, dx \qquad 226$$

$$g_1 = \frac{\bar{x}}{1-\rho} \qquad 213$$

$$g_2 = \frac{\overline{x^2}}{(1-\rho)^3} \qquad 214$$

$$\sigma_g^{\,2} = \frac{\sigma_b^{\,2} + \rho(\bar{x})^2}{(1-\rho)^3} \qquad 214$$

$$g_3 = \frac{\overline{x^3}}{(1-\rho)^4} + \frac{3\lambda(\overline{x^2})^2}{(1-\rho)^5} \qquad 214$$

$$g_4 = \frac{\overline{x^4}}{(1-\rho)^5} + \frac{10\lambda\overline{x^2}\,\overline{x^3}}{(1-\rho)^6} + \frac{15\lambda^2(\overline{x^2})^3}{(1-\rho)^7} \qquad 214$$

$$F(z) = zB^*[\lambda - \lambda F(z)] \qquad 217$$

$$P[N_{\text{bp}} = n] = \int_0^{\infty} \frac{(\lambda y)^{n-1}}{n!} e^{-\lambda y} b_{(n)}(y)\, dy \qquad 226$$

$$h_1 = \frac{1}{1-\rho} \qquad 217$$

$$h_2 = \frac{2\rho(1-\rho) + \lambda^2\overline{x^2}}{(1-\rho)^3} + \frac{1}{1-\rho} \qquad 218$$

$$\sigma_h^{\,2} = \frac{\rho(1-\rho) + \lambda^2\overline{x^2}}{(1-\rho)^3} \qquad 218$$

$$\frac{\partial F(w,t)}{\partial t} = \frac{\partial F(w,t)}{\partial w} - \lambda F(w,t) + \lambda\int_{x=0}^{w} B(w-x)\, d_xF(x,t) \qquad 227$$

(Takács integrodifferential equation)

$$F^{**}(r,s) = \frac{(r/\eta)e^{-\eta w_0} - e^{-r w_0}}{\lambda B^*(s) - \lambda + r - s} \qquad 229$$

$$Q(z) = \frac{(1-\rho)(1-z)B^*[\lambda - \lambda G(z)]}{B^*(\lambda - \lambda G(z)) - z} \qquad \text{(bulk arrival)} \qquad 235$$

M/G/∞

$$p_k = \frac{\rho^k}{k!} e^{-\rho} \qquad 234$$

$$T = \bar{x} \qquad 234$$

$$s(y) = b(y) \qquad 234$$

G/M/1

$$r_k = (1 - \sigma)\sigma^k \qquad k = 0, 1, 2, \ldots \qquad 251$$

$$\sigma = A^*(\mu - \mu\sigma) \qquad 251$$

$$W(y) = 1 - \sigma e^{-\mu(1-\sigma)y} \qquad y \geq 0 \qquad 252$$

$$W = \frac{\sigma}{\mu(1 - \sigma)} \qquad 252$$

G/M/m

$$q'_{n+1} = q_n' + 1 - v'_{n+1} \qquad 242$$

$$\sigma = A^*(m\mu - m\mu\sigma) \qquad 249$$

$$P[\text{queue size} = n \mid \text{arrival queues}] = (1 - \sigma)\sigma^n \qquad n \geq 0 \qquad 249$$

$$\mathbf{r} = J[R_0, R_1, \ldots, R_{m-2}, 1, \sigma, \sigma^2, \sigma^3, \ldots] \qquad 254$$

$$R_{k-1} = \frac{R_k - \sum_{i=k}^{m-2} R_i p_{ik} - \sum_{i=m-1}^{\infty} \sigma^{i+1-m} p_{ik}}{p_{k-1,k}} \qquad 254$$

$$p_{ij} = 0 \qquad \text{for} \qquad j > i + 1 \qquad 242$$

$$p_{ij} = \int_0^\infty \binom{i+1}{j} [1 - e^{-\mu t}]^{i+1-j} e^{-\mu t j} \, dA(t) \qquad j \leq i + 1 \leq m \qquad 244$$

$$\beta_n = p_{i,i+1-n} = \int_{t=0}^\infty \frac{(m\mu t)^n}{n!} e^{-m\mu t} \, dA(t) \qquad 0 \leq n \leq i + 1 - m, m \leq i$$

245

$$p_{ij} = \int_0^\infty \binom{m}{j} e^{-j\mu t} \times \left[\int_0^t \frac{(m\mu y)^{i-m}}{(i - m)!} (e^{-\mu y} - e^{-\mu t})^{m-j} m\mu \, dy\right] dA(t) \qquad j < m < i + 1$$

245

$$J = \frac{1}{\dfrac{1}{1-\sigma} + \sum_{k=0}^{m-2} R_k} \qquad 254$$

$$W = \frac{J\sigma}{m\mu(1-\sigma)^2} \qquad 256$$

$$w(y \mid \text{arrival queues}) = (1-\sigma)m\mu e^{-m\mu(1-\sigma)y} \qquad y \geq 0 \qquad 250$$

$$W(y) = 1 - \frac{\sigma}{1 + (1-\sigma)\sum_{k=0}^{m-2} R_k} e^{-m\mu(1-\sigma)y} \qquad y \geq 0 \qquad 255$$

G/G/1

$$w_{n+1} = (w_n + u_n)^+ \qquad 277$$

$$c(u) = a(-u) \circledast b(u) \qquad 281$$

$$W(y) = \begin{cases} \int_{-\infty}^{y} W(y-u)\, dC(u) & y \geq 0 \\ 0 & y < 0 \end{cases} \qquad \text{(Lindley's integral equation)} \qquad 283$$

$$A^*(-s)B^*(s) - 1 = \frac{\Psi_+(s)}{\Psi_-(s)} \qquad 286$$

$$\Phi_+(s) = \frac{1}{\Psi_+(s)} \lim_{s\to 0} \frac{\Psi_+(s)}{s} = \frac{W(0^+)}{\Psi_+(s)} \qquad 290$$

$$\Phi_+(s) = \frac{\Psi_-(0)(1-\rho)\bar{t}}{[A^*(-s)B^*(s) - 1]\Psi_-(s)} \qquad 290$$

$$W = \frac{\sigma_a^2 + \sigma_b^2 + (\bar{t})^2(1-\rho)^2}{2\bar{t}(1-\rho)} - \frac{\overline{I^2}}{2\bar{I}} \qquad 306$$

$$W = -\frac{\overline{u^2}}{2\bar{u}} - \frac{\overline{y^2}}{2\bar{y}} \qquad 305$$

$$\tilde{w} = \sup_{n\geq 0} U_n \qquad 279$$

$$w(y) = \pi(c(y) \circledast w(y)) \qquad 301$$

$$W^*(s) = \frac{a_0[1 - I^*(-s)]}{1 - C^*(s)} \qquad 307$$

$$W^*(s) = \frac{1-\sigma}{1 - \sigma\hat{I}^*(s)} \qquad 310$$

Index